Ecological Research Monographs

Series Editor

Yoh Iwasa, Department of Biology, Kyushu University, Fukuoka, Japan

The book series Ecological Research Monographs publishes refereed volumes on all aspects of ecology, including Animal ecology Population ecology Theoretical ecology Plant ecology Community ecology Statistical ecology Marine ecology Ecosystems Biodiversity Microbial ecology Landscape ecology Conservation Molecular ecology Behavioral ecology Urban ecology Physiological ecology Evolutionary ecology The series comprise books and edited collections by international experts in their fields.

More information about this series at https://link.springer.com/bookseries/8852

Tao Wang • Atsushi Tsunekawa • Xian Xue •
Yasunori Kurosaki
Editors

Combating Aeolian Desertification in Northeast Asia

Springer

Editors
Tao Wang
Northwest Institute of Eco-Environment and Resources
Chinese Academy of Sciences
Lanzhou, Gansu, China

Atsushi Tsunekawa
Arid Land Research Center
Tottori University
Tottori, Tottori, Japan

Xian Xue
Northwest Institute of Eco-Environment and Resources
Chinese Academy of Sciences
Lanzhou, Gansu, China

Yasunori Kurosaki
Arid Land Research Center
Tottori University
Tottori, Tottori, Japan

ISSN 2191-0707 ISSN 2191-0715 (electronic)
Ecological Research Monographs
ISBN 978-981-16-9027-3 ISBN 978-981-16-9028-0 (eBook)
https://doi.org/10.1007/978-981-16-9028-0

This Springer imprint is published by the registered company Springer Nature Singapore Pte Ltd.
The registered company address is: 152 Beach Road, #21-01/04 Gateway East, Singapore 189721, Singapore

Preface

Desertification, defined by the United Nations Convention to Combat Desertification (UNCCD) in 1994, means "land degradation in arid, semi-arid and dry sub-humid areas resulting from various factors, including climatic variations and human activities." Desertification is primarily caused by soil erosion that occurs due to wind and/or water erosion; deterioration of the physical, chemical, and biological or economic properties of soil; and long-term loss of natural vegetation. Globally, approximately 24% of the land is experiencing desertification, and 110 countries are at risk of desertification. The livelihood of 1.5 billion people relies on desertified land. The desertification-induced annual loss of arable land is 8000 km^2, and the annual economic cost is 42 billion USD (UNCCD 2009).

Aeolian desertification is a type of desertification caused by wind erosion through aeolian processes. This type of desertification is often accompanied by processes related to the occurrence of blown sand (deflation, ground surface coarsening, sand dune formation, etc.) in formerly nondesertified areas that serve as a baseline area (benchmark). We defined aeolian desertification as land degradation caused by wind erosion mainly resulting from human impacts in arid, semi-arid, and dry sub-humid regions of northern China and southern Mongolia.

Since 1980, Professor Seiei Toyama from Tottori University, Japan, has been devoted to research and efforts related to combating aeolian desertification and studying sandy land with his team and collaborating researchers from the Lanzhou Institute of Desert Research, the Chinese Academy of Sciences, and different stakeholders in Inner Mongolia, Ningxia, and Gansu provinces. In the past four decades, many universities and national research institutes in China and Japan have systematically and holistically collaborated to focus on aeolian desertification processes and control techniques. The collaborative activities have included large spatial field surveys, in situ field observations, laboratory analyses, theoretical compilations, development of control techniques, and student exchanges, and these activities have had good results. In 2004, a trilateral framework was developed for China-Japan-Mongolia scientists to work on aeolian desertification issues. The three sides reached an agreement in terms of academic exchanges; field surveys; data sharing; student, environmental manager, and practitioner exchanges; and field station establishment and improvement.

Based on the achievements of these collaborative efforts that were implemented to understand aeolian desertification processes and the techniques to combat, we compiled a book, *Combating Aeolian Desertification in Northeast Asia*, with the following conclusions and outcomes:

1. According to the actual situations, research, and practices in Northeast Asia, we provided a definition of aeolian desertification and its implications, which allowed us to clearly understand the research object and issues, spatiotemporal scope, and targets for combating aeolian desertification.
2. To understand the impacts of natural climatic variations and human activities on aeolian desertification, we provided results on the analysis of the variations in the natural environment, human activities, and temporal dynamics of aeolian desertification derived from a remote-sensing dataset.
3. In view of the comprehensive study on the process of aeolian desertification by both natural factors with wind erosion and human land use patterns and intensities, we concluded that human activities caused more than 80% of the aeolian desertification in Northern China over the last 60 years. Thus, improper land use is the primary driver of aeolian desertification, and only appropriate human activities will successfully combat aeolian desertification.
4. We argued that environmental policies and national ecological engineering projects are important for combating aeolian desertification; we demonstrated the efficacy of physical, biological, and comprehensive protection systems in terms of combating aeolian desertification, and we provided information on how to alleviate poverty and gain ecological benefits.
5. We identified successful practices for combating aeolian desertification in China and Mongolia that can serve as references to other countries that face the same aeolian desertification problems.
6. This work showed that sustainable land management is necessary to improve aeolian desertification control efforts.

The collaboration among Japan, China, and Mongolia was financially supported by the Japan Society for the Promotion of Science, Japan Science and Technology Agency, Japan International Cooperation Agency, Ministry of Science and Technology of China, Chinese Academy of Forestry, Chinese Academy of Science, National Science Foundation of China, and Mongolian Academy of Science. This support indicates the level of concern related to combating aeolian desertification among these three nations, and the collaboration among the three nations also showed that aeolian desertification is not only an environmental-social-ecological problem of an individual country but also a transboundary concern. Efforts and collaborations among Japan, China, and Mongolia are indispensable to addressing aeolian desertification.

We thank all the authors for submitting their valuable manuscripts to this book and all the scientists, staff, and students who participated in and contributed to the research introduced in this book. We are grateful to Springer Japan and its staff for their assistance and encouragement.

Lanzhou, China Tao Wang
Tottori, Japan Atsushi Tsunekawa
Lanzhou, China Xian Xue
Tottori, Japan Yasunori Kurosaki

Contents

About the Editors

Tao Wang is a professor at Northwest Institute of Eco-Environment and Resources, Chinese Academy of Sciences (CSA). He obtained his doctoral degree from Lanzhou Institute of Desert Research, CAS. He has been at the forefront of aeolian desertification and disaster research and control since 1983, and obtained encouraging results on the status, causes, blown sand physics and biology process, development, and reversal trend of aeolian desertification in Northern China. His work has helped the national policy maker and organ of authority to take notice of aeolian desertification and provided theoretical foundation and technique system for implementation of national layout and projects to combat aeolian desertification and disaster. He served the UNCCD for years as member of the Scientific Advisory Committee of Third Scientific Conference (SAC, UNCCD) and as member of Science-Policy Interface (SPI) of the UNCCD.

Atsushi Tsunekawa is a professor at Arid Land Research Center, Tottori University (Japan). He graduated from the Graduate School of Agricultural Sciences, University of Tokyo, after which he joined the National Institute for Environmental Studies, Japan. Currently, he serves as a professor at the Arid Land Research Center, Tottori University, and also serves as a head of Strategic Management Office of International Platform for Dryland Research and Education, Tottori University. His primary scholarly interests are developing sustainable land management (SLM) technologies and approaches to restore degraded land and improve farmers' livelihood, and monitoring and modeling of terrestrial ecosystems under climate change using remote sensing and GIS. He has been selected as a Science and Technology Correspondent from Japan to the Committee on Science and Technology of United Nations Convention to Combat Desertification.

Xian Xue is a professor at Northwest Institute of Eco-Environment and Resources (NIEER), Chinese Academy of Sciences. She obtained her PhD in physical geography from the Graduate School at the Chinese Academy of Sciences in 2002. She currently leads the Department of Desert and Desertification in the NIEER, the Dryland Salinization Station of NIEER, and the Salinization Research Station of Gansu Province. Her main research fields are desertification and restoration, and climate and environmental change in arid and cold regions. Her interest focuses on the dynamic process and mechanism of desertification and its rehabilitation. One of her current research is exploring the impact of global warming and human activity on land degradation in the arid and alpine ecosystems. Her other work is to explore restoration possibilities of degraded land in the drylands by integrating water management, sustainable agriculture, and biological techniques with her team.

Yasunori Kurosaki is a professor at Arid Land Research Center (ALRC), Tottori University, Japan. He obtained his PhD from Graduate School of Life and Environmental Sciences, University of Tsukuba. His main research theme is elucidation of the aeolian dust emission mechanisms. Prof. Kurosaki began his research at the Meteorological Research Institute, Japan Meteorological Agency, in 2001, and he has continued it at Chiba University, Georgia Institute of Technology, and Tottori University. He has elucidated the causes of aeolian dust emission that change with time and place in drylands of Northeastern Asia mainly by analyzing meteorological observatory data and GIS data such as satellite data. He has also studied it by field surveys in the Gobi Desert. His interest and the purpose of his research are to apply his elucidated mechanisms to improve numerical aeolian dust models and to prevent wind erosion and aeolian dust damage.

Part I
Overview of Aeolian Desertification

Chapter 1
Definition of Aeolian Desertification and Its Implications

Tao Wang

Abstract In this book, land degradation is defined as a negative trend in land conditions resulting from various factors, including climatic variations and human activities, and expressed as a long-term reduction or loss in biological productivity, ecological integrity, or value to humans. Land degradation is caused by the processes related to soil degradation (including soil erosion, salinization, and soil fertility loss) and ecosystem degradation (including reduction in vegetation cover, reduction in biomass, and decrease in biodiversity). Desertification is defined as a type of land degradation in arid, semiarid, and dry subhumid areas. Aeolian desertification is a type of desertification caused by wind erosion through aeolian processes. This type of desertification is often accompanied by processes related to the occurrence of blown sand (deflation, ground surface coarsening, sand dune formation, etc.) in formerly nondesertified areas that serve as a baseline area (benchmark), and aeolian desertification results from various factors, including overuse of land and imbalances in the fragile ecosystem under dry and windy climate conditions and loose sand surfaces. Topics related to aeolian desertification research mainly include its occurrence and developmental processes; its distribution, status, causes, and damage; its types and indicator systems; restoration options; remote sensing monitoring; and mapping.

Keywords Land degradation · Desertification · Aeolian desertification · Classification · Index system

---

T. Wang (✉)
Northwest Institute of Eco-Environment and Resources, Chinese Academy of Sciences, Lanzhou, Gansu, China

Key Laboratory of Desert and Desertification, Chinese Academy of Sciences, Lanzhou, Gansu, China

Drylands Salinization Research Station, Northwest Institute of Eco-Environment and Resources, Chinese Academy of Sciences, Lanzhou, Gansu, China
e-mail: wangtao@lzb.ac.cn

T. Wang et al. (eds.), *Combating Aeolian Desertification in Northeast Asia*, Ecological Research Monographs, https://doi.org/10.1007/978-981-16-9028-0_1

1.1 Concept of Aeolian Desertification and Its Development

1.1.1 Formulation of the Term "Desertification"

Aeolian desertification is one of the main types of desertification, and it is necessary to introduce the concept of desertification before explaining the definition and implications of aeolian desertification.

The term "desertification" was first presented by Aubreville, A, a French botanist, in his studies as long ago as 1949 (Aubreville 1949). In his study on the ecological problems in tropical African forest regions, he defined "desertification" as a process of environmental degradation from a forest to a grassland and then to a desert-like landscape after the destruction of forest vegetation in the region. Other researchers have been discussing the "desertification" phenomenon since the late 1950s (Zhu 1959; Le Houerou 1962; Hou 1973). However, this topic did not attract the attention of the international community until 1968–1973, when the serious and extended drought in West Africa intensified the process of land desertification in the Sahel region at the southern margin of the Sahara Desert. Desertification has received widespread attention all over the world, as confirmed by the creation of the United Nations Conference on Desertification in Nairobi in 1977 (UN Secretariat of the Conference on Desertification 1977). It was described in the UN Conference on Desertification (UNCD 1978) as: "Desertification is the diminution or destruction of the biological potential of the land, which can lead ultimately to desert-like conditions. It is an aspect of the widespread deterioration of ecosystems, and has diminished or destroyed the biological potential...The deterioration of productive ecosystems is an obvious and serious threat to human progress. In general, the quest for ever-greater productivity has intensified exploitation and has carried disturbance by man into less productive and more fragile lands. Overexploitation gives rise to degradation of vegetation, soil and water, the three elements which serve as the natural foundation for human existence. In exceptionally fragile ecosystems, such as those on the desert margins, the loss of biological productivity through the degradation of plant, animal, soil and water resources can easily become irreversible, and permanently reduce their capacity to support human life. Desertification is a self-accelerating process, and as it advances, rehabilitation costs rise exponentially. Action to combat desertification is required urgently before the costs of rehabilitation rise beyond practical possibility or before the opportunity to act is lost forever."

After summarizing the situation and development trends of land desertification since 1977, the United Nations Conference on Environment and Development (UNCED, held in Rio de Janeiro, Brazil in 1992) complementarily defined desertification as "land degradation in arid, semi-arid and dry sub-humid areas resulting from various factors, including climatic variations and human activities" (UNCED 1992). In the definition, the arid, semiarid, and dry subhumid areas are classified by aridityindex (AI) (UNEP 1992; Table 1.1).

Table 1.1 Climate zones classified by the aridity index

Climate zone	arid index (AI)
Arid	0.05<AI<0.20
Semiarid	0.20<AI<0.50
Dry subhumid	0.50<AI<0.65
Humid	>0.65

Note: AI = *P*/APE, where *P* is annual precipitation (mm) and APE is the revised potential evapotranspiration (mm)

Although a number of research articles, papers, and reports from many countries have shown growing interest in the desertification issue (Glantz 1977; Quintanilla 1981; Zhu and Liu 1981; Zonn 1981; Wang 1989, 1991, 1993; Wang and Wang 1989; Zhu and Wang 1990, 1992a, b, 1993), drought was neither the first indicator of desertification nor the only reason for scientific interest in it. Desertification is acknowledged to be a complex phenomenon requiring the expertise of researchers in disciplines such as climatology, meteorology, soil science, hydrology, range science, agronomy, geography, political science, economics, and anthropology. To date, the term "desertification" is still widely interpreted globally in many different ways by researchers in these and other disciplines, and it has also been defined from many national and institutional perspectives, each emphasizing different aspects of the phenomenon. Some reviews of desertification literature show great diversity (and confusion) among definitions (Carder 1981; Chen 1994). Hence, the United Nations Convention to Combat Desertification (UNCCD, signed at the Diplomatic Conference in Paris, France, in 1994) further clarified the three basic elements of the above definition of desertification: (a) desertification is the result of a combination of factors such as climate change (natural factors) and human activities; (b) desertification occurs mainly in arid, semiarid, and dry subhumid areas with fragile ecological environments; and (c) desertification is part of the global land degradation process (UNCCD 2000). The UNCCD also highlighted that it is important to note that desertification is not the natural process of deserts expanding into new regions; it is a form of land degradation caused primarily by human activity in vulnerable areas. All regions in which the ratio of the total annual precipitation to potential evapotranspiration (P/PET) ranges from 0.05 to 0.65 should be considered vulnerable to desertification, and such regions constitute approximately 40% of the global terrestrial area. Broadly, land degradation refers to the processes of soil degradation (including soil erosion, salinization, and soil fertility loss) and ecosystem degradation (including reduction of vegetation cover, reduction in biomass, and decrease of biodiversity), but desertification is only part of the process.

1.1.2 The Term "Aeolian Desertification" and Its Development

As mentioned above, the term "desertification" originated officially from the southern Saharan region of Africa, and as the UN Conference on Desertification described, its salient feature is that "desertification can lead ultimately to desert-like conditions." With the deterioration of the global ecological environment and the expansion of the countries involved in the discussion, the scope of the discussion has been extended from arid areas to dry subhumid areas, and the causes of land degradation discussed are not limited to wind erosion but have involved water erosion, secondary salinization, and other factors. Desertification no longer has its original narrow concept; it now involves problems that are constantly growing, and its meaning is constantly expanding. Therefore, the use of the term under different national backgrounds requires a clear conceptual scope (Zhu 1994, 1999; Zhu and Chen 1994; Ci 1995; Chen 2001, 2002).

China is suffering from serious desertification. As early as the late 1950s, Professor Zhu noted that owing to anthropogenic causes, some former nondesert land had changed into desert and approximately 530,000 ha of sandy grasslands in Shaanxi and Inner Mongolia were "man-made desert" (Zhu 1959). He also stressed that water is a weapon to conquer deserts, and we can use water to open oases in deserts. Thus, researchers in China conducted studies on aeolian sand movement, water balance, physiological and ecological characteristics of sand binders, and vegetal sand stabilization measures. In these studies, special attention was given to the environmental degradation and the spread of deserts caused by anthropogenic factors. After the UN Conference on Desertification in 1977, the priority in desert research in China gradually shifted to the study of desertification processes and their control, and the studied regions gradually shifted from arid and hyperarid zones to the semiarid and part of subhumid regions that had relatively better eco-environmental conditions and higher productivity but were facing desertification problems. According to the situation in China, desertification in China mainly includes desertification from wind erosion, water erosion, soil salinization, and freeze-thaw. In northern China, desertification mainly occurs from wind erosion, which is now referred to as "aeolian desertification" or "sandy desertification." Although there has been international consensus on the concept of desertification since the UNCCD, the concepts of "desertification" and "aeolian desertification" have been used in contrasting ways for a relatively a long time in China, resulting in very different research results, especially in terms of desertification areas or aeolian desertification areas (Li 1994; Wang 1994; Zhu 1994, 1999; Zhu and Chen 1994; Ci 1995; Zhu and Zhu 1999; Chen 2001, 2002; Yi 2005).

In this book, we combined the research by Zhu and Chen (Zhu and Chen 1994) and the UNCCD in 1994 and CCICCD (1994) and defined aeolian desertification as land degradation in arid, semiarid, and dry subhumid regions with the occurrence of blown sand processes (deflation, ground surface coarsening, sand dune formation, etc.) in formerly nondesert areas as the main indicator, and the occurrence of this type of desertification has resulted from various factors, including overuse of land and an imbalance in the fragile ecosystem under the dry and windy climate conditions and loose sand surface.

1.2 Scope of Research on Aeolian Desertification

After the UN Conference on Desertification in 1977, scientific research on deserts in China changed from being sandy desert-oriented to aeolian desertification-oriented. Since that time, the process of aeolian desertification and its control mechanisms have become important components of desert scientific research. At the same time, some important information on aeolian desertification that is different from that on sandy desertification has been clarified (Zhu and Liu 1981; Zhu and Wang 1992a; Wang et al. 1999):

1. Aeolian desertification in China mainly occurred in the human historical period, especially over the last 100 years; aeolian desertification is mainly distributed where the earth's surface is covered by loose sand materials and winds are frequent in arid, semiarid, and even part of subhumid zones. This desertification type occurs in limited areas and mostly in patches on farmlands and grasslands, with simple and small aeolian sand features. Sandy desert is a barren area where the ground is completely covered with sand dunes, plants are very rare, rain is scarce, and the air is dry; sandy deserts are generally aeolian landforms such as the Taklimakan Desert and the Badain Jaran Desert. Sandy deserts in China were mainly formed during different stages of the Quaternary period; they are the outcomes of the dry climate and abundant surface sand source, and they mainly occur in arid zones, with large areas and complex aeolian landforms.
2. Aeolian desertification develops or reverses in a short period under the same climatic conditions due to the negative or positive effects of human economic activities. Recent climatic conditions and their variation amplitudes are insufficient to cause a large expansion or shrinkage of sandy deserts. In the past several decades, aeolian desertification developed rapidly due to the influences of undue human economic activities (overcultivating, overgrazing, overcutting, and overusing water resources) rather than due to climate change. Sandy deserts were formed under the influences of various natural factors, while the occurrence of aeolian desertification is mainly due to anthropogenic factors in addition to natural factors. Sandy deserts varied with the changes in the Quaternary climate; in the dry-cold stage, they expanded and became mobile, but in the warm-wet stage, they shrank and became fixed.
3. Aeolian desertification is a gradual process, and wind acts as the power to shape the landscape of desertified land surfaces. Therefore, blowing sand, wind erosion, and the morphological characteristics of wind-deposited surfaces can be used as quantitative indicators of the degree of aeolian desertification development. Aeolian desertification intensity and its spatial extent are related to the degree

of drought and human and animal pressure on land. With the interaction of these factors and the action of wind force, aeolian desertification can spontaneously spread; it leads to sand dune encroachment, land biological productivity reduction, and available land resource loss, but it can also be reversed or self-restored, depending upon natural conditions (especially water conditions), landscape complexity, and human activity intensity.

In summary, studies on sandy deserts and aeolian desertification do not belong to the same category in terms of time, space, cause, developmental trend, restoration, and utilization. Research on aeolian desertification mainly includes the following aspects:

- Occurrence and development processes of aeolian desertification and their types in arid, semiarid, and dry subhumid zones in northern China, especially in regions with fragile eco-environments and frequent human activities
- Studies on aeolian desertification status, distribution, causes, damage, and indicator systems
- Establishment of restoration experiments and demonstration projects for different desertified land types, i.e., examples at the margin of an oasis, in the desert steppe zone, in the semiarid agropastoral ecotone, and in some other typical areas
- Remote sensing monitoring, field investigations, and aeolian desertification mapping

1.3 Index System of Aeolian Desertification Monitoring

Aeolian desertification is a complex process of land degradation. During the process, the physical and chemical properties of soils change with the variation in the landscapes caused by sand movement. Therefore, establishing a set of identifying indicators to monitor the developmental stages of aeolian desertification is important. Although a set of indicator systems of aeolian desertification has been established by the FAO and UNEP (FAO and UNEP 1984), a practical indicator system for regional aeolian desertification monitoring and classification is still needed because of large differences from place to place. It is generally accepted that aeolian desertification indicators related to natural, human, and socioeconomic factors should be identified. However, there is currently no recognized index system of aeolian desertification in the world due to a lack of consistency in the selection of indicators. There are several commonly used classification index systems in China:

1. An aeolian desertification index system based on symbols of the aeolian desertification developmental phase (Table 1.2)
2. An aeolian desertification index system based on ecology (Table 1.3)
3. An aeolian desertification index system based on landscape configuration changes (Table 1.4)

Table 1.2 Degree of aeolian desertification based on developmental symbols (Wang 2011)

Degree	Percentage of aeolian desertification land	Features of shapes
Aeolian desertification-prone	≤5	Patches of shifting sand sparsely scatter in farmland and around water wells and residential areas
Ongoing aeolian desertification	6–25	Patches of shifting sand or blown land, farmlands suffer wind erosion
Intensively developed aeolian desertification	26–50	Sheets of shifting sand dunes and blown shrub sand dunes, interlaced with fixed and semifixed sand dunes
Severe aeolian desertification	>51	Predominant shifting sand dunes distribute densely

Table 1.3 Degree of aeolian desertification based on ecological indicators (revised according to Wang 2011)

Degree	Vegetation cover (%)	Land productivity (%)	Ratio of output to input (%)	Biomass (t/ha·a)
Latent	>60 of original	>80 of original	>80 of original	3.0–4.5
Ongoing	59–30 of original	79–50 of original	79–60 of original	2.9–1.5
Intensively developing	29–10 of original	49–20 of original	59–30 of original	1.4–1.0
Severe	9–0 of original	19–0 of original	29–0 of original	0.9–0.0

4. An aeolian desertification index system based on different land types (Table 1.5)
5. An aeolian desertification index system based on dynamic status (Table 1.6)

The results from the different evaluation systems applied to the same region were not completely consistent, which can cause confusion to in terms of the theory that guides practice. To establish a universal index system of aeolian desertification monitoring, the indicators selected should be representative and applicable and a statistical quantum, represent environmental phenomena related to aeolian desertification processes, or represent an existing specific environmental condition (Wang et al. 1998). The indicators should (a) contain clear information that is easy to obtain from observations; (b) be sensitive to variations in aeolian desertification status; (c) be suitable for repeated use; and (d) be reviewed with a quantitative check. According to the characteristics of aeolian desertification in northern China, we have established a relatively universal indicator and classification system of aeolian desertification (Table 1.7), which considers surface feature variations the main factors and considers the changes in soil, vegetation, and ecosystems the supplemental factors. We supplement several indicators to evaluate the aeolian desertification degree with the development of aeolian desertification.

Table 1.4 Degree of aeolian desertification based on landscape configuration changes (Wang 2011)

Degree	Synthetic landscape symbols
Slight aeolian desertification	1. Blowouts appear on windward slopes, with shifting sand depositing on leeward slope; rate of vegetation cover is 30–60% of nonaeolian desertified land; areas with patches of shifting sand occupy 5–25% 2. Different scales of shrub sand mounds appear; shrubs grow abundantly and thickly 3. A thin layer of sand deposits on the Earth's surface, even with gravel outcrops 4. In spring, farmland is eroded by wind, with less than a 50% loss in humus, and output is 50–80% of original yields 5. Blowouts appear where fine soil is thick, with certain vegetation cover
Moderate aeolian desertification	1. The difference between a blown slope and slip slope is obvious; vegetation cover is 10–30% of nonaeolian desertified land; area of shifting sand accounts for 25–50% 2. The whole sand mound cannot be completely covered by shrubs, with shifting sands on the windward slope 3. Small patches of shifting sand appear in the loess area, with much coarser sand and gravel at the surface but still with sparse plant cover; vegetation cover is 10–30% of the nonaeolian desertified land 4. Productivity is decreased due to wind erosion, with more than a 50% loss in humus and less than 50% of original output 5. Blowouts are mostly bare; small steep ridges emerge at the ground surface
Severe aeolian desertification	1. The whole sandy desertified area is shifting sandy land-like, with more than 50% of shifting sand and sparse vegetation, and vegetation cover is less than 10% of the nonaeolian desertified land 2. Gravel desertified area appears gobi-like, vegetation cover is below 10% of the nonaeolian desertified land, and farmlands with gravel desertification occurrence are deserted 3. Humus layer is almost blown away completely, calcic horizon or soil parent material is outcropped, and most farmlands are deserted 4. Soil residues from wind erosion appear on the Earth's surface

In remote sensing monitoring, the percentage of eroded land or shifting sand areas and its changes in a certain period are considered the key indicator (it is the easiest factor to judge and use), and the other factors are considered supplementary indicators. This is because the change in eroded land or shifting sand area is a combined result of vegetation cover, biological production, soil properties, water content, etc. Our classification system mainly relies on direct information on ground vegetation cover, plant species, and microtopographic features (Fig. 1.1). Based on this indicator system of aeolian desertification monitoring, the four types of aeolian desertification in northern China have the following features:

1. Slight aeolian desertification: (a) the blowouts appear on the windward slopes of sand dunes, there are some accumulative shifting sands on leeward slopes, vegetation cover is 30–50% of the original cover, and patches of shifting sands

Table 1.5 Degree of aeolian desertification based on different land use types (Wang 2011)

Degree	A Reactivation of fixed sand dunes	B Aeolian desertification of shrubs	C Gravel aeolian desertification	D Badland aeolian desertification	E Aeolian desertification of farmland
1 Original status (aeolian desertification-prone)	1a Fixed sand dunes or oases, farmlands	1b Dry steppe or desert steppe, steppification desert	1c Desert steppe or steppification desert	1d Dry steppe or desert steppe, steppification desert	1e Dry farmlands
2 Slight aeolian desertification	2a Blowouts appear on the windward slope; patchy shifting sand occupies 5–25%, with more than 90% of original vegetation cover	2b Shrubs flourish; shifting sand deposits under shrubs	2c Gravel is becoming concentrated at the ground surface	2d Shallow blown pits merge on the ground surface but without steep ridges	2e There is deposited sand on farmlands in spring, with obvious traces of wind erosion
3 Moderate aeolian desertification	3a The difference between blown slope and slip slope is obvious; area of shifting sand accounts for 25–50% of the original vegetation cover	3b The whole sand mound cannot be completely covered by shrubs, with shifting sand's occurrence on the windward slope	3c Substantially coarse sand and gravel at the surface but still with sparse plant cover, and vegetation cover is more than 25% of the original vegetation cover. The landscape is gravel steppe	3d Most blowouts are bare; obvious small steep ridges emerge on the Earth's surface	3e Small patches of shifting sand appear on loessial farmlands; productivity is decreased due to wind erosion, with more than a 50% loss in humus
4 Severe aeolian desertification	4a Sandy land is semishifting; area of shifting sand exceeds 50%, with less than 50%	4b Large patches of shrubs are dead; vegetation cover is below 25% of the original vegetation cover, and	4c The Earth's surface is covered by gravel completely, with minimal sand in small holes among the gravel, and vegetation	4d Soil residues from wind erosion appear on the Earth's surface, with sparse vegetation at the interdune areas, and	4e Humus layer is almost blown away completely, calcic horizon or soil parent material is outcropped, and most

(continued)

Table 1.5 (continued)

Degree	A Reactivation of fixed sand dunes	B Aeolian desertification of shrubs	C Gravel aeolian desertification	D Badland aeolian desertification	E Aeolian desertification of farmland
	of the original vegetation cover	the area of shifting sand exceeds 50%	cover is 10–25% of the original vegetation cover	farmlands with gravel desertification are deserted	sandy desertified farm-lands are deserted
5 Very severe aeolian desert-ification land	5a Shifting sand dunes or sandy land, with the veg-etation cover being less than 10% of the original vegetation cover	5b Undulated shifting sandy land, with the vegetation cover being less than 10% of the original vegetation cover	5c Gobi; vegetation cover is less than 10% of original vegetation cover	5d Yardang landscape	5e Flat sandy land or gravel land; vegetation cover is less than 10% of the original vegetation cover

Table 1.6 Degree of aeolian desertification based on dynamic status (Wang 2011)

Degree	Rate of annual increase in aeolian desertification land (%)
Reversed aeolian desertification	Negative
Stable aeolian desertification	<0.25
Developing aeolian desertification	>0.25
Thereinto: Commonly developing	0.25–3.0
Intensively developing	>3.0

Table 1.7 Index system of aeolian desertification classification (Wang 2011)

Degree	Main indicators			
	Blown-sand area (%)	Annual expansion area (%)	Vegetation cover[a](%)	Annual reduction in biomass (%)
Slight (L)	<5	<1	>60	<1.5
Moderate (M)	5–25	1–2	60–30	1.5–3.5
Severe (S)	25–50	25	30–10	3.5–7.5
Very severe (VS)	>50	>5	10–0	>7.5

Degree	Supplementary indicators				
	Soil deflation thickness (cm)	Accumulative thickness (cm)	Soil deflation rate (t/ha·a)	Overload population (%)	Overload livestock (%)
Slight (L)	<5	<5	>0.5	−50 to −31	−50 to −31
Moderate (M)	5–10	5–10	0.5–1.0	−31 to 0	−31 to 0
Severe (S)	10–20	10–20	1.0–3.0	0–31	0–31
Very severe (VS)	>20	>20	>3.0	>31	>31

[a]Vegetation cover is calculated by the projection method, and the vegetation cover of the local primary landscape is regarded as 100%

occupy 25% of the area; (b) shrubs grow well, and sand mounds of different sizes appear around shrubs; (c) there is a thin layer of shifting sands on the land surface; (d) the ridges of the cultivated field are eroded, sands accumulate between ridges, and the humus layer loss in the soil is less than 50%; (e) crop yield is 50–80% of the initial stages of cultivation; and (f) shallow blowouts occur in the sandy area, but some vegetation still exists; in addition, the blowouts gradually transform without a visible steep bench.

2. Moderate aeolian desertification: (a) a distinct differentiation between eroded slopes and slip faces appears, vegetation cover is 15–30% of the original cover, and the area of shifting sand occupies 25–50%; (b) leaved shrubs cannot entirely

Fig. 1.1 Remote images (left) and field pictures (right) of different grades of aeolian desertification in the Horqin Region, Inner Mongolia, China (provided by Hanchen Duan). (**a**) Slight aeolian desertification. (**b**) Moderate aeolian desertification. (**c**) Severe aeolian desertification. (**d**) Very severe aeolian desertification

cover sand mounds, and there are shifting sands on the windward side of the sand mounds; (c) small patches of shifting sand occur on loessial farmland or land surface covered by coarse sands or gravel, but minimal vegetation still exists with a coverage of 10–30%; (d) there is obvious wind erosion on the cultivated land, less than 50% of the humus layer has been blown away, and crop yield is 50% of the initial stages of cultivation; and (e) blowouts are mostly exposed, and ridges are easy to distinguish.

3. Severe aeolian desertification: (a) sandy land is in a semifixed state, the area of shifting sands exceeds 50%, and vegetation coverage is less than 15% of the original cover; (b) gobi landscape occurs, and vegetation cover is less than 10%; (c) humus layer of the soil is eroded and almost blown away, calcic horizon is exposed, and most desertified croplands are abandoned; and (d) deflation mounds and pillars appear on the land surface.
4. Very severe aeolian desertification: (a) land loses its productivity completely; (b) a mobile sand dune landscape occurs in the sandy lands; (c) gobi landscape occurs in the gravel lands; and (d) yardangs occur in wind-eroded lands.

1.4 Conclusion

It is essential to scientifically define aeolian desertification. Different understandings will lead to considerable differences in the area and distribution of aeolian desertified land. More importantly, different perspectives will affect the places where aeolian desertification prevention measures are implemented and the choice of prevention measures. For example, treating natural dune fields as aeolian desertified land will lead to unsustainable land use methods such as large-scale afforestation in dune fields. A scientific and unified aeolian desertification discrimination index system is also essential. Achieving the global goal of Zero Land Degradation requires a scientific index system to evaluate whether the land is in the process of desertification development or reversal. Without a unified scientific standard, it is difficult to determine at a global scale whether zero land degradation has been achieved or the case is far from it. This chapter only briefly introduces the concepts and classification systems of aeolian desertification in China. A definition and evaluation system suitable for aeolian desertification on a global scale also requires the joint efforts and contribution of scientists across the world.

References

Aubreville A (1949) Climats, forêts et désertification de l'Afrique tropicale. Société d'Editions Géographiques, Maritimes et Coloniales, Paris

Carder DJ (1981) Desertification in Australia-a muddled concept. Search 12(7):218–221

Chen GT (1994) UN convention to combat desertification and definitions of desertification. J Desert Res 14(4):71(in Chinese with English abstract)

Chen GT (2001) Causes of the difference of statistic sandy desertification area in China. J Desert Res 21(2):209–212 (in Chinese with English abstract)

Chen GT (2002) Land desertification. Chemical Industry Press, Beijing (in Chinese)

China National Committee to Implement the UN Convention to Combat Desertification (CCICCD) (1994) United nations convention to combat desertification in those countries experiencing serious drought and/or desertification, particularly in Africa. China Science and Technology Press, Beijing

Ci LJ (1995) Desertification and sandy desertification. Sci Technol Bull (8):32 (in Chinese with English abstract)

FAO, UNEP (1984). Provisional methodology for assessment and mapping of desertification. Report in Rome

Glantz MH (ed) (1977) Desertification: environmental degradation in and around arid lands. Westview Press, Boulder

Hou RZ (1973) Changes of the Mu Us Desert from vicissitude of the ancient cities along the Hongliu River course. Cult Reli (1):35–42 (in Chinese with English abstract)

UNEP (1992) World Atlas of Desertification.

Le Houerou HN (1962) Les pâturages naturels de la Tunisie aride et désertique. Inst Sces Econ Appl Tunis, Paris

Li LK (1994) My view on the Chinese translation of "desertification". World Forest Res 7(2):87–88 (in Chinese with English abstract)

Quintanilla EG (1981) Regional aspects of desertification in Peru. In: Combating desertification through integrated development, UNEP/UNEPCOM International Scientific Symposium, Abstract of Papers, Tashkent, USSR, 114–115

UN Conference on Desertification(UNCD) (1978) Round-up, plan of action and resolutions. United Nations, New York

UN Secretariat of the Conference on Desertification (1977) Desertification: an overview. In: Desertification: its causes and consequences. Pergamon Press, New York

UNCCD (2000) Desertification, global change, and sustainable development. Desertification Control Bulletin, 20: Fact Sheet 2, 10

UNCED (1992) Earth Summit Agenda 21: programme of action for sustainable development. United Nations Department for Public Information, New York

Wang T (1989) A comparative study of desertification in typical areas in Northern China. J Desert Res, 9(1):113–136 (in Chinese with English abstract)
Wang T (1991) A case study on rangeland desertification in Bashang Region, Hebei Province, Progress Report on Part 2, Desertification Application Research Programme, World Laboratory, pp 163–171
Wang T (1993) Desertification in the North China- causes, processes, trends and harness. J Agric Meteorol 45(5):691–694
Wang LX (1994) Current situation and development trend of global desertification control. World Forest Res 7(1):10–17 (in Chinese with English abstract)
Wang T (2011) Deserts and aeolian desertification in China. ELSEVIER in Amsterdam and Science Press in Beijing
Wang KF, Wang T (1989) Characteristics of study to combat desertification in Pakistan. Desert Sea 1:1–3 (in Chinese with English abstract)
Wang T, Wu W, Wang XZ (1998) Remote sensing monitoring and assessing sandy desertification: an example from the sandy desertification region of northern China. Quat Sci (2):108–118 (in Chinese with English abstract)
Wang T, Zhao HL, Xiao HL (1999) Progress in the research of desertification of China. J Desert Res 19(4):299–311 (in Chinese with English abstract)
Yi LY (2005) The area of desertification region and sandy land in China shrunk for the first time. Pratacult Sci, (7):115 (in Chinese with English abstract)
Zhu KZ (1959) Rehabilitating desert is our historical mission. China Daily 1959-8-2 (in Chinese)
Zhu ZD (1994) The status and prospect of desertification in China. Acta Geograph Sin, 61(S1): 650–659 (in Chinese with English abstract)
Zhu ZD (1999) Spatial distribution of land desertification in China. In: Zhu ZD et al (eds) Sand desert, sandy desertification, desertification and control in China. China Environmental Science Press, Beijing (in Chinese)
Zhu ZD, Chen GT (1994) Sandy desertification in China. Science Press, Beijing (in Chinese)
Zhu ZD, Liu S (1981) Aeolian desertification processes and its control regionalization in North China. China Forestry Publishing House, Beijing (in Chinese)
Zhu ZD, Wang T (1990) An analysis on the trend of land desertification in northern China during the last decade based on examples from some typical areas. Acta Geograph Sin 57(4):430–440 (in Chinese with English abstract)
Zhu ZD, Wang T (1992a) Theory and practice on sandy desertification in China. Quat Sci (2): 97–106 (in Chinese with English abstract)
Zhu ZD, Wang T (1992b) Development trend of sandy desertification in northern China. Chin Nat Geograp (1):2–3 (in Chinese with English abstract)
Zhu ZD, Wang T (1993) The trends of desertification and its rehabilitation in China. UNEP Desertificat Cont Bull 22:27–30
Zhu JF, Zhu ZD (1999) Sandy desertification control in China. China Forestry Publishing House, Beijing (in Chinese)
Zonn IS (ed) (1981) USSR/UNEP projects to combat desertification. Moscow Centre of International Projects GKNT, p 33

Chapter 2
Environmental and Ecological Setting in Northeast Asia

Tao Wang and Mandakh Nyamtseren

Abstract Northeast Asia, especially northern China and southern Mongolia, has experienced severe aeolian desertification in the past several decades and is considered a primary source region of Asian sand-dust storms. Abundant sand, strong winds, and minimal plant coverage are the three aspects that cause an area to be prone to aeolian desertification and sand-dust storms. This chapter describes the process through which abundant sand occurs (from where and how), climatic systems and climate change, and land coverage and land use change due to climate and human activities in northern China and southern Mongolia based on previous studies, climatic data, and remote sensing.

Keywords Climate change · Land use and cover · Aeolian desertification · Northern China · Mongolia

2.1 Climate of Mongolia

Its geographical location, especially remoteness from seas and oceans, high elevations, and surrounding mountain systems predetermine the harsh and distinct continental climate in Mongolia. The climate of Mongolia, therefore, is characterized by a high variability in climatic indices both in terms of temporal and spatial distributions

T. Wang (✉)
Northwest Institute of Eco-Environment and Resources, Chinese Academy of Sciences, Lanzhou, Gansu, China

Key Laboratory of Desert and Desertification, Chinese Academy of Sciences, Lanzhou, Gansu, China

Drylands Salinization Research Station, Northwest Institute of Eco-Environment and Resources, Chinese Academy of Sciences, Lanzhou, Gansu, China
e-mail: wangtao@lzb.ac.cn

M. Nyamtseren
The Institute of Geography and Geoecology, Mongolia Academy of Sciences, Ulaanbaatar, Mongolia

T. Wang et al. (eds.), *Combating Aeolian Desertification in Northeast Asia*, Ecological Research Monographs, https://doi.org/10.1007/978-981-16-9028-0_2

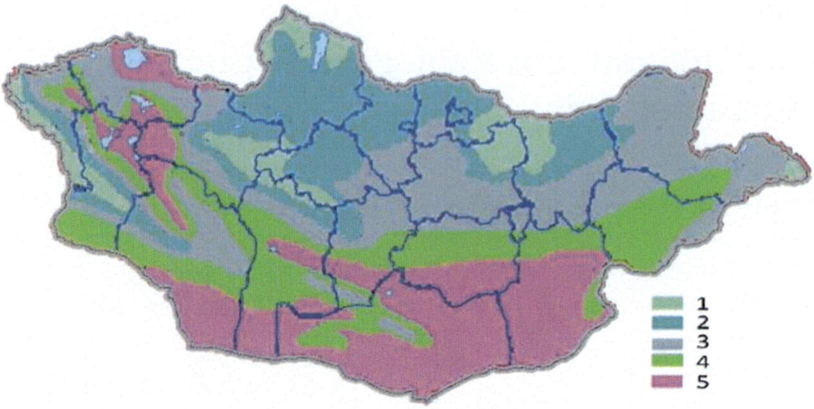

Fig. 2.1 Climatic regions of Mongolia (Jambaajamts 1989). Note: (1) Extra humid, cold region; (2) humid, temperate region; (3) subhumid, cool region; (4) semiarid, warm region; (5) arid, hot region

with high amplitudes of diurnal and seasonal temperatures and relatively low atmospheric precipitation. Over the majority of the land, the climatic index distribution follows latitudinal and altitudinal variations, and only in the easternmost corner in the Khingan Mountains are small areas of longitudinal changes observed due to the impact of the monsoon climate. According to the climate classification (Fig. 2.1), approximately 40% of the total territory is classified as arid and semiarid, and only 10–15% of the territory is classified as extra humid (Jambaajamts 1989).

Mean annual air temperatures differ by region; for instance, in the Altai, Khangai, Khentei, and Khuvsgol mountainous regions, the air temperatures vary from −2 to −4 °C, and over the large depressions and river basins, they decrease to a range of −6 to −8 °C in the southern regions; the mean annual temperatures can exceed +2 °C (Fig. 2.2). The recorded highest annual mean air temperature is 8.5 °C at the Ekhiingol station. The mean air temperatures in winter range from approximately −22 to −25 °C on average for the country. The lowest temperatures can be observed in the mountainous regions of the Altai, Khangai, Khuvsgol, and Khentei mountains and range from −30 to −34 °C. On the steppes and high plains, winter temperatures are normally between −20 and −25 °C, and it is relatively warm to the south, where temperatures range from −15 to −20 °C. The mean temperatures in summer are approximately 15 °C in the mountain regions, 15–20 °C on the plains and in the depressions, and 20–25 °C in the eastern and southern parts of the country. The long-term minimum temperature recorded in Zuungobi, Uvs aimag, was −55.3 °C, and the maximum temperature documented in Darkhan city, Darkhan-Uul aimag, was 44 ° C.

The spatial distribution of precipitation in Mongolia is very specific due to the complex impact of geographical factors, i.e., location, landforms, and surface roughness. Typically, precipitation decreases from north to south and from east to west (Fig. 2.3); however, the heterogeneity of the landforms influences the general

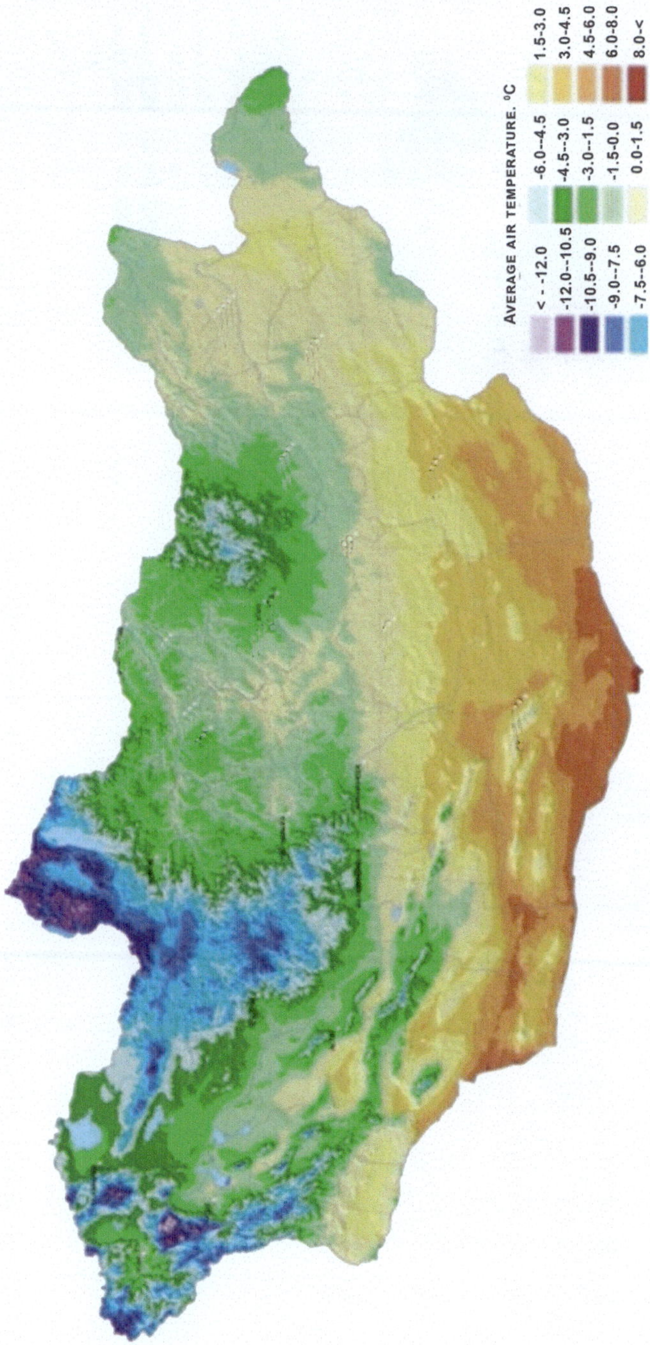

Fig. 2.2 The mean annual temperatures over Mongolia, 1961–1990 (Dagvadorj et al. 2014)

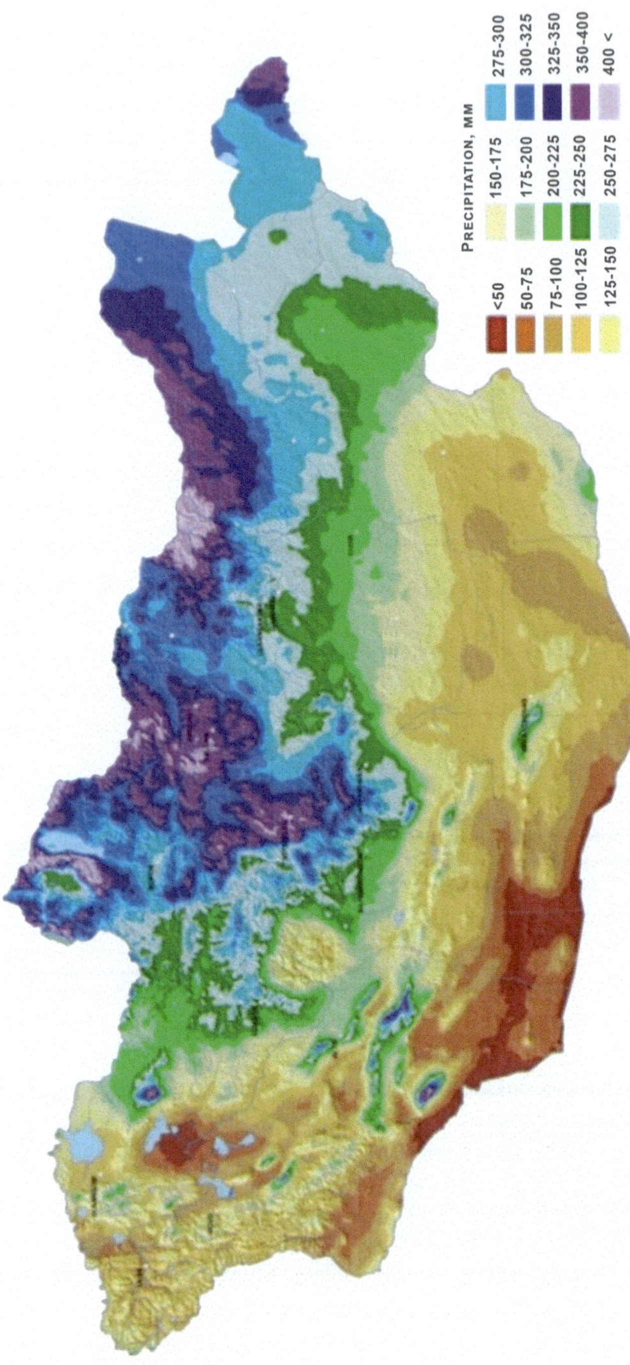

Fig. 2.3 Spatial distribution of annual precipitation, 1961–1990 (Dagvadorj et al. 2014)

distribution of precipitation. The annual total precipitation exceeds 400 mm only in the high mountains of the Khangai, Khubsugul, and Khentei mountain regions, and all other mountainous parts of the country, including the northeastern high plains, receive 300–400 mm of precipitation annually. The majority of the territory, especially the middle part of the country occupied by grasslands, receives approximately 250–300 mm of precipitation over the year, which decreases to 100–150 mm/year in the southern semidesert and desert regions. Due to the rain shadow effect of the Altai Mountain, the territory to the south has a minimum amount of precipitation, ranging between 50 and 55 mm/year.

On a temporal scale, approximately 70–80% of the total precipitation occurs during the warm season (April to September), of which 50% falls only within July and August. Special attention should be paid to winter snowfall because it defines vegetation onset. The average duration of snow cover in the territory of the country ranges between 100 and 150 days; however, in the middle and southern parts of the country, snow cover occurs for only 50–100 days and be absent some years.

The major indicator of climate suitability is the level of the aridity index (AI), which is calculated as the ratio of rainfall to potential evapotranspiration (UNEP 1992). According to the aridity index, 22.7% of the entire territory can be classified as having a hyperarid and arid climate. Almost 60% of the land is considered to have a dry and semiarid climate with an aridity index less than 0.2. A very small proportion of land located in the northernmost and easternmost regions of the country has aridity values higher than 0.5, which is assumed to be humid and very humid.

Wind is another characteristic of the climate in Mongolia. Generally, steppe and desert steppe regions of the country are considered regions with high-intensity winds. The average wind speed in the northern part of Mongolia ranges between 2 and 3 m/s and increases to the south, where the mean annual wind speed is between 6 and 8 m/s (Jambaajamts 1989). According to national meteorological observation records, more than 90 days in a year have windy days with speeds exceeding 4 m/s. On a temporal scale, the wind regime in Mongolia has two maximums and two minimums over the year. The spring maximum is in April–June when the average wind speed ranges from 2.1 to 3.0 m/s in the north and from 4 to 6 m/s in the south. The second maximum can be observed from September to October with an average wind speed of 2–4 m/s. During these seasons, sand-dust storms may occur anywhere in the country. According to recent studies, in the mountainous part of the country, sand-dust storms occur on approximately 10 days, which increases to 50 days on the plains. This number exceeds 90 days in the regions of Trans-Altai in the southwest and Great Lake Depression in the west (Chung et al. 2004) (Fig. 2.4).

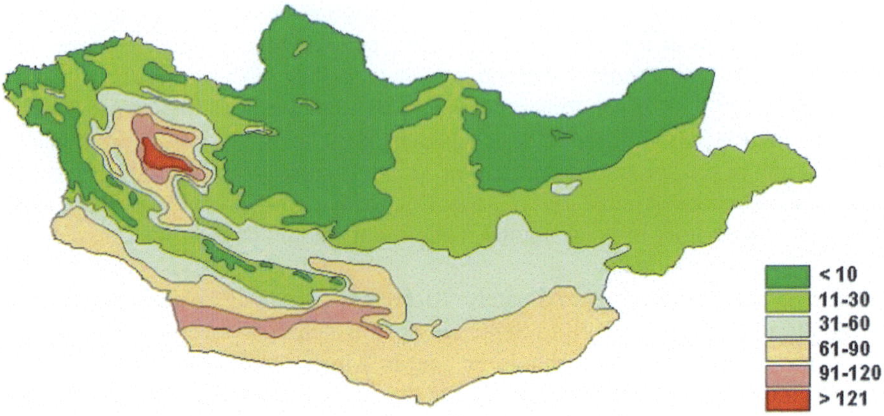

Fig. 2.4 Number of days with sand and dust storms (1960–2008) (Chung et al. 2004)

2.1.1 Observed Climate Change and Its Future Scenarios in Mongolia

The rate of global warming over the past 50 years has increased the fastest in recorded history (IPCC 2014). The observed changes in climate, especially warming, in Central Asia and Mongolia are at their highest levels. The observed near-surface temperatures and their annual mean over Mongolia have increased by 2.24 °C since the start of meteorological observations at a country level (Dagvadorj et al. 2014). The intensity of warming is higher in mountainous regions of the country than in other areas and relatively low in the steppe and Gobi regions; however, the country temperature means still exceed global warming averages. According to studies, the hottest 10 years in history were observed after the 2000s, specifically in 2001, 2009, and 2015, and these years are characterized by having the longest duration of heatwaves leading to extended drought throughout the country (Fig. 2.5).

In terms of precipitation, its change is not significant (Figs. 2.6 and 2.7). The observed changes in precipitation showed an approximate 7% decrease over the entire period of meteorological observations. The significant decrease in precipitation observed in Mongolia occurred in its central regions; however, the amount of precipitation is increasing in the Trans-Altai Desert region (Dagvadorj et al. 2014). In addition to changes in the amount of precipitation, notable changes occurred in rainfall patterns. The number of days with torrential rains has increased, leading to an increased impact of water erosion on land (Khudulmur and Tsogtbaatar 2013).

The aforementioned observed changes in climatic indices indicate that the territory of Mongolia has experienced relatively rapid changes in aridity. The latter is likely to threaten the livelihood and incomes of many households, which predominantly depend on healthy and productive grazing lands. The increased aridity will increase risks and reduce the resilience of communities to environmental changes,

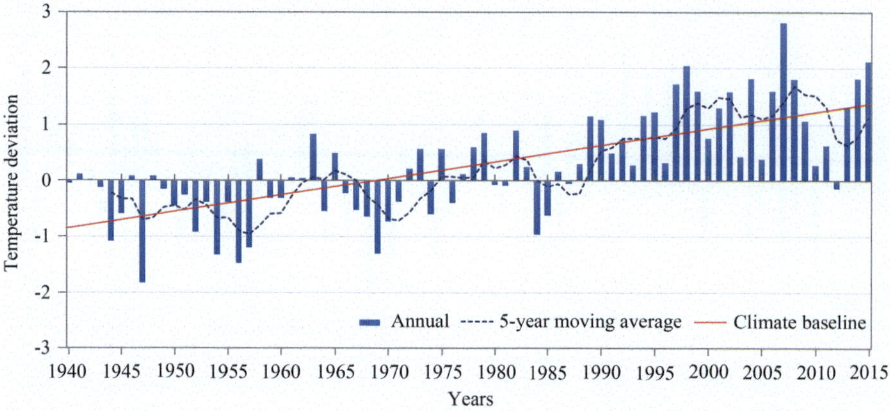

Fig. 2.5 Deviation in the mean annual temperatures compared to the climatic baseline (1961–1990) (Batjargal 2018)

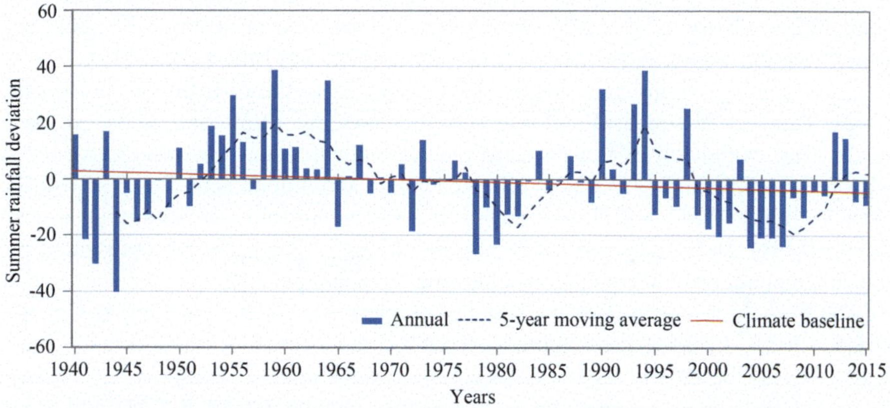

Fig. 2.6 Deviation in summer rainfall compared to the climatic baseline (1961–1990)

diminishing adaptive capacities in both natural and socioeconomic spheres. Almost 70% of the territory has experienced a decrease in the aridity index, and in other areas, the climate has become more arid, especially in the Central Khalkha Plain, Selenge-Orkhon River Basin, Mongol Daurian steppe, and Great Lake depression (Fig. 2.8). According to the station-based analysis, 21 stations out of 61 analyzed had a decreasing trend with relative changes varying between 13% and 57% (Nyamtseren et al. 2018).

At a spatial scale, an observed increase in aridity occurs in the middle and northern parts of the country, which contain steppe and dry steppe ecosystems. With increasing aridity in these regions, the risks related to water deficiency might increase significantly, which is likely to impact the largest water use sectors such as urban development, agriculture, and livestock. In comparison to other climates of the

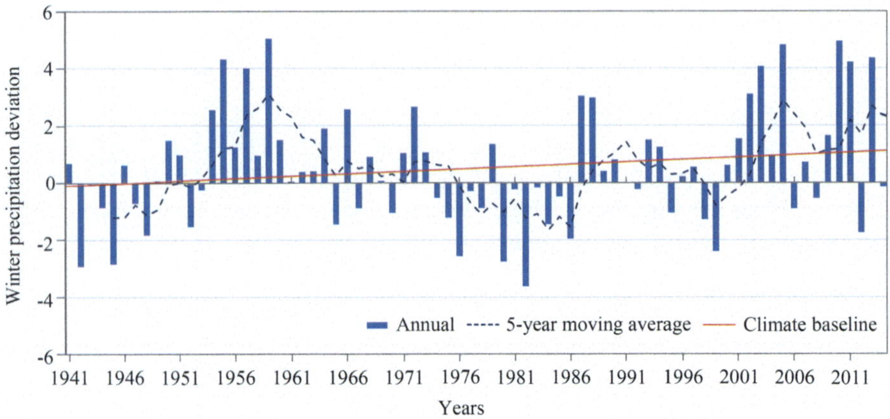

Fig. 2.7 Deviation in winter precipitation compared to the climatic baseline (1961–1990)

world, arid and semiarid regions are more sensitive to water resource variability and availability (Zhang et al. 2010). Moreover, the increase in aridity will definitely exacerbate the extent and level of aeolian desertification (Tabari et al. 2014).

Consequently, significant changes have occurred in the distribution of climatic zones. The area of semiarid and arid land increased by 18.2% and 25.3%, respectively. The expansion of the semiarid climate indicates that the vast grassland of Mongolia is threatened from both climate change and human activities. The overall dryness of the land will lead to an increased occurrence of natural disasters, especially for the middle part of Mongolia, which will mainly be represented by the frequent occurrence of sand-dust storms.

The Fifth Assessment Report of the Intergovernmental Panel on Climate Change (AR5, IPCC) included greenhouse gas (GHG) representative concentration pathways (RCPs) based on future socioeconomic development trends (IPCC 2014). These RCPs were accounted for in the global climate model (GCM) and were used to produce quantitative future climate change projections. Based on the results, impact and risk assessments of different sectors have been performed, and finally, adaptation actions have been developed to reduce vulnerabilities and risks. Many international centers, institutions, and universities are involved in running GCMs worldwide. Historical simulations from 1860 to 2005 and future projections from 2006 to 2100 have been performed by approximately 40 GCMs at 28 centers under different GHG emission scenarios globally. There are four GHG emission scenarios, which increase the radiation budget by 2.6, 4.5, 6.0, and 8.5 w/m^2, and they correspond to each RCP (Taylor et al. 2012). To assess climate change for Mongolia, ten GCM scenarios were selected based on the assessment of model simulation ability for the baseline climate period 1986–2005. The ranking was performed by multicriteria analysis using pattern correlation and bias between model monthly output and gridded observational data (Harris et al. 2013). The domain covers Mongolia, from 41.5 to 52° latitude and 87.5- to 120.0° longitude. In future

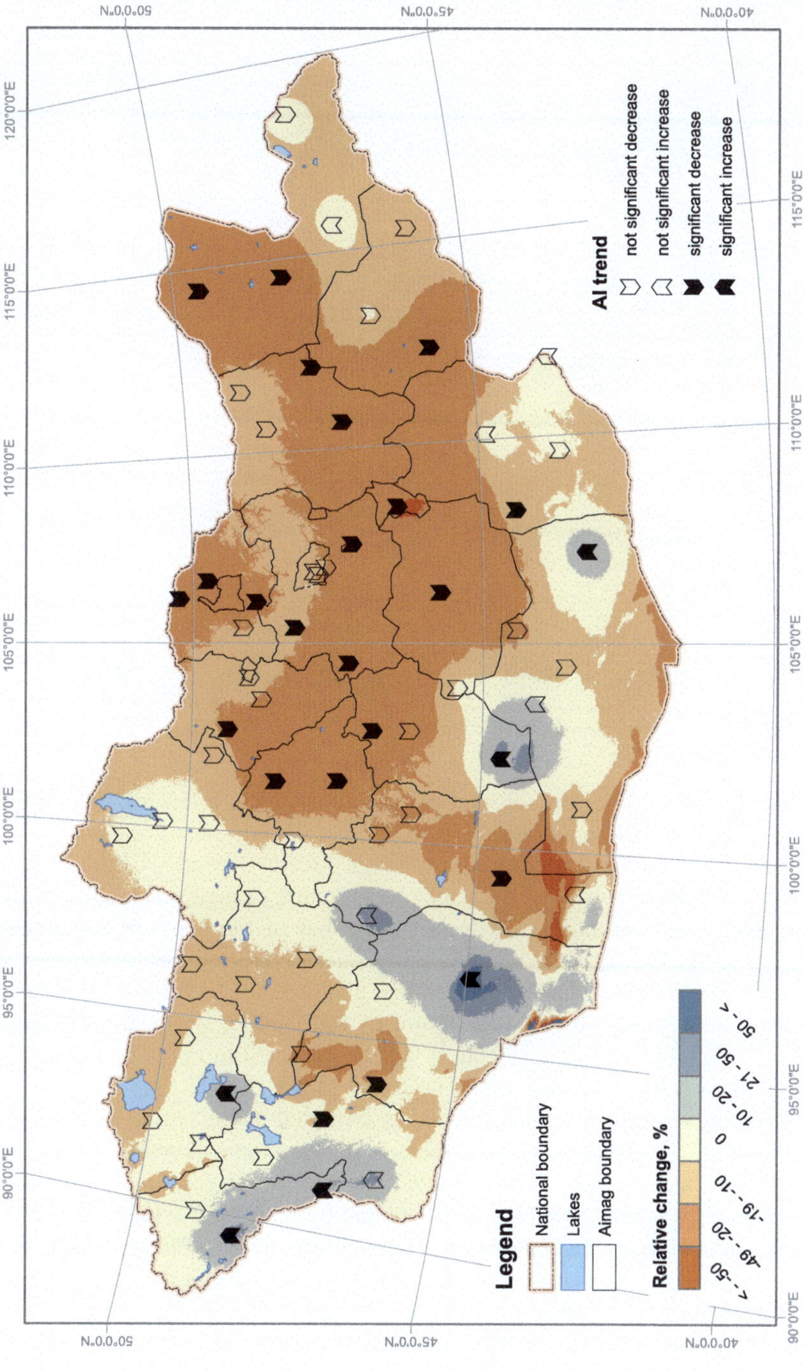

Fig. 2.8 Changes in the aridity index

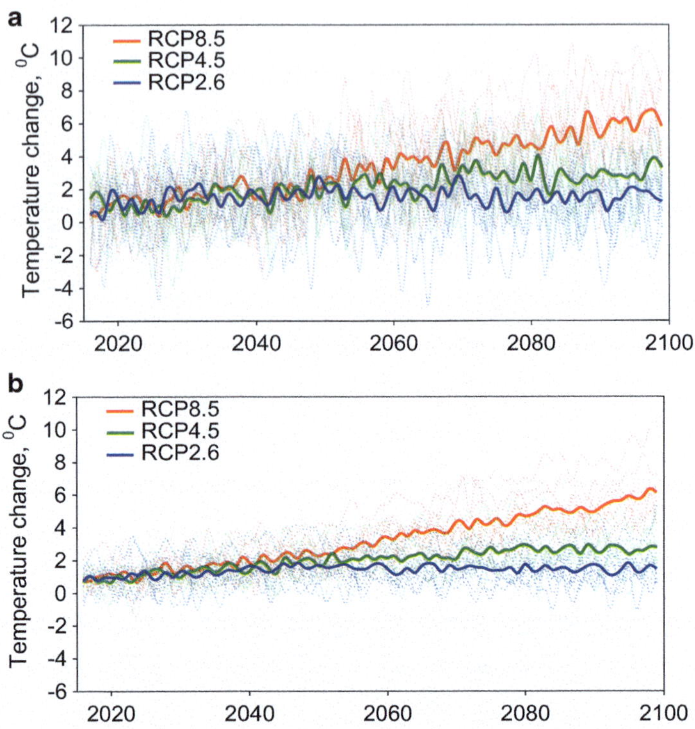

Fig. 2.9 (**a**) Winter and (**b**) summer temperature change, 2016–2100 (Batjargal 2018)

projections, the time slice was selected as near future from 2016 to 2035 and far future from 2081 to 2100 under different RCPs.

The results showed that the temperature change directly depended on the intensity of the GHG emissions. However, the winter temperature change was relatively low, and the interannual variability was higher than that of the summer temperature change (Fig. 2.9). The intensities of the temperature changes were similar for all RCP scenarios until the first half of this century, and then, the scenarios had different results, increasing yearly. In the near future of 2016–2035, seasonal temperature change will range only 2.0–2.3 °C, but it will be expected to range by 2.4–6.3 °C depending on each RCP scenario in the far future of 2081–2100.

For precipitation change, winter snow is expected to increase, and summer rainfall has not significantly changed; there is only a slight increase of less than 10% for all scenarios (Fig. 2.10). Winter snow will increase by 10.1–14.0% depending on each scenario in the near future and by 15.5–50.2% in the far future.

In light of the aforementioned changes in the future, it is interesting to note how climatic zones will evolve. The results of future projections using the ECHAM5 and HadGEM scenarios show that in the early twenty-first century, areas of arid, semiarid, dry, and subhumid zones will still expand, which we are currently experiencing. Even though the numbers vary between the two different scenarios,

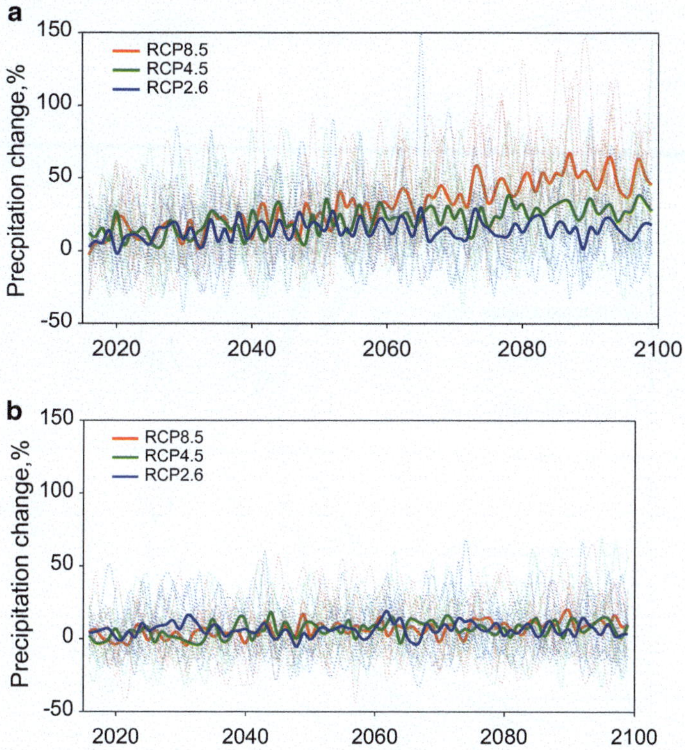

Fig. 2.10 (**a**) Winter and (**b**) summer precipitation change, 2016–2100 (Batjargal 2018)

it is evident that the current steppe and dry steppe zones will shrink in the near future. Furthermore, the hyperarid zone is likely to cover approximately 3.3% of the total territory, the arid zone is likely to cover 14.6%, the semiarid zone is likely to cover approximately 28.1%, the dry-subhumid zone is likely to cover approximately 14.9%, and the humid region is likely to cover approximately 39.1% (Mandakh 2017). The predicted trend in the expansion of dry conditions towards the north is expected to continue in the middle of the century. Compared to the early twenty-first-century conditions, both scenarios predict that the area of the hyperarid zone will slightly shrink, which is an interesting scenario overall for desert land. Although such relatively positive changes might occur in the south of the country, the homogenous dry climate condition will likely expand to the north, threatening relatively productive steppe, forest-steppe, and even forest ecosystems in the country.

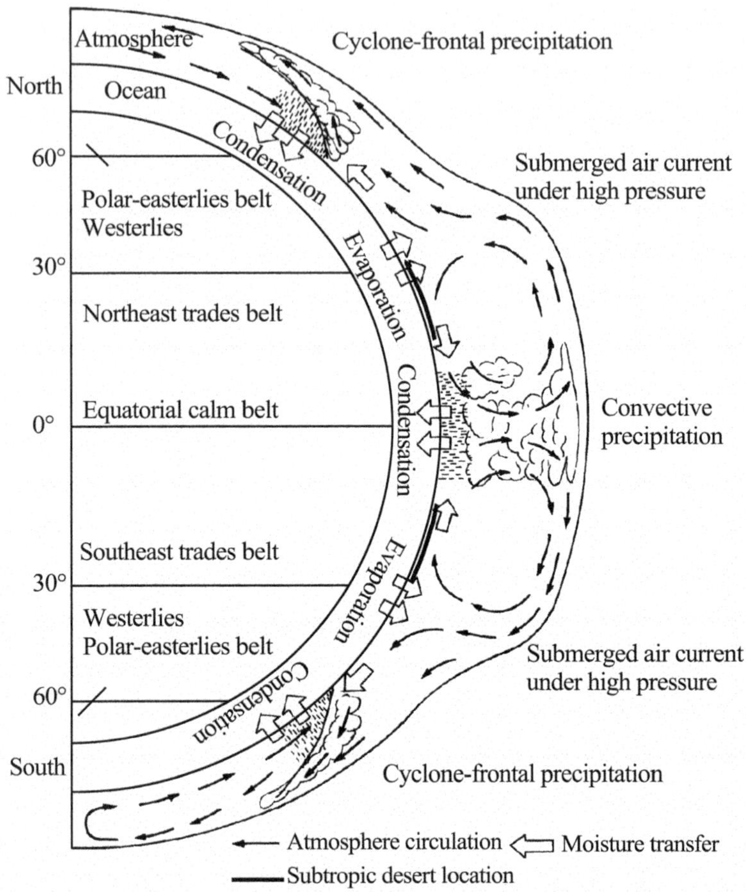

Fig. 2.11 Schema of planetary atmosphere circulation

2.2 Climate in Northern China

2.2.1 Atmospheric Circulation Regimes Over Northern China

Arid climate is a necessary condition for the occurrence of aeolian desertification (Zhu et al. 1980). Only in this specific climate can humans cause aeolian desertification by excessive economic activities. A lack of precipitation or rarely occurring precipitation that consistently occur is the essential reason for arid climates, and the mechanism of airflows is the most important condition for precipitation.

As shown by data (Fig. 2.11), deserts and gobi areas of the world are concentrated in zones between 15° and 25° latitudes southward and northward from the equator and are controlled by subtropical anticyclone belts with planetary airflows depressing steadily. However, airflow depression is the determining factor of arid climates, and desert surfaces are not the only factor. Outside subtropical zones, deserts, gobi areas, and many aeolian desertified lands are found elsewhere at higher latitudes. In northern China, there are vast sandy areas located in the warm-temperate zone, which are connected with those in Central Asia and the People's Republic of Mongolia, forming the largest temperate desert zone of the world.

Far away from oceans and surrounded by mountains and plateaus, northern China is controlled by a temperate continental climate with intensive sunshine, infrequent precipitation, large temperature differences, and a high sand-blown frequency. There are three atmospheric circulation systems dominating the evolution of the sandy areas in China: the East Asian monsoon, the north branch of the Westerly in East Asia, and the Qinghai-Tibetan Plateau monsoon.

2.2.1.1 East Asian Monsoon

The East Asian monsoon is induced by the large difference in thermal properties between the Pacific Ocean and the Eurasian continent. The land-sea thermal contrast induces a zonal pressure gradient between the Asian continental low (East Asian subtropical front) and the oceanic western Pacific subtropical high (Wang and Lin 2002). There is a distinct seasonal reversal in the pressure gradient, and the annual cycle of the East Asian monsoon can be divided into warm-wet summer and cold-dry winter monsoons (Fig. 2.12).

From a sea-land position, China is located in the east of the world's largest continent Eurasia and on the west coast of the Pacific Ocean. It is at a convergence area between the Pacific subtropical high and the Siberia-Mongolia high. In summer, the Pacific subtropical high gradually moves northward and stretches westward, eventually to approximately 30°N latitude, where it stays. The warm-wet air current coming from the southwest or southeast brings abundant precipitation to mainland China. In the East Asian summer monsoon-affected area of China, the precipitation

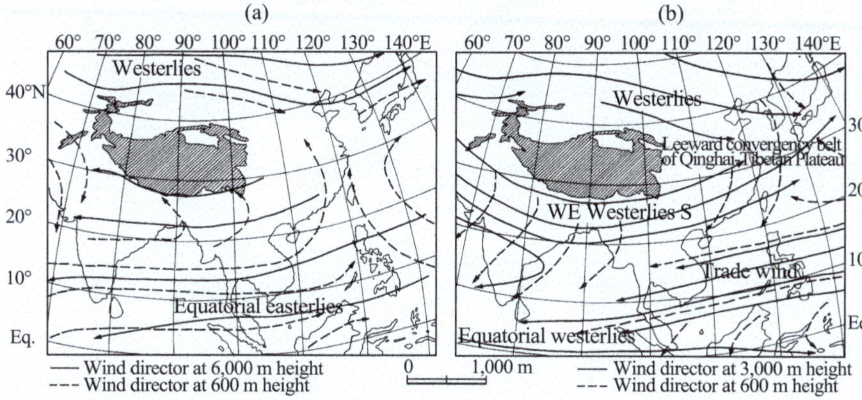

Fig. 2.12 Diagram for air flow scenarios of the East Asian monsoon (Wang 2011)

amount and duration decline from south to north following the monsoon track. Northern China benefits from the humid air current at the end of summer with a small amount of rainfall.

In winter, the Siberia-Mongolia high ridge always stretches out to northern China with strong cold air flowing southward, causing extremely cold weather (Tao 1959; Ding and Krishnamurti 1987; Ding et al. 1991; Gong and Wang 1999; Wang and Ding 2006). The Siberia-Mongolia high generally forms in October of each year and becomes the strongest in January of the next year. During this period, the sandy areas in northern China are predominantly affected by dry and cold fronts, most often with cold weather and arid and strong winds. Then, the Siberia-Mongolia high gradually weakens until April, and then, the high pressure center disappears in the summer and is replaced by a thermal low.

Thus, the kind of climate background in northern China presenting with a small amount of summer precipitation and a long-term winter drought becomes the essential preconditions for land desertification.

2.2.1.2 North Branch of the Westerly in East Asia

In comparison to that of the monsoon, the effect of the westerly on the climate in East Asia is not as great, but in winter, its northern branch (the westerly from west to east is separated into the southern and northern branches forced by the high Qinghai-Tibetan Plateau), as an anticyclone, can move relatively freely in areas with broken terrain (Fig. 2.13), forming a constantly stable high ridge over the courses of the branches. The underlying climate is cold and dry, so there are vast barren areas covering stony mounds and black gobi. Eastern Xinjiang and western Gansu

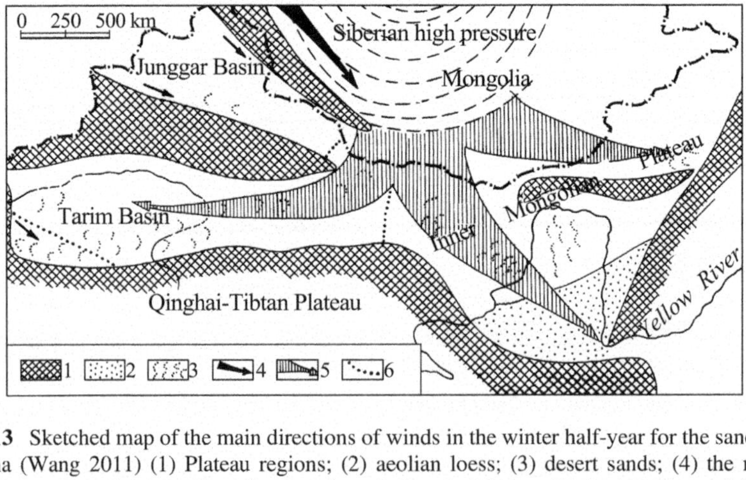

Fig. 2.13 Sketched map of the main directions of winds in the winter half-year for the sandy areas in China (Wang 2011) (1) Plateau regions; (2) aeolian loess; (3) desert sands; (4) the northern branch across the plateaus; (5) main direction of the wind flows; (6) the boundary of the main directions of wind flows

provinces are even regarded as the driest places on the Eurasian continent, and some scholars call them "the extremely dry pole of the Eurasian continent."

2.2.1.3 Qinghai-Tibetan Plateau Monsoon

Due to thermal differences between the Qinghai-Tibetan Plateau and the surrounding atmosphere, the prevailing wind direction on the plateau is almost opposite in winter and summer (a cold high in winter and a hot low in summer); this phenomenon is called the "plateau monsoon." The time, position, intensity, and height of the plateau monsoon will influence the winter and summer monsoons, which in turn will affect the climate in China.

First, the plateau monsoon increases the monsoon thickness in the lower troposphere layer both in winter and summer and greatly strengthens the winter monsoon and summer monsoon. In winter, the plateau monsoon results in a much drier climate in the western region of the sandy areas in China; in summer, it promotes the northward expansion of the summer monsoon, resulting in increased transportation of energy and water vapor. However, while the ground air rises because the plateau surface warms rapidly under strong radiation in summer, this ground air is continuously compensated by the air source around the plateau; in addition, the air column high over the plateau becomes increasingly dense and thick, and consequently, the world's most powerful warmest high-pressure center forms over the plateau airspace at 300 hPa (it is called the "Tibetan high"). The Tibetan high strengthens and stabilizes the northern branch of the westerly.

Second, the variation in the Tibetan high position exerts different influences on the climate of its eastern areas. If the position of the Tibetan high is to the south, then northern China will experience drought, and central China will be rainy; if its position is to the west, then it will be rainy in the middle and lower reaches of the Yangtze River and in eastern Sichuan and Guizhou provinces, while less rain will occur in western Sichuan and northern China.

Another manifestation of the dynamic action of the Qinghai-Tibetan Plateau is to act as a barrier to the atmospheric circulations in East Asia. The plateau not only prevents the eastward movement of the weather system from the west but also directly blocks the exchange of cold and warm airflows in the lower troposphere between the north and south in western China. Consequently, an extremely arid desert climate occurs in northwestern China, mostly controlled by the arid cold high in winter and minimally affected by the humid summer monsoon.

2.2.2 Climate Characteristics in Northern China

The sandy area in northern China is vast and has obvious regional differences, but all the sandy areas occur based on a background of infrequent precipitation, dry air, exposed ground surfaces, large daily and annual temperature variations, minimal

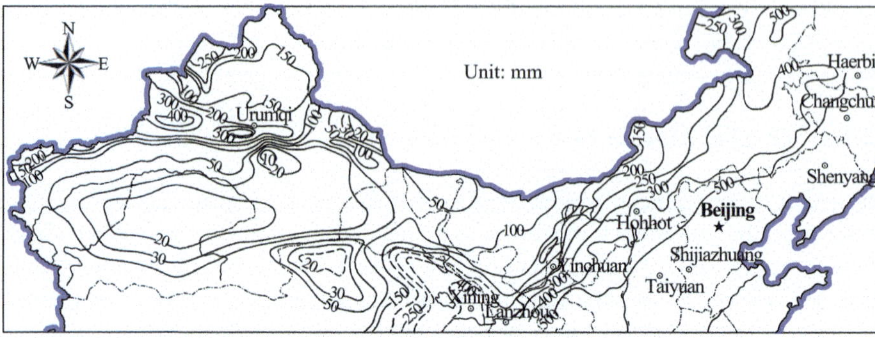

Fig. 2.14 Mean precipitation distribution map of sandy zones in China (Wang 2011)

clouds, relatively long sunshine, and strong and frequent wind flows. The climate in these areas has remarkable characteristics compared with other arid desert areas in the world as described in the following sections.

2.2.2.1 Infrequent Precipitation with Great Variability

Precipitation in the sandy area in northern China is much lower than the average precipitation at the same latitude in the Northern Hemisphere. Annual precipitation in most sandy areas in northern China is less than 400 mm, and even in alpine areas, annual precipitation rarely exceeds 600 mm (Fig. 2.14). Except for some sandy areas in the eastern semiarid region that are affected by the East Asian summer monsoon in midsummer, which results in an annual precipitation of 200–400 mm, the annual precipitation in western arid areas is less than 200 mm. In the western sandy basins intersected by high mountains and plateaus, the precipitation decreases from the periphery of the area to the center, and the hinterland is the most arid area. According to historical records, the annual multiyear mean precipitation in Ruoqiang County (86°45′–93°45′E, 36°–41°23′N) in the southeastern margin of the Tarim Basin is 17.4 mm, Tuokexun County (87°14′05″–89°11′08″E, 41°21′14″–43°18′11″N) in the western Turpan Basin has the lowest annual precipitation record in the country at 3.9 mm over 10 years, and the annual precipitation at Lenghu town (92°88′–94°30′E, 37°46′–39°18′N) in the northwestern Qaidam Basin is 17.6 mm. The triangular zone connecting Tuokexun, Ruoqiang, and Lenghu is called "the extremely arid spot in the Eurasia continent," and precipitation increases outward from this triangular zone.

In addition to infrequent precipitation, the seasonal and annual variations are also great in the sandy areas in China, and the less precipitation there is, the greater the precipitation variation rate is. In the western sandy areas, the annual mean change rate of precipitation is greater than 40%, and a record of no effective rainfall for 12 consecutive years was documented in Ruoqiang County in the southeastern margin of the Tarim Basin. The seasonal distribution of precipitation is extremely uneven; the precipitation in winter accounts for less than 10% of the annual total, and that in summer accounts for more than 70% of the total. Moreover, summer rainfall is usually concentrated over a few days, with almost no effect on plant growth. However, it is worth noting that unlike tropical desert areas, there is a small amount of snow in the desert areas of China owing to low winter temperatures. This limited precipitation can also bring great benefits to vegetation in desert areas. The snow melts in spring of the following year and moistens the sand, allowing a large number of plants to germinate, grow, blossom, bear fruit, and complete their life cycles in 20–30 days. The vegetation coverage can increase by as much as 20% or more, and vegetation plays an important role in the fixation of sand dunes.

In the eastern sandy areas, the annual mean change rate of precipitation is 25–40%. Situated at the weakened end of the East Asia monsoon, the seasonal change in precipitation in the eastern sandy areas is alternatingly controlled by the Pacific Ocean subtropical high and Siberia-Mongolia high, which are prone to drought or unexpected floods.

2.2.2.2 Large Temperature Variations

The annual mean air temperature in the sandy areas in China varies within 0–10 °C, and it decreases with increasing latitude and altitude (Fig. 2.15). The isotherms, controlled by high mountains and plateaus, appear as closed circles or semiclosed belts. In ranges from eastern Inner Mongolia through the Helan Mountains to the plains on both sides of the Tianshan Mountains, the days with air temperatures ≥0 °C and the accumulated temperature with a daily averaged value ≥0 °C vary within 180–240 days and 2200–4500 °C, respectively; the active accumulated temperature (≥10 °C) changes from approximately 2000 to 4000 °C. In the driest Turpan Basin, the ≥0 °C temperature occurs for 275 days, the ≥0 °C accumulated temperature reaches 5500 °C, and the ≥10 °C active accumulated temperature reaches 5400 °C.

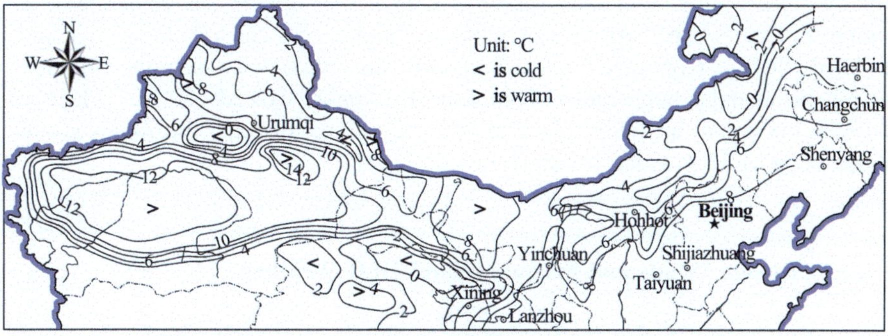

Fig. 2.15 Mean temperature distribution map of the sandy zones in China (Wang 2011)

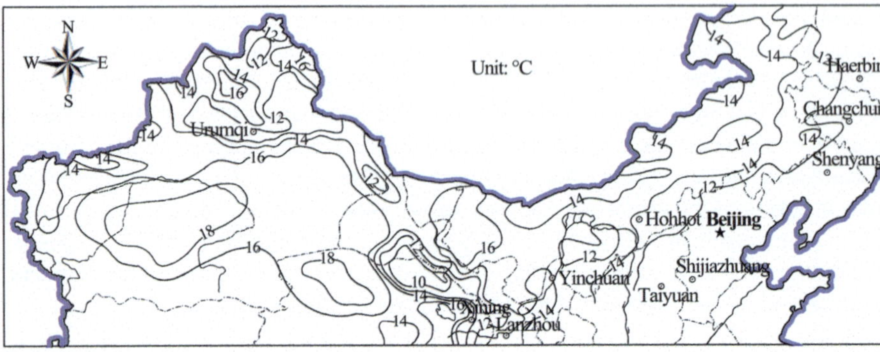

Fig. 2.16 Daily range of the mean temperature distribution map for sandy zones in China (Wang 2011)

The large seasonal variation in air temperature is a notable difference between the sandy areas and tropical/subtropical areas in China. Chinese desert regions are cold in winter and hot in summer, varying dramatically. The minimum monthly mean temperature varies from –10 to −20 °C ranging from the western Northeast Plain to the Odors Plateau and to northern Xinjiang, with the coldest temperature recorded in China at −51.5 °C in Fuyun located in the northeastern margin of the Junggar Basin. The maximum monthly mean temperature is approximately 24–26 °C in the Hexi Corridor, Junggar Basin, and Tarim Basin, with the hottest record of 33 °C in the Turpan Basin. Affected by the continental climate, the annual variation range of the mean air temperature in the sandy areas of China is generally 30–50 °C; for example, there is an annual range of 36.3 °C in the Taklimakan Desert hinterland to over 40 °C in the Badain Jaran Desert to 41.3 °C in the Turpan Basin to far exceeding 18.9 °C in the tropical Aswan Desert. Both spring and autumn in desert areas are short, with generally "only two distinct seasons." In contrast, the daily change in air temperature seems to vary throughout the four seasons. The daily temperature differences are generally in the range of 10–20 °C or more, with a maximum of 35–40 °C (Fig. 2.16). The proverb "to wear a fur coat in the morning but wear only one single thin shirt at noon and to eat watermelon around a heating stove in the evening" vividly depicts the violent daily variation in air temperature in the temperate desert regions of China. The Turpan Basin, with an extremely dry and hot climate in China, is called "the Fire Continent." In the Turpan Basin, although it is near the latitude of 43°N, the average temperature from June to August exceeds 30 °C, where an extreme maximum air temperature of 48.9 °C and an extreme maximum ground temperature of 82.3 °C have been recorded.

2.2.2.3 Strong Winds and Frequent Sand-Dust Weather

In the desert areas of China, there are many extremely aperiodic weather events, such as extreme cold events, cold fronts, cyclones, and intense showers, which are much more common than those in tropical deserts. During the extremely cold winter, the arid regions in northwestern China are near the center of the Siberian high, which is the most powerful anti-cyclone in the world; thus, chilly north winds blow strongly, and the weather is extremely arid and cold. Both in spring and autumn, cold and warm airflows vigorously alternate over desert areas, demonstrating abrupt changes between warm and cold weather, especially when there are many windstorms. In summer, the desert areas are controlled by a low pressure, with more vertical airflows and hot tornadoes, in the piedmont with a foehn. The wind speed is usually the highest in spring and early summer in the desert areas of China, as it is affected by the increased transit of cold fronts and upper troughs and frequent confrontation of cold and warm airflows.

The spatial distribution of wind speed in the sandy areas of China shows that wind speed is stronger in the west than in the east and stronger in the north than in the south, and it is influenced by the barometric distribution and atmospheric circulation; affected by terrain, higher wind speed zones are found on steppes (the eastern sandy areas), deeply weathered residual hills, and flat and the vast gobi areas, and lower wind speed zones are found on the relatively lowlands. Strong wind centers mostly appear along the northwestern borderline of the country, especially in some intermountain passes and canyons with extraordinarily strong windy areas. There are many well-known strong wind areas within the northwestern territory of China, such as those that follow:

- Known as the "No.1 gale pass," the Alataw intermountain pass is situated at the eastern end of the Junggar Basin. There is an average of 164 days in a year with wind speeds greater than 17.2 m/s (wind force: level 8), with a maximum speed of 55 m/s and an annual mean speed of 7 m/s. The nearby Ebinur Lake is also known as a "wind lake."
- Known as "continental wind stores," the Turpan Basin is located in the central part of Xinjiang (41°12′–43°40′N, 87°16′–91°55′E), where more than 95% of the year has strong winds. There is an average of 100 days in a year with wind speeds greater than 17.2 m/s, and wind speeds at 60 m/s were the record of a century.
- Known as the "world's wind stores," Guazhou County of Gansu Province is located at the western end of the Hexi Corridor (94°45–97°00′E, 39°52′–41°53′N). The northeastern wind and the northwestern wind intersect in this region, and there is an average of 80 days (the most 105 days) in a year with wind speeds greater than 17.2 m/s. This scenario is referred to as "one wind a year, from spring to winter" by the local people.
- Known as the "one hundred kilometers wind zone," the Qijiaojing section along the Lanzhou-Xinjiang Railway refers to the 123 km range from the Hongqikan Station eastward to the Thirteen Room Station. Because of the "tunnel effect" of the local terrain, the airflow accelerates through the passageway, and the wind speed increases rapidly. There is an average of 320 days in a year with wind speeds greater than 17.2 m/s, and winds above 32.6 m/s (wind force: level 12) are also common.

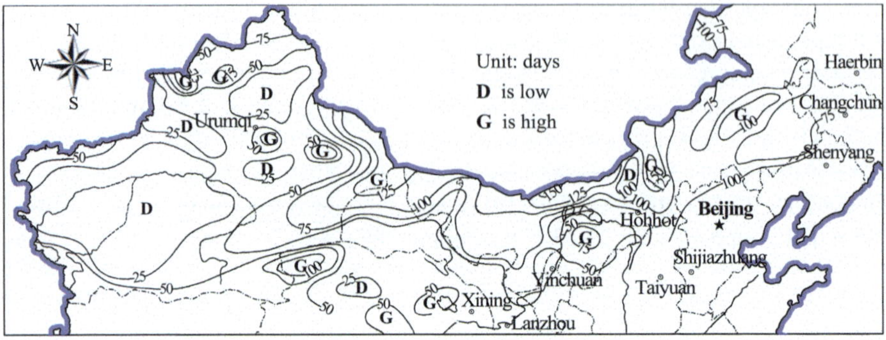

Fig. 2.17 Map of the annual frequency of sand-dust storms in the sandy zones in China (Wang 2011)

There is a large difference in the seasonal distribution of winds between the east and the west in the sandy regions. In the west, particularly in the region with wind gorges, although there are more strong winds, they are concentrated in the spring and have small change amplitudes. In the eastern sandy areas, there are strong winds in winter, spring, and autumn; the range of wind magnitudes is significantly large; and the wind is fairly weak in summer.

Sand-blown activities are closely related to both wind force and ground surface. In the western sandy areas, a large amount of dried and loose sand materials are prone to blow, thus forming sand-blown weather; in the east, sand-blown events are much less common than those in the west due to relatively vegetated surfaces.

There are generally 20–100 days with sand-blown events in the sandy areas of China (Fig. 2.17), among which as many as 35–60 days are sandstorms. The sand-blown event frequency and proportion of the three weather types, including sand-blowing, dust-floating, and sandstorm, are closely related to the local airflow field and terrain (Geng 1986). The northwestern wind system is the prevailing wind direction in sandy areas of China, demonstrating that most of the sand areas face the east-south direction. However, the multidirectional sand-blown activities in the western sandy area are apparent in comparison with the unidirectional activities in the eastern sandy area. For example, in the Tarim Basin, the airflow field is the convergence of the north branch of the westerly with that from the Mongolia high and that from the Plateau monsoon. They alternatingly play a leading role with season change. The basin is frequently shrouded in floating dust, which lasts for more than 100 days. However, the floating dust settles and is deposited locally and does not affect areas outside the basin. In the Hexi Corridor, as an important passage of the northern branch of the westerly, sand blown weather occurs year-round, sand blown events occur on 60 days in the March–June windy seasons, and the highest wind speed is often greater than 20 m/s. The western corridor is the area with the most concentrated sandstorm events in China. There are vast gobi ground surfaces, and the ground temperature rises sharply under strong sunlight, causing strong thermal convection; winds roll sand/dust, forming a strong sandstorm. East of the corridor bypasses the Badain Jaran Desert and the Tengger Desert, with over 130 days a year being cloaked in blowing sand. Generally, the duration of sandstorms is not long, but the damage is great. Strong winds deeply erode the ground surfaces, and then, the settled sands occupy farmlands, bury water conservancy facilities and transportation facilities, and cause a series of secondary disasters. The northern side of the Yinshan Mountains, located in the middle of Inner Mongolia, is also a part of the "wind corridor" of the northern branch of the westerly. Strong winds pass over the bare and loose land surfaces, resulting in sandstorms. In the last few years, the sandstorms that threaten the Beijing-Tianjin area have mainly originated from this area.

2.3 Land Cover and Land Use in Northern China

2.3.1 Basic Terrain Structure

Chinese geomorphology is divided into the north and the south by the Kunlun Mountains-Qilian Mountains-Qinling Mountains. Deserts and desertified lands in China are mainly distributed in the north. The geomorphological units in the north generally contain two great plateaus (the Inner Mongolia Plateau and Ordos Plateau), two major basins (the Junggar Basin and Tarim Basin), and their surrounding mountains. The two great plateaus and two major basins in northern China are the main areas that contain deserts, gobi, and desertified lands in China (Zhao 1990).

2.3.1.1 Two Great Plateaus

The Inner Mongolia Plateau and the Ordos Plateau, taking the Yinshan Mountains as the "backbone" boundary, are the southeastern part of the Mongolian Plateau in the middle of Asia.

The Inner Mongolian Plateau is a very broad area and is located to the west of the Greater Hinggan Mountains and to the north of the Yinshan Mountains, with an extent of more than 2000 km in the northeast-southwest direction and an elevation of 600–1500 m from the northwest to southeast. The Inner Mongolian Plateau can be further divided into the Hulun Buir Plateau (in the undulated middle part distributed with strips of sandy lands), the Xilin Gol Plateau (with the Hunshandake sandy land in the south), the Ulan Qab Plateau, and the Alxa Plateau (with the Badain Jaran Desert and the Tengger Desert at its southern rim and the Ulan Buh Desert at the eastern margin) from east to the west according to the geomorphological combination.

The Ordos Plateau, with an elevation range of 1200–1600 m, is almost a square high land. Its western, northern, and eastern areas are surrounded by the Yellow River, and the southern area is connected to the Jinshan (Shanxi-Shaanxi) Loess Plateau. The land surface is undulating and covered by aeolian sand, loess, and lacustrine/alluvial gravel, with the most extensive aeolian landforms. The Hobq Desert is in the north, and the Mu Us sandy land is in the south.

2.3.1.2 Two Major Basins

The western part of the northwestern arid area in China is occupied by two large inland basins, namely, the Tarim Basin in southern Xinjiang and the Junggar Basin in the north, and they are separated by the Tianshan Mountains.

The Tarim Basin is the largest inland basin in China. It is located between the Tianshan Mountains and the Kunlun Mountains, with a greatest width of 520 km from north to south and a maximum length of 1400 km from east to west. The basin's topography is high in the west and low in the east, inclined slightly northward. The Tarim River is located in the north of the basin, and the water flows to the east. The vast territory to the south from the Tarim River is the famous Taklimakan Desert. With an area of approximately 330,000 km^2, it is the largest desert in China and the second largest mobile desert in the world. The desert covers the alluvial/pluvial plains at the foot of the Kunlun Mountains and the ancient alluvial plain of the Tarim River. The area of the deserts to the north of the Tarim River is limited, and these deserts are distributed discontinuously in the front of the alluvial fans of the Tianshan Mountains. The natural landscape of the Tarim Basin has circularly distributed characteristics: its margin is gravel gobi connected to mountains, the center is endless desert, and the transition part contains alluvial fans and alluvial plains where oases are distributed.

The Junggar Basin is located between the Altay Mountains and the Tianshan Mountains, the Junggar western mountainous area is to the west with a series of low mountains at 2000–3000 m above sea level, and the eastward extension of the Altay Mountains is to the east. The basin is inclined westward, slightly higher in the north than in the south. Ebinur Lake in the southwest is the lowest point of the Junggar Basin. The basin is approximately 450 m from north to south and approximately 700 m from east to west, with an area of 180,000 km^2, and 30% of the area is occupied by desert. There are several low altitude gaps on the west side of the basin, such as the Ertix River valley, Emin River valley, and the Alataw intermountain pass, and through the gaps, the moist westerly airflow brings precipitation to the basin and its surrounding mountains. The landforms of the Junggar Basin can be generally divided into three plain areas: (a) the plain at the southern piedmont of the Altay Mountains extends south to the edge of the Gurbantunggut Desert. The main characteristics of this plain area are that there are thin Cenozoic strata and large areas of wind eroded lowlands with thin soil layers. (b) The plain at northern piedmont of the Tianshan Mountains extends north to the southern margin of the Gurbantunggut Desert. The soil layer gradually thickens from the south to the north, the gravel gobi in the south is mainly used as the spring and autumn grasslands, and the fine soil plain in the north is an important irrigated agricultural area. (c) The Gurbantunggut Desert in the middle of the Junggar Basin is the third largest desert in China, dominated by fixed and semifixed dunes, with mobile dunes accounting for approximately 3% of the area. The annual precipitation in the desert areas is 100–150 mm, and there is steady snow cover in winter. The vegetation coverage is much higher than that on the desert in southern Xinjiang. It is approximately 20% on semifixed dunes and 40–50% on fixed dunes, and pasture plants grow well in the interdune depressions.

2.3.2 Hydrology and Water Resource Features

Water resources are basic natural resources, strategic economic resources, and the controlling elements of ecology and the environment. The hydrological cycle and water resources in the arid areas of China are generally characterized by the following factors:

1. Each basin has its own runoff forming area, its own water system, and its own tail lake or direct ending into the desert.
2. The water resource formation area is separated from the consumption and utilization area, with rivers as a link to connect the mountain area with the basin plain; the water cycle is considered the main body, and it is closely related to biology and ecosystems, forming an independent and complete inland local water circulation system (Chen 2010).
3. Water resources are sensitive to global warming, and their temporal and spatial distributions are obviously uneven. Both surface water and groundwater originate from mountainous areas and convert frequently to the other. Water resources coexist in many forms, such as glaciers, snow cover, groundwater, lakes, and surface runoff.

2.3.2.1 Runoff Depth Distribution in Northern China

Northern China has obvious regional and vertical variations (Fig. 2.18). There are four zones: (a) the zone with no runoff due to infrequent precipitation and strong evaporation and infiltration, mainly including the Taklimakan Desert, Gurbantunggut Desert, Badain Jaran Desert, Tengger Desert, Qaidam Basin, Turpan Basin, Hexi Corridor, and the border of Gansu and Xinjiang; (b) the dry zone with an annual runoff depth of less than 10 mm, covering most areas of the piedmont plains; (c) the low runoff zone with an annual runoff depth of 10–50 mm, located in the middle of the Song-Liao Plain, the upper reaches of the Liaohe River, the south edge of Inner Mongolian Plateau, most parts of the Loess Plateau, and the low mountains and hills at the northern and western parts of the Qinghai-Tibetan Plateau; and (d) the transition zone with an annual runoff depth in the range of 50 to 200 mm, including parts of the Songnen Plain, Sanjiang Plain, the North China Plain, the plain at the lower reaches of the Liaohe River, most parts of Shaanxi and Shanxi provinces, the

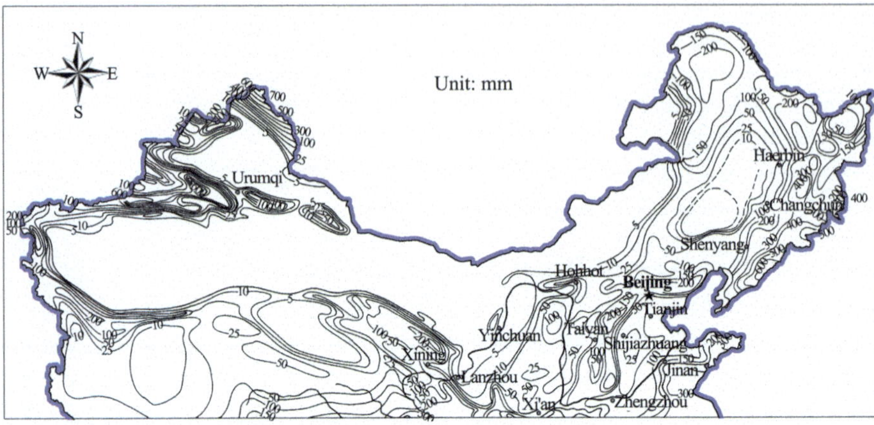

Fig. 2.18 The annual runoff depth in northern China (Wang 2011)

middle of the Qinghai-Tibetan Plateau, and the mountainous areas of the Qilian Mountains, Tianshan Mountains, and the western part of Xinjiang.

2.3.2.2 Rivers in Northern China

Rivers in Arid Regions

Except for the Ertix River that flows to the Arctic Ocean, the rivers in the arid region of northwest China have developed in closed basins, and most rivers will eventually become tail lakes or disappear in the desert naturally or under the influence of human activities. In the mountainous area, the river network density is generally as high as 0.5 km/km^2; in the plain area, its density is less than 0.1 km/km^2. The rivers in the northern inland arid areas have the following unique hydrological characteristics:

(a) The small and medium rivers are numerous, but the total water volume is small; there are fewer large rivers, but they occupy the majority of the water resources (Table 2.1). There are 676 inland rivers, of which only 20 have annual runoff of $\geq 10 \times 10^8$ m^3, accounting for nearly 3% of the total, but the water yield accounts for 62.2% of the total runoff.
(b) The inland rivers are mainly recharged with ice and snow melt or have certain ice and snow meltwater supply components. In the runoff formation area of the inland river basin in China, there is a total glacier area of 35,390 km^2, which is mostly concentrated in the Kunlun (20.6%), Tianshan (15.5%), and Karakorum (10.5%) Mountains and on the Pamir Plateau (4.5%) and is less concentrated in the Qilian (3.25%) and Altay Mountains (0.47%).
(c) River runoff is concentrated in the warm season and has minimal interannual variation due to the regulation of the alpine glacier "white reservoir" and the subalpine forest shrubs "green reservoir." The large and medium rivers, mainly

Table 2.1 Statistics of inland rivers and runoff volume in the arid areas of northwestern China

Region	Annual runoff $\geq 10 \times 10^8$ m^3			Annual runoff $\geq 5 \times 10^8$ m^3			Annual runoff $\geq 1 \times 10^8$ m^3			Total		
	River number	Runoff ($\times 10^8$ m^3)	Percentage (%)	River number	Runoff ($\times 10^8$ m^3)	Percentage (%)	River number	Runoff ($\times 10^8$ m^3)	Percentage (%)	River number	Runoff ($\times 10^8$ m^3)	Percentage (%)
Northern Xinjiang	9	267.58	67.1	13	295.21	74.0	45	363.11	91.0	387	398.81	100
Southern Xinjiang	9	263.97	68.7	15	303.89	79.2	39	353.49	92.2	183	383.53	100
Qaidam	1	10.68	23.3	2	18.05	39.4	10	34.96	76.3	51	45.80	100
Hexi Corridor	1	15.97	23.3	3	33.23	47.5	15	59.90	85.6	55	69.96	100
Total	20	558.20	62.2	33	650.47	72.4	109	811.46	90.4	676	898.11	100

Note: This table is based on Tang et al. (1992)

recharged with ice and snow melt, have a late flood season and a concentration of summer water, and the discharge accounts for 70% of the whole year from June to September. The mixed recharge rivers with rain and snow melt are characterized by the summer flood just coming after the spring flood, and the maximum runoff occurs in May–August or June–September, accounting for 60–80% of the runoff of the whole year.

(d) Each inland river basin, with the surface and groundwater linked, forms an independent and unified inland ecosystem in which there is a linked and restricted relationship between the upstream and downstream.

Rivers in Semiarid Regions

The Inner Mongolia Plateau is 30.0×10^4 m^2, with the smallest river network density in China. The rivers there are relatively short. The rivers originate from the surrounding mountains or hills; when entering the plain area, the rivers gradually become winding rivers without obvious channels. Last, the rivers either disappear into grasslands and deserts or overflow the river channels during flood periods, forming many terminal small lakes and wetlands. The rivers often dry up in the drought season (mainly in winter and spring). Rainwater is the main supply of river runoff, which generally accounts for more than 60% of the annual runoff.

2.3.2.3 Lakes in Northern China

In the inland arid area of northern China (area of annual precipitation $\leq$200 mm), there are more than 400 lakes with an area of more than 1 km^2, and their total area is approximately 1.70×10^4 km^2. Among them, there are more than 80 lakes with freshwater or brackish water with salinities of 1–5 g/L, which are estimated to have a capacity of 300×10^8 m^3 for water storage. These lakes are situated in the upper reaches of rivers and cover an area of 0.35×10^4 km^2, which is less than 10% of the total area of freshwater lakes ($\geq$1 km^2) in China. There are two larger freshwater lakes ($\geq$1000 km^2): Bosten Lake in southern Xinjiang and Hulun Lake in Inner Mongolia. Bosten Lake, Kaidu River being its main supply, has a salinity level of approximately 1.3 g/L. The water quality of Hulun Lake, mainly recharged by surface runoff and groundwater recharge in addition to atmospheric precipitation, is dependent on an increase (appears to be a freshwater lake) or decrease (appears to be a brackish water lake) in the lake water volume.

Most lakes in arid northwestern China have developed as salty or saline lakes due to minimal surface runoff replenishment and strong evaporation. The lakes developed in the sandy desert regions are small and shallow, so lake water easily evaporates and with accumulated salt remaining in it. These small lakes restore some of the water area in the rainy season but dry again in the drought season. Some lakes become underground lakes due to encroachment of blown sand. The available lake water resources are extremely poor, which makes it difficult to meet the needs of industrial and agricultural production and people's lives.

2.3.2.4 Groundwater in Northern China

The abundant loose sediments in Cenozoic tectonic basins are not only the material base of desert formation but also good storage sites for interstitial water. Therefore, the groundwater in arid basins of northwestern China is rich. In a natural state, before flowing into the plain region, most small rivers are transformed into seasonal flood channels through evaporative loss and infiltration into underground runoff, with the exception of some large rivers that still have a certain runoff flow into the plain directly; some runoff from medium rivers infiltrates the alluvial fan in front of a mountain to form underflow, and the underflow is exposed again in the form of spring water at the leading edge of the alluvial fan and converges into river runoff. After entering a plain area, surface water recharges groundwater and replenishes soil moisture to supply plants and crop growth through natural infiltration or irrigation. Groundwater recharge in plain areas mainly depends on leakage from natural/artificial rivers and farmland irrigation. However, in the basin center, groundwater is often mineralized because of strong evaporation and inappropriate development and utilization. A large amount of river water being blocked or diverted increases evaporation consumption, decreases the downstream runoff component, and changes the water resource distribution pattern and the water cycle degree; the overexploitation of groundwater changes its transportation route and circulation rate, affecting the participation of groundwater in ecological processes and the dynamic equilibrium characteristics of groundwater (Hu et al. 2002). There are unique characteristics of the groundwater regimes in different groundwater basins in northern China (Wang 2011).

In the Tarim Basin, there are approximately 149 rivers that are fed by precipitation and snow melt. Sloping gravel plains, fine earthy plains, sandy desert plains, and salt desert plains exist from the mountain foot to the center of the basin. The thicknesses of the sand/gravel layer in the piedmont, the clay/sand layer on the clay plain, and the sandy layer on the sandy desert plain are approximately 50–300 m, 800–1000 m, and 300–500 m, respectively. The Tarim Basin has a groundwater resource of 292.7×10^8 m^3, of which 203.22×10^8 m^3 is in the piedmont. The fresh groundwater amount is 121.6×10^8 m^3 (including the part that also serves as surface water). In the Taklimakan Desert and its marginal regions, groundwater resources have been widely developed in recent years and are mainly used as injection water for oil field exploitation and greening water for the establishment of wind-sand prevention systems. As a result of overexploitation, the groundwater level continuously drops, and the groundwater mineralization degree has increased.

In the Junggar Basin, the sediment is 2000–6000 m thick in the Ebinur Lake region, 800–3000 m thick near Karamay, and 2000–4000 m thick in the Karameili region, providing an ideal storage site for interstitial water. There is shallow groundwater and deep confined water in the middle and lower parts of the alluvial fans in front of the Tianshan Mountains, where the aquifer consists of pebbles, gravel, and sand, and the water table is deeper than 50 m. In the western part of the Shihezi region, there are six layers of confined water within a depth of 200 m. In the central part of the basin, there is shallow phreatic water with a water table ranging from 1–4 m in most of the areas to 5–10 m in the northern part of the desert. Under the plain in the northern piedmont of the Tianshan Mountains, there is an 80×10^8 m^3 groundwater resource, of which approximately 80% is cycled as surface water. At present, the main mechanisms of groundwater consumption in most regions are still overflow of a spring, water evaporation, and vegetation transpiration. However, with the rapid increase in agricultural and industrial water consumption on the plains, the groundwater level in some places has largely decreased, and the ecological environment has deteriorated, especially in the lower reaches of the Urumqi River.

The Qaidam Basin is widely covered by loose Quaternary sediments. There are four types of groundwater, namely, mountain fissure water, phreatic water in alluvial/diluvial sand-gravel layers, phreatic water in diluvial lacustrine layers, and confined water under lacustrine plains. The groundwater resource is approximately 40.7×10^8 m^3, and its development degree is not high at present.

In the Hexi Corridor, there are four large basins: the Yumen, Jiuquan, Zhangye, and Wuwei basins, where there are approximately 56 rivers, with the main rivers including the Shiyang River, the Heihe River, and the Shule River. The Hexi Corridor has deep and loose Cenozoic sediments. The aquifer is mainly composed of sand and gravel, and the corridor plain is the most important groundwater aquifer. Groundwater is mainly recharged by the infiltration of the river courses, canal systems, and farmlands in alluvial fan zones and the direct replenishment of lateral runoff in mountainous areas. The Hexi Corridor has groundwater resources of 44.7×10^8 m^3, of which infiltration from river courses, canal systems, and farmlands accounts for 89%, and only 5×10^8 m^3 is an unduplicated natural supply of water resources. Overexploitation of groundwater has resulted in a sharp decrease in the water table in the Hexi Corridor, especially at the lower reaches of the Shiyang River, where the exploitation rate of groundwater has reached 70%.

On the western Inner Mongolian Plateau (the Alxa Plateau), there is a substantial amount of phreatic water in the deep and loose Quaternary sediments on the plain, confined water in the Tertiary and Cretaceous layers, and fissure water in the bedrock. The groundwater on the plain region mainly originates from the seepage of the Ejin River (also the lower reaches of the Heihe River). In the Badain Jaran Desert and the Tengger Deserts, no runoff forms, but rainstorm infiltration and a small amount of condensation water in deserts are also favorable to the replenishment of groundwater. On the eastern Inner Mongolian Plateau, with an annual precipitation of 200–350 mm, there are some short inland rivers and several inland lakes. The total annual runoff is approximately 10×10^8 m^3. Bedrock fissure water exists in lower hills, and phreatic and confined water exists in valleys between hills. Natural groundwater resources are 27.7×10^8 m^3 and are mainly distributed in basins and ancient river courses.

On the Ordos Plateau, the Hobq Desert is to the north, the Mu Us sandy land is to the south, hilly gullies are to the east, and the middle and western areas contain mound and erosion depressions. There are extensive no-flow areas, few rivers, and many saline lakes. The groundwater level in depressions between sand dunes is below 2–3 m. The Hetao Plain is the largest Yellow River diversion irrigation area in China. The irrigation area is 50×10^4 hm^2, and the annual channel water from the Yellow River is 56×10^8 m^3. The groundwater is mainly recharged by the infiltration of farmlands. However, the topography is not conducive to the discharge of groundwater, land salinization is severe, and at least 50% of irrigated farmlands have suffered from salinization. The annual replenishment rate of groundwater is approximately 41.42×10^8 m^3, and extractable resources are 26.8×10^8 m^3.

On the western Liaohe River Plain, where the Horqin sandy land occurs, there is a total annual runoff of 22×10^8 m^3. The groundwater, recharged by the western Liao River, is plentiful and fresh. Due to vegetation degradation and serious water erosion, the upper reaches of the western Liaohe River are the main sediment-producing region; the Horqin sandy land is covered by a thick layer of Quaternary sand deposits. The aquifer thickness increases from approximately 5 m at the northern margin to nearly 150 m at the southern edge of the sandy land.

2.3.3 Vegetation Components and Their Distribution

Desert and aeolian desertification areas in northern China cover arid, semiarid, subhumid, and humid climatic zones, which also include various natural zones such as desert, grassland, and forest. Therefore, there is complex and diversified vegetation in these areas.

The plants in sandy desert regions of China belong to the Holarctic flora kingdom. The most important subregion is the Asian desert plant subregion, followed by the Eurasian grassland plant subregion (Wu and Wu 1999). The main flora are (a) components of Central Asia, including xerophytes, strong xerophytes, and hyperxerophytes. Most edificators in Chinese desert regions, especially in its middle and western parts, belong to central Asian components. (b) Components of the Tethys-sea region are the species of Tethys-sea flora that have an important position in the vegetation of the middle and western desert regions of China. They are either edificators or dominant species in some deserts, desert steppes, and sandy lands. (c) Components of Black Sea-Kazakhstan-Mongolia and Dawuli-Mongolia belong to the components of Eurasian steppes, mainly distributed over the Hulun Buir Desert, the Horqin Desert, the Hunshandake Desert, the Mu Us sandy land, and their surrounding areas. They include xerophytes and mid-xerophytes, dry mesophytes, and halophytes. (d) Components of East Asia are the plants in semiarid, subhumid, and humid regions of China and are primary and secondary forests and shrubs, and they belong to the main components of the East Asia flora, with mesophytes as principal features. (e) Endemic components are endemic plant species that are low in number; however, most components are the dominant species, and some endemic species are ancient relict species.

Sandy desert areas of China exhibit both zonal and azonal distribution features. First, they show an obvious horizontally distributed pattern from east to west: the temperate deciduous and broad-leaf forest (in the Huang-Huai-Hai sandy lands), the middle-temperate forest grassland (in the Song-Nen sandy lands), the typical middle-temperate steppe (in the Hulun Buir, Horqin, Hunshandake sandy lands, and their surrounding areas), the middle-temperate desert steppe (in the Ulanqab-Wulate grasslands), the typical warm-temperate steppe (in the Hobq Desert and the Mu Us sandy land), the warm-temperate desert steppe (in western Hobq Desert and the Hedong sandy lands in Ningxia), the temperate desert (west to the line from Helan Mountains-Wushao Mountains-Qilian Mountains), the warm-temperate desert (in the Kumtag Desert, the Taklimakan Desert, and the surrounding areas), and the high-cold vegetation (in deserts and sandy lands of Qinghai-Tibetan Plateau). Second, these areas also show a distinct, vertically distributed pattern from low to high altitudes: the mountainous steppe above the basal vegetation zone, the xeromorphic grass-steppe and deciduous broad-leaf bushes on sunny slopes, the forest vegetation on shady slopes in middle-high mountainous regions, and the typical meadow in high-mountainous regions. The vertical zonality of vegetation is still influenced by local climate in addition to mountain altitude because the mountains are situated in different natural zones.

The azonal distribution of vegetation has two characteristics. One is that vegetation is distributed under different water conditions; some hygrophytes, hydrophytes, and desert riparian vegetation are distributed along banks and beaches of inland rivers and lakes. The other characteristic is that they are distributed in different soil types; the physical and chemical properties of soils in sandy lands are greatly different from those in zonal soils. Therefore, in comparison to zonal vegetation, azonal vegetation has the following specific features: (a) the components and structure of plants are simple and easily influenced by seasonal precipitation and easily renewed after implementing artificial control measures; (b) in comparison to surrounding zonal vegetation, vegetation in the arid and semiarid sandy lands tends to become mesophytes, and the life forms of the edificators are higher than those in the surrounding zonal vegetation; (c) the vegetation in sandy lands in subhumid to humid regions tends to become xerophytic compared with surrounding zonal vegetation; and (d) plants only inhabit the interdune depressions where groundwater rises to near the surface, whereas the surrounding desert or gobi regions are bare.

2.3.4 Soil Characteristics in Northern China

According to the *Soil Classification System of China*, Sandic Entisols are the main soil types in desert and aeolian desertified regions of China. Sandic Entisols refer to soils that possess sandy sediment characteristics. The Sandic Entisols are formed under a background of arid climate and aeolian sands as their parent materials. In the

Table 2.2 References between system classification and genetic classification of some major soils in desert and aeolian desertification regions of China

Soil genetic classification (1978–1992)	Soil system classification (2000)
Aeolian sandy soil	Freezing/dump/arid/semiarid/wet Sandic Entisols
Oasis soil, anthropogenic alluvial soil, anthropogenic desert soil	Weak salt/mellow/water tillage/speckle/general anthropogenic alluvial/arid tillage/artificial soil
Chestnut soil	Calcic/gley/semiarid/embryonic soil
Brown calcic soil	Calcic/gley normal arid soil
Gray brown soil	Calcic/gley normal arid soil
Gray desert soil	Calcic/saline/sliming/gley normal arid soil
Gray brown desert soil	Gypseous/gley normal arid soil
Brown desert soil	Gypseous/saline normal arid soil
Takyr soil	Gley normal arid soil
Saline soil	Arid/dump normal saline-sodic soil
Marshy soil	Organic/black/perch gley soil
Meadow soil	Gley/black/tint/speckle/dump embryonic soil

previous soil classification, these soils correspond to aeolian sandy soils (Xiong and Li 1987; Chen et al. 1998; Chen and Li 1999). In the present research, shifting sands are also Sandic Entisols because they have vital movements (including microbes) and soil fertility in shifting sands. To avoid misunderstandings from different soil classification systems, Table 2.2 shows some different information about soil system classification and soil genetic classification.

Based on water content or temperature, Sandic Entisols in northern China can be divided into four types as follows:

(a) Cold Sandic Entisols. The soil temperature is low or even below 0 °C. They are mainly distributed over the Qinghai-Tibetan Plateau, especially over those regions where the elevation is over 4000 m a.s.l. and the annual mean temperature below 0 °C. Cold Sandic Entisols mainly occur in the high terraces of lakes, where the parent material comes from aeolian sand and the silt and clay content is usually less than 100 g/kg. In these regions, the zonal vegetation is mainly high-cold steppe and high-cold desert steppe.
(b) Arid Sandic Entisols. The water content of the soil is low. This soil is mainly distributed over the northwestern Ordos Plateau, in the middle and eastern part of the Hexi Corridor, and in the Junggar and Tarim basins, where the annual precipitation is in the range of 50–200 mm, the annual mean temperature is 3.5–10.0 °C, and the dryness is approximately 4–8.
(c) Semiarid Sandic Entisols. The water content of the soil is moderate. These soils are widely distributed in the Hulun Buir Desert, the river paleochannels of the Huang-Huai-Hai Plain.
(d) Humid Sandic Entisols. The water content is relatively high, and at least one layer ($\geq$10 cm) within a 50 cm depth has the characteristics of oxidation and

reduction. They are mainly distributed in the interdune depressions where there are high groundwater levels.

In addition to the principal Sandic Entisols, there are some other common soil types in northern China. They are distributed with the Sandic Entisols, and most of these soils appear on the rims of deserts and mainly include (a) irrigation-warping aridic soil, which is widely distributed in the arid and semiarid irrigated farming regions, with an obvious irrigation-warping diagnostic horizon of $\geq$50 cm and a thickness as high as 2 m as a result of long-term cultivation with irrigation; (b) common aridic soil, which represents the soil occurring in aridic regimes, with an aridic epipedon but not a cold soil temperature condition; its counterpart in soil genetic classification includes sierozem, desert gray soil, gray-brown desert soil, brown desert soil, and takyr; (c) halogenic soil, which is also called salt-affected soil or saline-sodic soil in the soil genetic classification system, where the soil profile generally contains saline or alkaline horizons; it occurs at low-flat basins, semienclosed depressions, river deltas, and dry deltas in extensive arid, semiarid, and desert regions of China; (d) gleisoil, which is a marshy soil in the soil genetic classification system; the soil profile contains at least one $\geq$10 cm gleization-featured layer, and it often occurs in various lowlands, such as a depression at a rim of an alluvial fan, lake depression, river confluence, and paleochannel; (e) udorthent, which a relatively newly named soil type that corresponds to the chestnut soil in the soil genetic classification system; this soil mainly occurs in semiarid hills and high plains with the annual precipitation at approximately 250–400 mm and the steppe vegetation; and (f) aquic embryonic soil, which corresponds to the meadow soil in the soil genetic classification system. The soil profile has at least one oxidation-reduction layer, and this type of soil is mainly distributed in the depressions and littoral lands of rivers and lakes.

2.4 Land Cover and Land Use in Mongolia

Stretching 1259 km from north to south and 2392 km from east to west, Mongolia's landforms and natural landscapes distinctly change with latitude, longitude, and altitude due to the surrounding mountain ranges (Dorjgotov 2009).

Based on its terrain composition, Mongolia is a mountainous country. Its terrain consists of mountains, hills, and elevated, denuded plains that form the three main, historically formed regional tiers of its surface. Mountains (1500–3000 m) account for more than 40% of its total area, hills and hummocks (1000–1500 m) account for 40% of its total area, and denuded plains account for approximately 15% of its total area. If we exclude the high mountains (3000 m), which account for only 3.6% of the total area, then the vast majority of the country is located within altitudes ranging from 1000 to 3000 m; i.e., in Mongolia, the prevailing terrain type is mid-mountain and low-mountain (hilly). There are few accumulative plains in Mongolia and no lowlands. The average height of its surface above sea level is 1580 m. Within

Table 2.3 Natural zones and belts of Mongolia

Name of the zone or belt	Area (km^2)	Percentage
High mountain belt	56394.0	3.6
Mountain taiga belt	70492.5	4.5
Forest-steppe zone	238108.0	15.2
Steppe zone	535743.0	34.2
Gobi zone	366561.0	23.4
Desert zone	299201.5	19.1
Total	1566500.0	100.0

Mongolia, mountains occur mainly in the northern and western parts of its territory, while denuded plains occur the southeastern part; hills and hummocks are equally present throughout the territory but are primarily developed to the south from the watershed of the desert areas. Generally, it can be assumed that the northwestern part of Mongolia is mountainous and the southeastern part is a complex of hills, hummocks, and flat plains. The latter area is mainly formed due to its lithology, which in many ways predetermined the morphological differentiation of these plains and hills. The main orographic elements of Mongolia are the mountain system of the Mongolian Gobi Altai, the Khangai and Khentei highlands, the Khubsugul riftogenic highlands, the Orkhon-Selenge erosional mid-mountains, the basin of large lakes, Valley of Lakes, Dzungarian Gobi (Barun-Khurai depression), Trans-Altai Gobi, the Central Gobi peneplain, and the southeaster and eastern Gobi plain. These structures, given their close correspondence to geological structure, are also the largest morphologic and structural elements of the terrain (Dash and Mandakh 2018).

The geography of this country, e.g., location, terrain, and distribution of climatic indices, predefined the existence of both latitudinal zones and altitudinal belts in the territory of Mongolia. According to Dash (2013), seven natural zones and belts can be distinguished as shown in Table 2.3 and Fig. 2.19.

2.4.1 *Land Cover and Its Change*

To monitor land cover and its change to evaluate both the cause and impact of global change on ecosystems, Mongolia has developed its own land cover system in 1995 using available remote sensing techniques and data. Land cover mapping practices in Mongolia are based on annual NDVI time series phenology characteristics. Based on the classification of isoclusters, the national land cover classification was divided into 19 classes, as shown in Table 2.4 and Fig. 2.20 (source: National Remote Sensing Center).

According to the land cover change analysis using a confusion matrix, 27758.97 km^2 of the forest, 33651.83 km^2 of the grassland, 3163.84 km^2 of the cropland, and 4183.64 km^2 of the wetland are considered degraded or affected by desertification (Table 2.5).

Fig. 2.19 Natural zones of Mongolia

Table 2.4 Land cover/land use classification system

Level 1	Level 2/ESA CCL	National LC/LU classes
Forestland	Forest tree cover	Evergreen forest Mixed forest
Grassland	Pasture and natural grassland Shrubland, bushland, and heathland Sparsely vegetated areas Natural vegetation associations and mosaics	High mountain steppe Steppe Shrubland Dry steppe
Cropland	Medium to large fields of rain-fed herbaceous cropland Medium to large fields of irrigated herbaceous cropland Permanent crops, agricultural plantations Agricultural associations and mosaics	Cropland fallow Permanent croplands
Wetlands	Open wetlands	Riparian meadow
Settlements	Urban and associated developed areas	Urban
Other land	Barren land Permanent snow and glaciers	Glacier High mountain taiga Desert steppe Barren land Semidesert Desert Sand Abandoned cropland
Water bodies	Inland water bodies, coastal water bodies, seas	Lakes and rivers

The majority of the desertification is associated with deforestation and vegetation loss. Relatively high degradation rates occurred in wetlands. Urbanization caused degradation to occur in an area of 330.24 km^2.

2.4.2 *Land Use and Land Use Change in Mongolia*

Almost 90% of the total territory of Mongolia is located in arid, semiarid, dry, and subhumid climatic regions, which are areas prone to aeolian desertification. Geographically, the country is located in a transition area between Siberian taiga forests and Central Asian deserts, and this area defines the sensitivity of the country to both climate change and changes related to social development strategies, land use, and natural resource use (Batjargal 1997). The warming process occurring in Mongolia, the drastic increase in livestock, unsustainable mining industry development, and other direct and indirect impacts are the critical factors affecting desertification and exacerbating land degradation. According to a recent nationwide assessment, 76.8% of the land area has been degraded, and of these lands, severely degraded lands account for 22.9%. The total desertified area accounts for 64.7% of the country (Fig. 2.21).

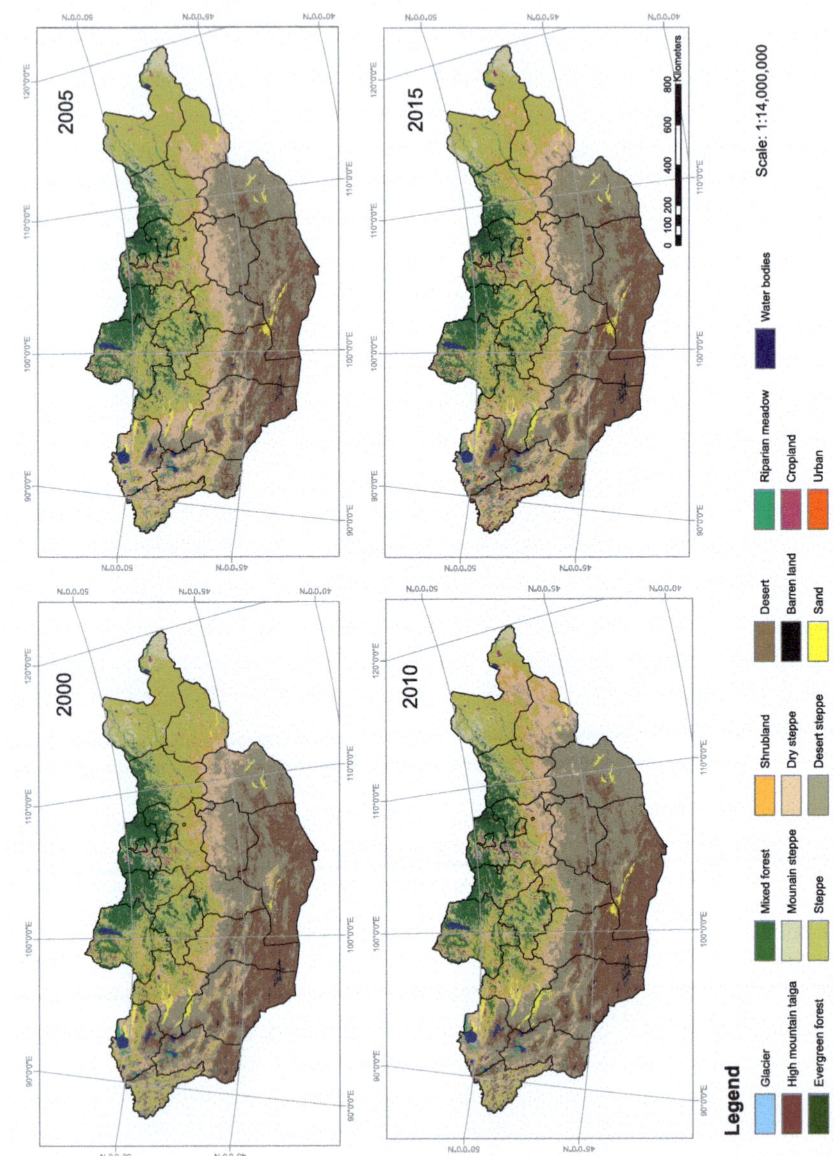

Fig. 2.20 Land cover dynamics of Mongolia

Table 2.5 Land area by type of land cover change (km^2)

Land cover in 2015									
Land cover in 2000		Tree-covered areas	Grasslands	Croplands	Wetlands	Artificial areas	Other lands	Water bodies	Total
	Tree-covered areas	105,989.44	24,796.86	389.18	2542.91	1.80	28.22	26.06	133,774.47
	Grasslands	26,759.27	670,583.33	1260.51	12,086.64	287.31	21,277.88	923.11	733,178.05
	Croplands	35.07	1340.00	7304.09	30.57	0.13	1793.15	3.04	10,506.04
	Wetlands	219.89	3835.71	102.35	9059.42	0.37	25.32	80.99	13,324.05
	Artificial areas	0.00	9.26	0.00	1.71	141.30	9.95	0.00	162.23
	Other lands	4.19	118,087.10	799.32	425.23	40.63	540,374.16	1081.91	660,812.54
	Water bodies	2.33	196.23	0.00	30.21	0.00	150.08	12,524.40	12,903.25
	Total	133,010.19	818,848.48	9855.46	24,176.70	471.54	563,658.77	14,639.50	1,564,660.64

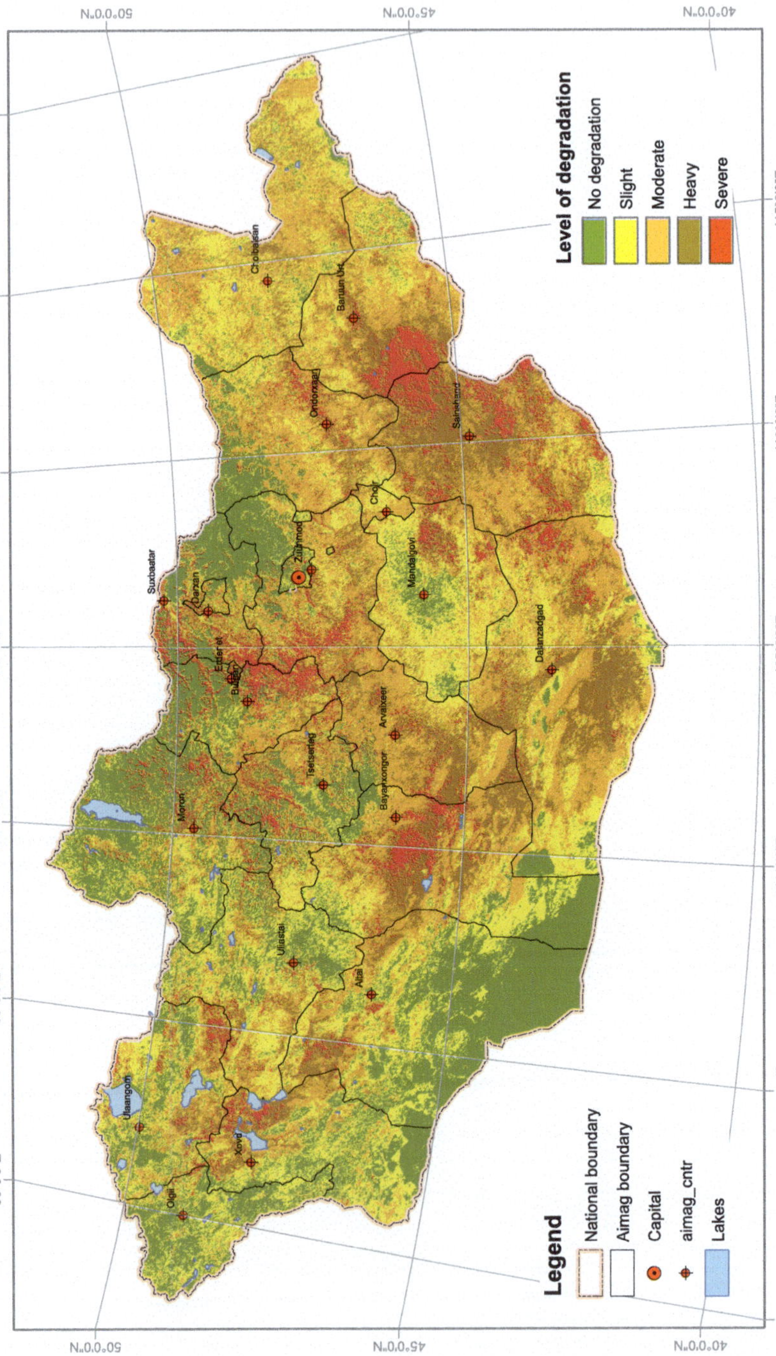

Fig. 2.21 Land degradation and desertification status in 2015 (Mandakh 2017)

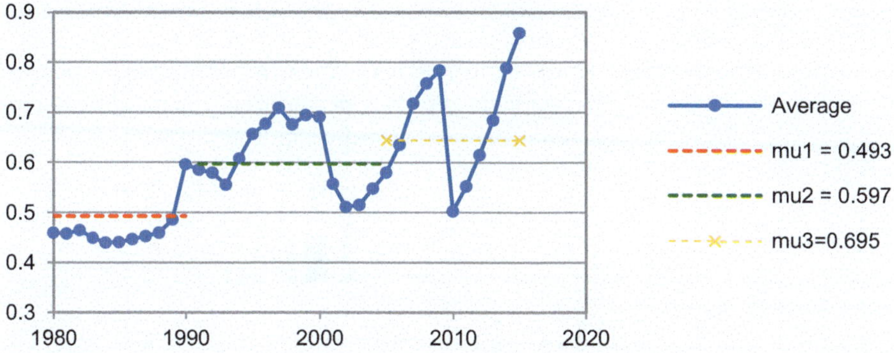

Fig. 2.22 Change in sheep forage units per ha (Mandakh 2017)

According to the national land use classification system of Mongolia, approximately 70% of the total land area is used for grazing, 0.2% is used for extractive industries, and 0.3% is used for roads and transportation. The main land use types that have an adverse effect on land health are uncontrolled grazing and mining activities.

Mongolia has a long tradition of raising livestock. Pastoral nomadism is the prevailing form of land use. Currently, approximately one-third of Mongolia's population lives as nomads and carries out livestock husbandry (Dagvadorj et al. 2009). Currently, there are approximately 52 million heads of livestock, with approximately 23.3 million sheep and 22.0 million goats in addition to several other kinds of livestock (National Statistical Office (NSO) 2015). Total number of livestock in Mongolia is estimated on the basis of sheep forage units (National Statistical Office (NSO) 2013). Studies on changes in the livestock number per unit of grazing area show that the number of livestock per 100 ha pasture was 40 sheep forage units in 1980–2000, and then, this value increased to 60–70 sheep forage units by 2000–2015 (Fig. 2.22).

According to Tserendash (2006), the total carrying capacity of grassland pastures is 80–90 million sheep head units or 50–60 sheep units per unit ha of pasture. From 1980 to 1990, the number of livestock was consistent with the given capacity. However, livestock heads have steadily exceeded pasture capacity, even after several droughts and zud (Mongolian for "severe winter conditions") that have been observed since 1991.

Regarding the spatial distribution of livestock, a sharp increase in livestock has been observed in the eastern and central regions. The number of livestock per unit area of pasture increased by 50–70 sheep forage units per year in Darkhan, Orkhon, Tov, and Arkhangai provinces and around Ulaanbaatar city over the last 15 years. In other regions, such an increase is also pronounced, and the increase has been estimated to be approximately 10–30 sheep forage units per year (Fig. 2.23).

In terms of spatial distribution, approximately 32% of the total livestock of Mongolia graze on the forest steppe and steppe, 29% graze on the mountain steppe, 15% graze on the eastern steppe, 14% graze in the semidesert, and 9% graze on the

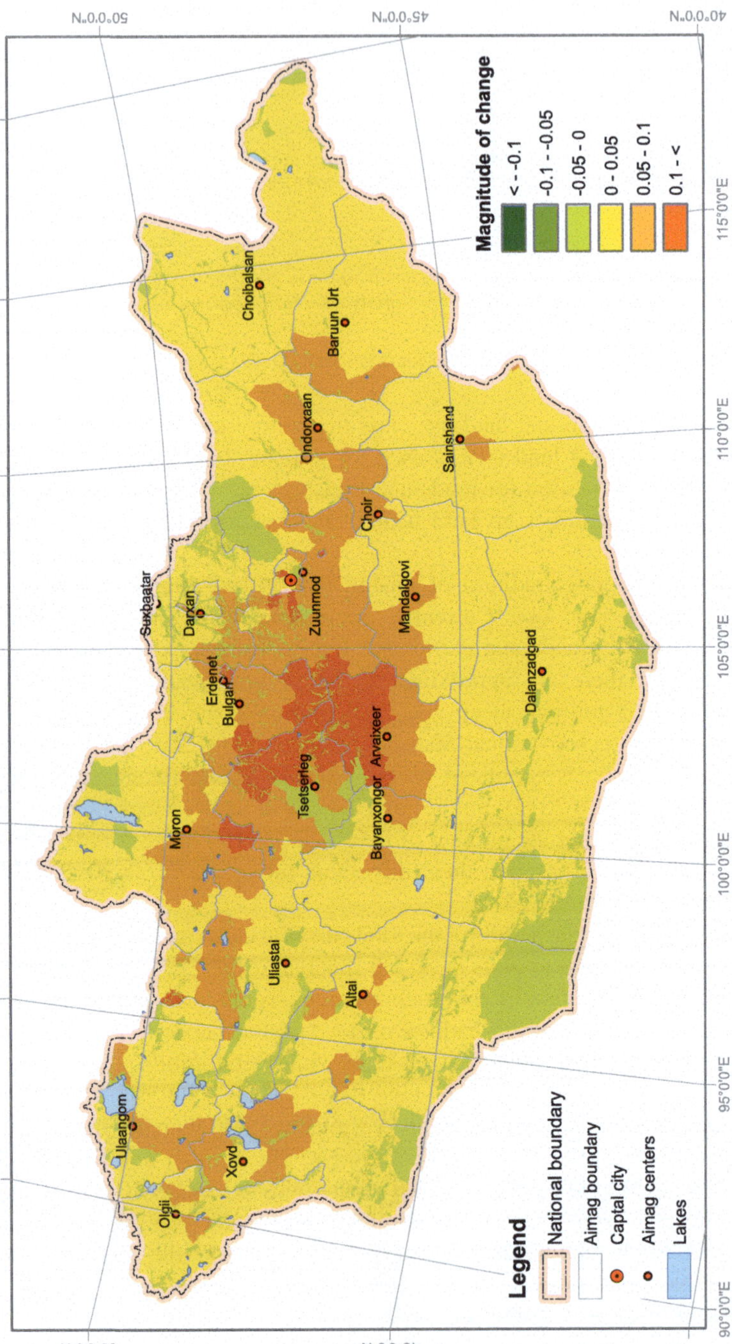

Fig. 2.23 Magnitude of livestock change over the territory (Mandakh 2017)

dry steppes of Mongolia. Livestock density is relatively sparse throughout the nation. High-density areas are concentrated around economically viable cities such as Khovd, Erdenet, Darkhan, and Ulaanbaatar. Overall, the dynamics of livestock density remain the same as that in the 2000s, except for the formation of new hotspots in the Gobi Desert that might be linked to the development of the mining industry.

Mining is another type of land use that is considered to highly contribute to land degradation and aeolian desertification because most of the operating mining sites employ open-pit technology. Degradation due to the extractive industry occurs independently of the other drivers. As the economic development policy of the country since the 1990s has mainly been directed to achieve growth through developing the mineral industry, the Government of Mongolia has granted 3580 licenses covering 13.9 million has of area (approximately 8.9% of the total territory); 0.9% of these licenses were exploitation licenses, and 8.0% were exploration licenses (www.mrpam.gov.mn). Even though the mining companies were required to restore the land during and after their operations, another threat from this industry is unrestored open pits, which directly contribute to land degradation. According to the Ministry of Environment, by 2018, 27405.46 ha of land formerly used for mining were considered degraded, of which 8871.8 ha needed to be restored to sustain ecosystem integrity (Saran 2018).

References

Batjargal Z (1997) Desertification in Mongolia. RALA Rep 200:107–113

Batjargal Z (ed) (2018) Mongolia – Third National communication under the UNFCCC. Ministry of Environment and Tourism, Ulaanbaatar

Chen X (2010) Physical geography of arid land in China. Science Press, Beijing (in Chinese)

Chen LH, Li FX (1999) Sandy neo-soils. In: Gong ZT et al (eds) Classification of soil system in China: theory, methods and practices. Science Press, Beijing (in Chinese)

Chen LH, Li FX, Di XM et al (1998) Aeolian soils in China. Science Press, Beijing (in Chinese)

Chung YS, Kim HS, Natsagdorj L et al (2004) On yellow sand occurred during 1997-2000. J Meteorol Soc 4:305–316

Dagvadorj D, Khuldorj B, Aldover RZ (eds) (2009) Mongolia assessment report on climate change. Ministry of Environment and tourism. Ulaanbaatar

Dagvadorj D, Batjargal Z, Natsagdorj L (eds) (2014) Mongolian second assessment report on climate change. Ministry of Environment and Tourism. Ulaanbaatar

Dash D (2013) Mongol ornii fizik-gazarzuin muzjlal (physical geography regionalization of Mongolia). In: Dash D (ed) Ladshaft ecologiin asuudluud. Admon, Ulaanbaatar

Dash D, Mandakh N (2018) Geomorphology of Mongolia. In: Undarmaa J et al (eds) Rangeland ecosystems of Mongolia. Munkhiin Useg, Ulaanbaatar

Ding YH, Krishnamurti TN (1987) Heat budget of the Siberian high and winter monsoon. Mon Weather Rev 115:2428–2449

Ding YH, Wen SG, Li YJ (1991) A study of dynamic structures of the Siberian high in winter. Acta Meteor Sin 49(4):428–439 (in Chinese with English abstract)

Dorjgotov D (ed) (2009) Mongol Ulsiin Undesnii atlas (National Atlas of Mongolia). The Institute of Geography, Ulaanbaatar

Geng KH (1986) Climate in sandy zones in China. Science Press, Beijing (in Chinese)

Gong DY, Wang SW (1999) Long-term variability of the Siberian high and the possible connection to global warming. Acta Geograph Sin 54(2):125–133 (in Chinese with English abstract)
Harris I, Jones PD, Osborn TJ et al (2013) Updated high-resolution grids of monthly climatic observations – the CRU TS3.10 dataset. Int J Climatol 34(3):623–642. https://doi.org/10.1002/joc.3711
Hu RJ, Fan ZL, Wang YJ et al (2002) Groundwater resources and their characteristics in arid lands of Northwestern China. J Nat Res 17(3):321–326 (in Chinese with English abstract)
IPCC (2014) Climate change 2014 impacts, adaptation, and vulnerability. Cambridge University Press, Cambridge
Jambaajamts B (1989) Mongol ornii uur amisgal (Climate of Mongolia). Ulsiin Hevlel, Ulaanbaatar
Khudulmur S, Tsogtbaatar J (2013) Desertification atlas of Mongolia. Admon, Ulaanbaatar
Mandakh N (2017) Mongol ornii tsuljilt, gazriin doroitol (desertification and land degradation in Mongolia). In: Environment of Mongolia, vol 1. Ministry of Environment and Tourism, Ulaanbaatar
National Statistical Office (NSO) (2013) Statistikiin bulletin (National statistics bulletin). Ulaanbaatar
National Statistical Office (NSO) (2015) Statistikiin bulletin (National statistics bulletin). Ulaanbaatar
Nyamtseren M, Feng Q, Deo R (2018) A comparative study of temperature and precipitation-based aridity indices and their trends in Mongolia. Int J Env Res 12(6):887–899
Saran B (2018) Evdersen gazriin toollogiin ur dun (results of the accounting of disturbed land). In: Enhbat A et al (eds) The state of environment of Mongolia. Ministry of Environment and Tourism, Ulaanbaatar
Tabari H, Talaee PH, Nadoushani SSM et al (2014) A survey of temperature and precipitation based aridity indices in Iran. Quatern Int 345:158–166
Tang QC, Qu YG, Zhou YC (1992) Hydrology and water resource use in arid region of China. Science Press, Beijing (in Chinese)
Tao SY (1959) Study on East Asia cold snap in China in decades. Acta Meteor Sin 30(3):226–230 (in Chinese with English abstract)
Taylor KE, Stouffer RJ, Meehl GA (2012) An overview of CMIP5 and the experiment design. Bull Am Met Soc 93(4):485–498
Tserendash S (2006) Belcheer ashiglah onoliin undeslel (theoretical background to the pasture use). Admon, Ulaanbaatar
UNEP (1992) World atlas of desertification. Edward Arnold, London
Wang T (2011) Deserts and aeolian desertification in China. ELSEVIER in Amsterdam and Science Press, Beijing
Wang ZY, Ding YH (2006) Climate change of the cold wave frequency of China in the last 53 years and the possible reasons. Chin J Atmos Sci 30(6):1068–1076 (in Chinese with English abstract)
Wang B, Lin H (2002) Rainy season of the Asian–Pacific summer monsoon. J Climatol 15:386–398
Wu ZY, Wu SG (1999) China flora division. In: Liu GM (ed) Natural geographic atlas of China. SinoMaps Press, Beijing (in Chinese)
Xiong Y, Li QK (1987) Soils in China, 2nd edn. Science Press, Beijing (in Chinese)
Zhang Q, Xu CY, Tao H et al (2010) Climate changes and their impacts on water resources in the arid regions: a case study of the Tarim River basin, China. Stoch Env Res Risk A 24:349–358
Zhao SQ (1990) Arid zone in China. Science Press, Beijing (in Chinese)
Zhu ZD, Wu Z, Liu S (1980) Sandy deserts in China. Science Press, Beijing (in Chinese)

Chapter 3
Spatial Distribution and Temporal Dynamics of Aeolian Desertification

Tao Wang and Mandakh Nyamtseren

Abstract This chapter describes the spatial distribution of aeolian desertification and its temporal change in northern China and Mongolia using remote sensing images and the aeolian desertification index. In northern China, aeolian desertification mainly occurs in semiarid and subhumid agropastoral ecotones, arid and semiarid pure pastoral regions, and arid oases in inland river basins. The area of aeolian desertification in northern China rapidly increased from 13.7×10^4 km^2 in the 1950s to 38.6×10^4 km^2 in 2000 and then slowly decreased to 37.6×10^4 km^2 in the 2010s. Human activities are the main driving forces of aeolian desertification and its restoration in northern China. Aeolian sands cover approximately 2% of the total territory of Mongolia and are distributed through aeolian processes. In terms of aeolian desertification, the latest studies indicate that approximately 68% of degraded lands are caused by wind erosion, i.e., deflation, transportation, and accumulation of sand. Aeolian desertification is mainly linked to the rapid decline in vegetation cover, decreases in surface roughness, and increases in the aridity of the climate.

Keywords Aeolian desertified land · Distribution · Mongolia · Northern China · Spatial and temporal

---

T. Wang (✉)
Northwest Institute of Eco-Environment and Resources, Chinese Academy of Sciences, Lanzhou, Gansu, China

Key Laboratory of Desert and Desertification, Chinese Academy of Sciences, Lanzhou, Gansu, China

Drylands Salinization Research Station, Northwest Institute of Eco-Environment and Resources, Chinese Academy of Sciences, Lanzhou, Gansu, China
e-mail: wangtao@lzb.ac.cn

M. Nyamtseren
The Institute of Geography and Geoecology, Mongolia Academy of Sciences, Ulaanbaatar, Mongolia

T. Wang et al. (eds.), *Combating Aeolian Desertification in Northeast Asia*, Ecological Research Monographs, https://doi.org/10.1007/978-981-16-9028-0_3

3.1 Introduction

Aeolian desertification not only leads to the loss of land resources, reduces land productivity, reduces biodiversity, and influences the global climate but also seriously affects the social economy in terms of food availability, economic and political stability, peace between different regions, sustainable development, and human health. Aeolian desertification is not a regional issue limited to the developing countries suffering from it but a global economic, social, and environmental issue. Most areas in Northeast Asia are seriously affected by aeolian desertification. Many desertified lands are concentrated in northern China and southern Mongolia. Analysis of the spatial-temporal evolution of aeolian desertification based on maps of desertification in different periods can provide distribution characteristics and evolutionary trends. Based on the research results, it should be reasonable to propose a key control target area and control recommendations for aeolian desertification and provide a decision tool for determining aeolian desertification control and regional ecological environment restoration and reconstruction.

3.2 Aeolian Desertification in Northern China Over the Past Five Decades

The history of aeolian desertification can be traced to the origination of human culture. Drought is the basic cause of land desertification, but unsustainable human behavior is essentially the main cause of aeolian desertification. In China, only 5.5% of the 38.6×10^4 km^2 of aeolian desertified land occurred due to the advancement of sand dunes (Wang et al. 2004a). Aeolian desertified land excludes deserts, gobi, salt deserts, and cold deserts formed in geological periods. The study of aeolian desertification in China started in the late 1950s, and prior to 1977, this type of desertification was mainly studied to investigate its distribution and methods to control sand damage to railways. The United Nations Conference on Desertification in 1977 raised global concerns about desertification. Since 1978, the study of desertification in China has been characterized by research on desertification status, development processes, change trends and predictions, and control measures (Zhu 1989).

3.2.1 The Spatial Distribution of Aeolian Desertification in Northern China

Aeolian desertified land in China is mainly distributed in the northwestern and the northern parts of China, covering 18 provinces, autonomous regions, and municipalities (Fig. 3.1). The total aeolian desertified area in northern China was 37.6×10^4 km^2 in 2010 (Wang 2014). Chinese researchers started studying aeolian

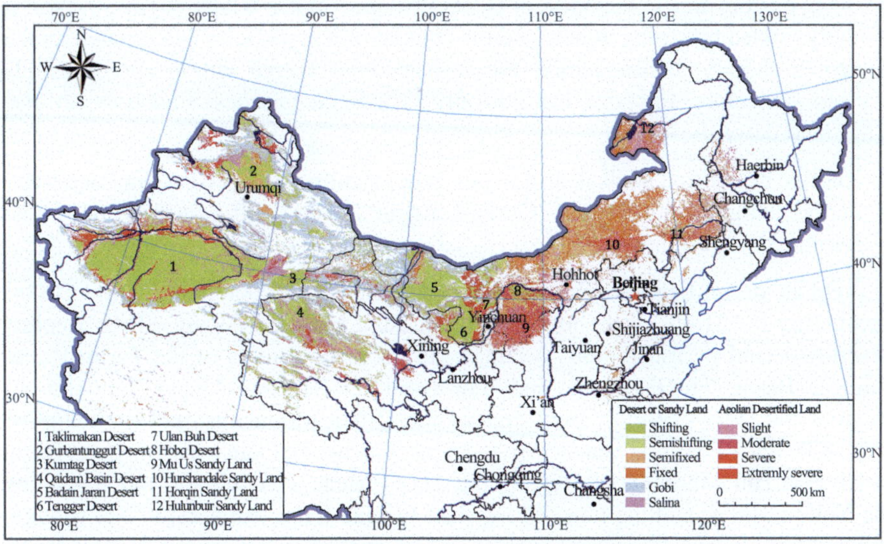

Fig. 3.1 Map of desert and aeolian desertified land in northern China (Wang 2011). Note: The international boundary is based on standard China Map No. GS (2016) 1570

desertification in the late 1970s and summarized the following spatial distribution characteristics (Zhu and Liu 1981; Zhu et al. 1981a, b; Zhu and Chen 1994; Zhu 1999; Wang 2002, 2011, 2014; Wang et al. 2003, 2004a, b):

1. Aeolian desertified lands are concentrated in semiarid regions.

In the 2000s, aeolian desertified land covered 201,326 km^2 in semiarid regions, accounting for 52.2% of the total desertified land in China, according to data from Landsat TM Imagery (Wang 2011). The area influenced by aeolian desertification in the semiarid zone mainly covers the following geographic units: the Hulun Buir steppe, Horqin steppe, Xilin Gol steppe, Hunshandake steppe, Qahar steppe, Ulanqab steppe, Erdos steppe, Mu Us sandy land, and sandy land on the eastern side of the Yellow River in Ningxia; the Bashang area in Hebei Province; the Qianshan area and Tumd Plain in Ulanqab Meng of Inner Mongolia; and northwestern Shanxi Province. On the one hand, there are natural factors, such as the synchronous occurrence of dry seasons and windy seasons, thick and loose sandy deposits, and the erodibility of loose sediments in pastures and agropastoral ecotones, that promote the development of aeolian desertification in semiarid areas. On the other hand, the population density in these areas is relatively high, with an average of 50–100 person/km^2, and the resulting overcultivation, overgrazing, and overcutting due to population pressure accelerate the process of land degradation in such fragile environments. For example, 85% of farmlands are cultivated by reclaiming undulating fixed sand dunes or sandy lands in the Houshan area to the north of the Yinshan Mountains in Inner Mongolia and on the Horqin steppe and the

Bashang pasture area to the north of Hebei Province. When winter begins, the soil is severely eroded by strong winds, and the content of fine particles and organic matter decreases sharply; consequently, the soil becomes barren within approximately 3 years. Therefore, this area has not only serious aeolian desertification but also rapid land desertification.

2. Aeolian desertified lands in arid regions are patchily distributed around oases and in the marginal area of the Gurbantunggut Desert with extensive fixed and semifixed sand dunes.

Aeolian desertified land in arid regions covered 122,000 km^2 in the 2000s, accounting for 31.4% of the total desertified land in China (Wang 2011). Aeolian desertified land in arid regions occur in the geographic units of the Tarim Basin, Junggar Basin, Tu-Ha Basin, and Yili Basin in Xinjiang Uygur Autonomous Region; the Hexi Corridor, front area of the Qilian Mountains, and the marginal area of Tengger Desert in Gansu Province; the Yinchuang Plain and Zhongwei Basin in Ningxia Hui Autonomous Region; the Alxa Plateau, Hetao Plain, and area along the Yellow River; and the Inner Mongolian Plateau (Houshan area) in Nei Mongol [Inner Mongolia] Autonomous Region. Most of the aeolian desertified lands are distributed around desert oases. Northwestern China has a concentrated distribution deserts with 90% of deserts in China, and these mainly include the Taklimakan Desert, the Badain Jaran Desert, the Gurbantunggut Desert, the Tengger Desert, the Kumtag Desert and, the Ulan Buh Desert. There are many oases either supported by ground water in pluvial fans at the edge of deserts or fed by runoff of rivers flowing into desert hinterlands. In arid regions, populations settle near oases and areas along rivers with available water, where the population density is generally up to 200–400 person/km^2. Frequent human development activities on the fixed and semifixed dunes around the oases result in sand dune encroachment and sandstorm damage. In the long history of humans, humans have searched for lush grass plants and adequate water sources, so desertified lands exist around modern oases and in the vicinity of ancient oases. In the ancient oasis regions, moving barchan dunes and chains and shrub-coppice dunes developed on the aeolian desertified land throughout history; in addition, there are also ruins of city walls, castles, houses, passageways, religious towers, and temples in these areas.

Moreover, some fixed and semifixed dunes in desert areas are also developed due to petroleum exploitation. A typical example is the aeolian desertification strongly developed in the marginal area of the Gurbantunggut Desert. A large number of fixed and semifixed dunes near oil extraction points and along the transportation lines are destroyed and shift. At the southwestern edge of the Gurbantunggut Desert, the width of developed sand dunes has increased to 5–10 km, stretching more than 350 km.

3. Aeolian desertified lands are mainly scattered in pluvial fans and along old riverbed banks in semihumid regions.

The characteristics of aeolian desertification in semihumid areas are obviously different from those in arid and semiarid regions. In semihumid areas, the climate is mainly controlled by the East Asian Monsoon, and land aeolian desertification commonly occurs due to soil wind erosion on farmland and has obvious seasonal changes. The aeolian desertified lands in these areas are concentrated in floodplains and deltas, and more than 80 counties experience aeolian desertification in their ancient deltas at the lower reaches of the Yellow River; the other aeolian desertified land occurs on the piedmont alluvial plains or in the floodplains of the Yongding River and the Chaobai River, the Luan River delta and lower valley, and the Songhuajiang Plain and the Nenjiang Plain, covering provinces of Heilongjiang Province, Jilin Province, Beijing City, Tianjin City, Hebei Province, Henan Province, Shandong Province, Jiangsu Province, Shaanxi Province, and An'hui Province in northern China. In the semihumid regions, the area of aeolian desertified land covers 24,660 km^2, accounting for 6.4% of the total desertified land in China during the 2000s (Wang 2011).

In summary, aeolian desertified land is the product of the interaction of unsustainable human activities and fragile ecological environments. This type of land is not limited to regions with drought indices in the range of 0.50–0.65 but is widely distributed across all kinds of natural systems in China. According to statistical data in 2010, there are 37.6×10^4 km^2 of aeolian desertified land in northern China, and this type of land is mainly distributed in the middle of Inner Mongolia (the middle and western parts of the semiarid zone and desert steppe zone), where reactivated fixed dunes and shifting sand spread; in regions of the Otindag sandy land, Horqin sandy land, Bashang area of Hebei Province, and Houshan area of Inner Mongolia (the farming-grazing mixed regions, rainfed farming regions in the eastern semiarid zone, and part of the subhumid zone), where wind erosion and sand sheets are striking features; and in western Inner Mongolia, the Hexi Corridor, and the lower reaches of the Tarim River (the oasis margin in arid zone and the lower reaches of inland rivers), where fixed dunes are the main feature (Wang 2014).

3.2.2 The Temporal Dynamics of Aeolian Desertification in Northern China Since the 1950s

Aeolian desertification is a dynamic process, and the distribution scope and characteristics of these surficial landforms were variable during different periods. To study the modern process of this aeolian desertification that is closely related to human activities, the scope of aeolian desertified lands in northern China since the 1950s was assessed based on aerial photos (1:50,000) obtained from the mid-1950s, mid-1960s, mid-1970s, and mid-1980s, with Landsat TM images obtained in 2000 and 2010. It was concluded that the aeolian desertified land area generally continued to expand before 2000 and has been decreasing since 2000. The total amount of aeolian desertified land in northern China was 13.7×10^4 km^2 in 1955, and it increased to 17.6×10^4 km^2 in 1975, 33.4×10^4 km^2 in 1987, and 38.6×10^4 km^2 in

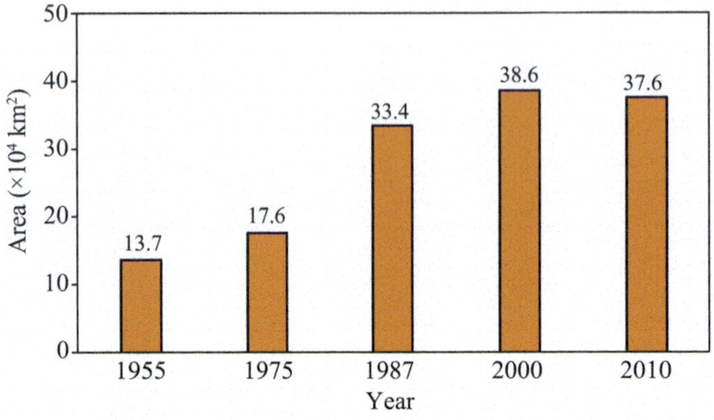

Fig. 3.2 Aeolian desertification area in China

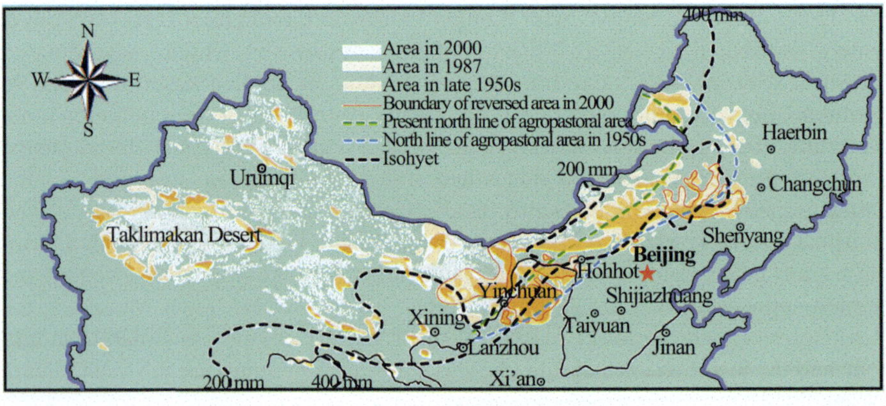

Fig. 3.3 Spatial distributions of aeolian desertified land in northern China in the 1950s, 1987, and 2000 (Wang et al. 2008; due to restoration policies, aeolian desertification was reversed in some areas in northern China by the end of the 1990s). Human activities also pushed the agropastoral boundary northward by approximately 200 km from the 1950s to 2000

2000 and decreased to 37.6×10^4 km^2 in 2010 (Fig. 3.2), with an expansion rate of 1560 km^2/a in the late 1950s to 1975, 2100 km^2/a in 1975–1988, and 3600 km^2/a in 1988–2000 and a decreasing rate of 1375 km^2/a in 2000–2010 (Wang 2002, 2014; Wang et al. 2003, 2004a, b).

Regionally, the dynamic changes in aeolian desertification are significantly different in the arid western part and the semiarid eastern part of China. Aeolian desertification has developed mainly in three regions of northern China in the past five decades (Fig. 3.3): (a) the agropastoral ecotone of the semiarid and subhumid regions account for 40.5% of the total aeolian desertified land of northern China; (b) the sandy grassland zone or pure pastoral semiarid region accounts for 36.5%; and (c) the arid oasis edge and lower reaches of inland rivers account for 23%.

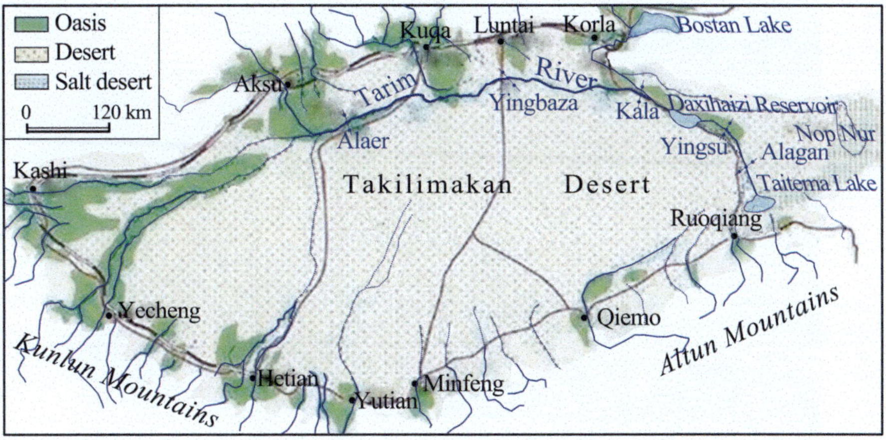

Fig. 3.4 The Tarim River system and oasis distribution

Remote sensing data also show that approximately 10% of aeolian desertified land maintained a stable state or exhibited a reversed trend. Since the start of the desertification control project in 1991, some relatively complete ecological protection systems have been established in arid areas, and the development of aeolian desertification has been reversed to some extent (Zhong 2003). Taking the Tarim River Basin as a typical example, the evolution of the Tarim River can well reflect the process of aeolian desertification along a river. The Tarim River is the longest continental river in China, with a length of 1280 km as its main stem (from the Xiaojiake River station to the terminal Taitema Lake, Fig. 3.4). By the 1960s, the Tarim River was still a perennial river, and boats could sail along the river. In the 1950s–1960s, water consumption in the middle and upper river reaches rose sharply due to large-scale land reclamation, resulting in the lower reaches being cut off after the 1960s, and its original terminal lake, Lop Nur, rapidly dried. After 1972, the lower reaches of the Tarim River were completely cut off. At the same time, aeolian desertification gradually developed from the lower reach to the upper reach, and the oasis shrank toward the riverhead. Starting in 2010, transporting water for ecological purposes into the lower reaches of the Tarim River effectively alleviated serious ecological degradation in the Tarim River Basin. Until November 11, 2019, the action of water transportation was carried out 20 times, with a cumulative transportation of more than 81.0×10^8 m^3 of water for ecological purposes. Therefore, 50.0×10^8 m^3 of water was transported into the degraded *Populus euphratica* forest area, ecologically improving these areas (Fig. 3.5). The area of vegetation restoration and improvement in the downstream area was 2285 km^2, including 362 km^2 of new vegetation cover and 854 km^2 of restored sandy land, and the plant species increased from 17 to 46 (http://www.tahe.gov.cn/xwzx/lylw/mInEFz.htm).

Modern aeolian desertification in the agropastoral ecotone in the semiarid region has expanded since the middle twentieth century due to overcultivation, overgrazing, overcutting, etc., and the dynamics mainly manifested in the following three ways:

Fig. 3.5 Photo of the green corridor in the lower reaches of the Tarim River (photo by George Steinmetz)

first, the expansion of aeolian desertified lands on the Ordos and Horqin grasslands resulted in the integration of patchily distributed aeolian desertified lands, and the proportion of moving sand dunes on the Horqin grassland increased from 2% to 3.74% in the 1980s. Second, aeolian desertification tended to develop toward the Ulanqab steppe and the Bashang grassland, and the area of lands affected by modern aeolian desertification increased from 25% in the 1950s to 40% in the 1980s as a result of large-scale reclamation of grasslands. Third, 42.3% of the total pasture area experienced aeolian desertification because of overgrazing. At the turn of the twenty-first century, the rapid development of aeolian desertification in most agropastoral interconnected areas was controlled and even reversed, benefiting from the "West Development" plan implemented by the Chinese government.

3.3 Aeolian Desertification in Mongolia

3.3.1 Distribution and Origin of Aeolian Sands

Aeolian sands occupy an area of approximately 40,000 km^2 or 2.5% of the total territory of Mongolia. The distribution of sands over the territory of Mongolia is very scattered and is not confined to a particular geographic location; however, the impact of geology is obvious. An interesting hypothesis on the dynamics of aeolian sand was proposed by Fedorovich (1950), who assumed that the sand originated due to weathering of ancient sediments and volcanic rocks that were constantly transported by surface water to lakes and from the shores of lakes to adjacent territories by wind.

This might have occurred within a short time during the postglacial period in the western part of Mongolia. The distribution of sands, for instance, in Central Mongolia, was linked with dryness during the preglacial period.

In the territory of Mongolia, we can delineate three different regions of aeolian sand: (1) western; (2) Gobi Desert; and (3) northern. All three zones differ from each other in terms of sand depth, accumulation type, and modern surface forms (Fig. 3.6).

The majority of sands are located in western Mongolia. The researchers mainly linked this distribution to the late Pleistocene when a major transformation in ancient lakes occurred (Dgebuadze 2013). Most researchers consider that the glaciation in the Altai and Khangai Mountains corresponded to the pluvial and interglacial periods and to arid climate in the western depressions. According to Kuznetsov and Murzaev (1963), in Central Asia during the Pleistocene glaciation, a special period of lake development occurred. Devyatkin (1981) also noted that the peak of the pluvial period in these areas coincided with the end of the interglacial period and the first half of the glaciation. During the maximum and second half of the glaciation, the climate conditions were more arid, which was facilitated by the large masses of water that were "preserved" in the glaciers. A slight but smaller increase in humidity was associated with a period of degradation of mountain glaciers. The interglacial period was characterized by a general increase in temperature, an increase in evaporation, and, consequently, a decrease in river flow and regression of lake basins, which is ongoing to date (Devyatkin 1981).

In general, the landforms of the basins in the western part of the country and the distribution of corresponding sediments, especially sand, indicate significant distribution ranges in the past. However, it is not always possible to determine the exact time of the maximum lake transgression in certain regions of Mongolia during the Pleistocene. This is especially difficult for the early Pleistocene because the sediments were mostly either not separated from the middle Pleistocene or not traced at all. At the same time, there is no reason to believe that at the beginning of the Pleistocene, there were no lakes in the Basin of the Great Lakes, the Transaltai depressions, the Valley of Lakes, and other large depressions in the northern and central parts of Mongolia.

Undoubtedly, lakes existed here, but there is a lack of surveys related to sedimentation. The sediments, in some cases, may have been washed away by erosion and denudation processes and, in some places, probably buried under younger strata. The existence of lakes in these regions of Mongolia in the early Pleistocene may be proven by indirect data. For instance, the northern foots of the Khan-Khukhii Mountain range have thick alluvial and alluvial fan deposits from the lower Pleistocene, whose height can reach up to 25 m in the Uvs lake depression. As noted by Devyatkin et al. (1989), their composition and position in the relief are comparable to those of the Eopleistocene.

Lower Pleistocene deposits of alluvial and alluvial fan origins were also found in the valleys and basins of the Selenga, Kherlen, Tuul, Onon, and other rivers, as well as near the Tori lakes, but they do not indicate the nature of the lake basins that may have existed at that time. Undoubtedly, there were small lakes in the early

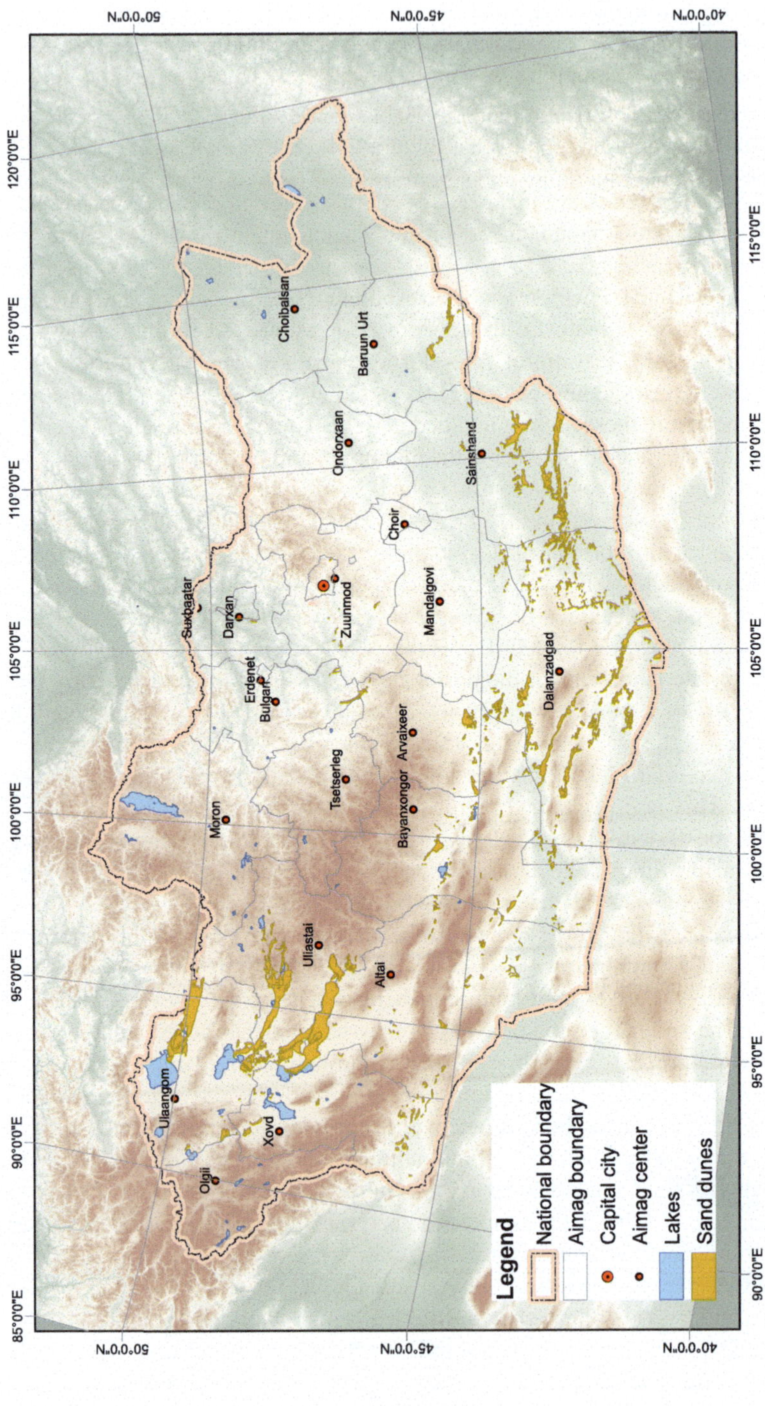

Fig. 3.6 Aeolian sand distribution of Mongolia (Baasan 2003)

Pleistocene in the south and in the Gobi regions of the country, but almost nothing is known about the structure and thickness of deposits of the lower Pleistocene, except dense gray cover ("upper Gobi") conglomerates possibly related to this age.

Substantially more complete information is available for the middle Pleistocene, during which lake basins reached maximum sizes in the Great Lakes Depression, in the Valley of Lakes, in the Transaltai depressions, on the northeastern plains, and in the other lowlands of Mongolia. In the middle of the Pleistocene, glaciation covered most of the mountain structures of Mongolia, primarily the Mongolian and Gobi Altai, Khubsugul, Khangai, and Khentei Mountains. Traces of glaciation in the mountains of Mongolia have been noted by many researchers (Berkey and Morris 1927; Obruchev 1947; Murzaev 1952; Marinov 1957; Sinitsyn 1959; Kuznetsov and Murzaev 1963; Nikolaeva and Shuvalov 1967; Marinov 1970; Devyatkin 1970; Selivanov 1972; Devyatkin et al. 1978; Devyatkin 1981).

Numerous rivers that originated from the aforementioned mountains carried large amounts of water into the Great Lakes Depression, the Valley of Lakes, and other basins of Mongolia, which caused a sharp increase in the level of lake reservoirs. The traces of ancient lake development are perfectly preserved in most of the major depressions of Mongolia. In terms of the Great Lake Depression, its northern part, where the largest rivers, the Khangai and Altai, flow in, the leading processes were lacustrine and alluvial accumulation, while in the south, alluvial fan and slope sedimentation dominated (Devyatkin et al. 1989). Sedimentation is the central part of the depression and is linked to lacustrine and lacustrine-takir sediments. In the subsequent late glacial and postglacial periods, lacustrine sediments of the Great Lake Depression were subject to aeolian processes, which resulted in the formation of the largest area of aeolian sands in Mongolia.

In conclusion, the formation of aeolian sands in the territory of Mongolia was homogeneous even though their spatial distributions may differ from each other. All paleogeographic studies show the spatiotemporal dynamics of paleolakes, and the late glaciation period with highly arid interglaciation periods predefined the modern distribution of sands in Mongolia. These predominantly lacustrine, alluvial, and sometimes alluvial-alluvial fan-originated sands are highly disturbed by aeolian processes in modern times, which should not be linked to current wind erosion-related deflation observed in many parts of the country. The latter is linked to soil erodibility, land use/land cover, and climate change, especially the land-climate nexus.

3.3.2 Aeolian Desertification in Mongolia

Aeolian desertification is a form of land degradation predominantly defined as a product of wind erosion in drylands (Wang 2004). This topic has not been well researched in detail over Mongolia. According to research conducted during the 1990s by T. Baasan (2003), sandy land covering approximately 2.5% of the total territory is the main source of aeolian desertification in the country. The dynamics of sand movement are estimated to be 1.6 m/year for the western part of the country and 0.9 m/year for the eastern part (Baasan 2003). According to the first national aeolian desertification assessment, implemented by the joint Mongolia-Turkmen expedition, the area of wind-blown sands in Mongolia is expanding, and the major factors of such critical processes can be attributed to the degradation of fixed and semifixed sand massifs located in the Great Lake Depression, the territory between the Altai and Khangai Mountains. The estimated area of land affected by aeolian desertification is 19,945 hectares (Baasan et al. 1992). Considering these research results, it can be said that approximately 1–2% of the territory potentially has a high risk of aeolian desertification due to its geological structure, soil erodibility, and fragile vegetation cover. However, the current scenario in relation to changes in climate, land cover, and land use is different from that in during previous researchers' studies.

The assumption of previous researchers was that only the sandy lands in drylands should be considered a hotspot of aeolian desertification. Considering the soil cover of Mongolia, however, most of the soil types, especially on elevated plains, along large river basins are highly susceptible to wind and other disturbances. According to Mandakh et al. (2016), approximately 60% of the territory covered by soils is highly susceptible to wind erosion, of which the Great Lake depression, Valley of Lakes, and eastern Gobi region should be considered hotspots (Fig. 3.7). Thus, the soil cover in the abovementioned regions is a direct factor in aeolian desertification due to high erodibility and low vegetation cover for fixing topsoil.

In all other parts of the country, the soil erodibility factor ranges between 1.5 and 2.0 tons ha^{-1}; however, due to increased aridity and wind gusts, the areas affected by wind erosion are increasing. The dynamics of wind-induced degradation have become increasingly critical in the middle parts of the country and have been caused by changes in climate indices as well as by changes in land use, especially due to increased livestock, expansion of extractive industries, and increased use of unpaved roads. According to the latest assessment of desertification in Mongolia (Khudulmur and Tsogtbaatar 2013), wind-induced degradation accounts for approximately 60% of the total land, of which increased degradation occurs in the eastern parts, in Dornogobi, Dundgobi, Gobisumber, and Sukhbaatar aimags (Fig. 3.8).

The spatial distribution of aeolian desertification in Mongolia can be summarized as follows:

1. Aeolian desertification in semiarid and arid regions: The degradation is mainly defined by the arid climate and strong winds. Spatially, aeolian desertification occurs in the territories of Khovd, Gobi-Altai, Zavkhan, Bayanhongor, Dornogobi, and Sukhbaatar aimags that have natural sand dunes and sandy sediment. The main mechanism of degradation is overgrazing on highly susceptible lands caused by the increased occupation of barren soils that are easily denuded by winds.
2. Aeolian desertification along old, large riverbeds: This type of degradation is pronounced in the basins of the Orkhon, Selenge, Kherlen, and Tuul Rivers. With an increased number of rainfed arable land areas as well as overgrazing, the onset

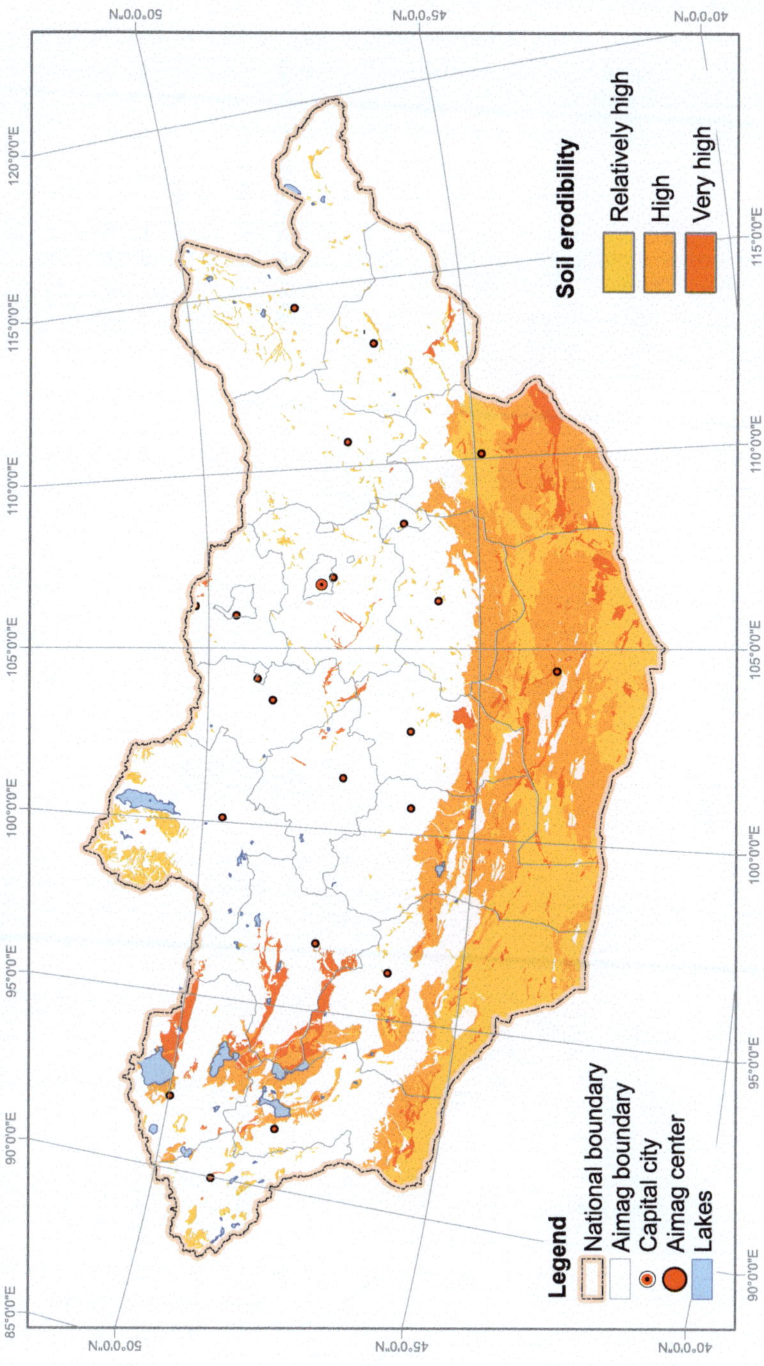

Fig. 3.7 Risk levels for soil erodibility (modified from Mandakh et al. 2016)

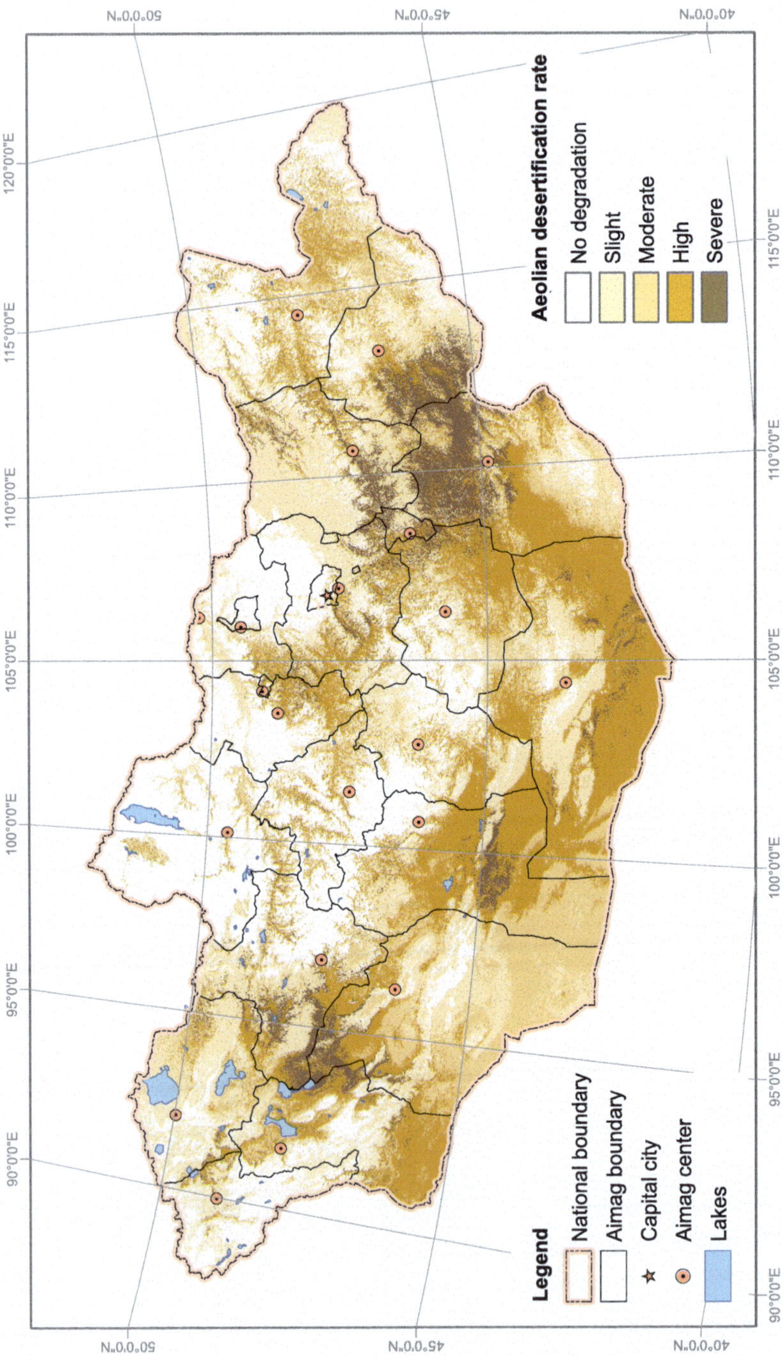

Fig. 3.8 Spatial distribution and rate of aeolian desertification in Mongolia (Mandakh 2017)

of vegetation cover is prolonged, leading to an increase in windblown topsoil, especially during the spring season.

3. Aeolian desertification with sporadic distribution: This type of degradation occurs in the areas of the eastern Mongolia steppe. Aeolian desertification in this region is highly interlinked with drought. Unfortunately, the frequency of drought in this region is increasing, causing a decline in the aboveground biomass, which decreases the surface roughness, thus leading to an increased risk of aeolian desertification.

References

Baasan T (2003) Mongol ornii els (Sands of Mongolia). Ministry of Environment, Ulaanbaatar

Baasan T, Dash D, Enkhtaivan D et al (1992) Huurai, ganduu nutgiin tsuljiltiin sudalgaanii ur dun (Results of desertification research in drylands of Mongolia). In: Dayan delhiin uurchlult – gobi tsul olon ulsiin baga hurliin huraangui, Ulaanbaatar, 22 July 1992

Berkey CP, Morris FK (1927) Geology of Mongolia, reconnaissance report based on the investigations of the years 1922–1926. American Museum of Natural History, New York

Devyatkin EV (1970) Geologiya Kainozoya Zapadnoi Mongolii (Cainozoic geology in Western Mongolia). In: Geology of Mesozoic and Cainozoic era in Western Mongolia, vol. 2. Moscow, p 44

Devyatkin EV (1981) Kainozoi Vnutrennei Azii (stratigrafiya, geohronologiya, korrelyatsiya) (Cainozoic in Inner Asia (stratigraphy, geochronology and correlation)). Nauka, Moscow

Devyatkin EV, Malaeva EM, Murzaeva VE et al (1978) Pluvialinie pleistocenovie basseini Kotlovini Bolishih Ozer Zapadnoi Mongolii (Pluvial Pleistocene baisns in Great lake depressions of Western Mongolia). Izvestiya AN SSSR Ser Geography 5:89–99

Devyatkin EV, Malaeva EM, Murzaeva VE (1989) Hirgis-Nurskaya, Hara-Us-Nurskaya I Hara-Nurskya vpadini (Depressions of Khyargas, Khar-Us and Khar Lakes). In: SSMGE: Late Cainozoic of Mongolia, vol. 47. Moscow, p 106

Dgebuadze YY (ed) (2013) Limnology and paleolimnology of Mongolia. Russia-Mongolian Complex Biological Expedition, Moscow

Fedorovich BA (1950) Proishojdenie i razvitie peschanih tolsh pustin Azii (Formation and development of sands in Asian deserts). In: SSMGE: Materiali po chetvertichnomu period SSSR, vol. 2. Moscow, p 221

Khudulmur S, Tsogtbaatar J (eds) (2013) Desertification atlas of Mongolia. Admon, Ulaanbaatar

Kuznetsov NT, Murzaev EM (1963) Nablyudeniya na ozerah Djungarii (Monitoring of lakes in Dzungaria). Lectures in memory of L.S. Berg, Moscow-Leningrad

Mandakh N (2017) Mongol ornii tsuljilt, gazriin doroitol (Desertification and land degradation in Mongolia). In: Environment of Mongolia, vol 1. Ministry of Environment and Tourism, Ulaanbaatar

Mandakh N, Tsogtbaatar J, Dash D et al (2016) Spatial assessment of soil wind erosion using WEQ approach in Mongolia. J Geogr Sci 26(4):473–483

Marinov NA (1957) Stratigraphiya Mongoliskoi Narodnoi Respubliki (Stratigraphy of Peoples Republic of Mongolia). Nauka, Moscow

Marinov S (1970) Novie materiali o drevnem oledenenii Mongolii (New data on ancient glaciation in Mongolia). In: Geology of Mesozoic and Cainozoic era in Western Mongolia, vol. 2. Moscow, p 170

Murzaev EM (1952) Mongoliskaya Narodnaya Respublika (Peoples Republic of Mongolia). Nauka, Moscow

Nikolaeva TV, Shuvalov VF (1967) Novie Dannie o pleistocenovom oledenenii Hangaya I Gobiiskogo Altaya (New data on Pleistocene glaciation in Hangai and Gobi-Altai mountains). Vestnik LGU 6:34–39

Obruchev VA (1947) Vpadini Tsenralinoi Azii I nauchnie sokrovisha, ojidayushie izucheniya (Depressions of Central Asia and scientific treasures for research). Izvestiya AN SSSR, Seriya Geol 5:56–68

Selivanov EI (1972) Neotektonika I geomorphologiya Mongoliskoi Narodnoi Respubliki (eotectonics and geomorphology of Peoples Republic of Mongolia). Nauka, Moscow

Sinitsyn VM (1959) Tsentralinaya Aziya (Central Aisa). Nauka, Moscow

Wang T (2002) Progress in desertification research of China. In: Integration and regional research to combat desertification. Proceedings of the 16th Global Environment, Tsukuba, Japan, pp. 47–57

Wang T (2004) Progress in sandy desertification research in China. J Geogr Sci 14(4):387–400

Wang T (2011) Deserts and aeolian desertification in China. ELSEVIER in Amsterdam and Science Press in Beijing

Wang T (2014) Aeolian desertification and its control in Northern China. Int Soil Water Conservat Res 2(4):34–41

Wang T, Wu W, Xue X et al (2003) Analysis of the evolution trends of desertification in Northern China. J Desert Res 23(3):230–235 (in Chinese with English abstract)

Wang T, Wu W, Xue X et al (2004a) Study of spatial distribution of sandy desertification in North China in recent 10 years. Sci China (D) 47:78–88

Wang T, Wu W, Xue X et al (2004b) Spatial-temporal changes of sandy desertified land during last 5 decades in northern China. Acta Geograph Sin 59(2):203–212 (in Chinese with English abstract)

Wang T, Xue X, Luo YQ et al (2008) Human causes of aeolian desertification in Northern China. Sciences in cold and arid regions, initial Issue: 1–13

Zhong D (2003) Modern desert evolution in China. SinoMaps Press, Beijing (in Chinese)

Zhu ZD (1989) Advance in desert research in China. J Desert Res 9(1):1–13 (in Chinese with English abstract)

Zhu ZD (1999) Spatial distribution of land desertification in China. In: Sand desert, sandy desertification, desertification and control in China. China Environmental Science Press, Beijing (in Chinese)

Zhu ZD, Chen GT (1994) Sandy desertification in China. Science Press, Beijing (in Chinese)

Zhu ZD, Liu S (1981) Desertification processes and its control regionalization in North China. China Forestry Publishing House, Beijing (in Chinese)

Zhu ZD, Chen ZP, Wu Z et al (1981a) Study of aeolian sand landforms in Taklimakan Desert. Science Press, Beijing (in Chinese)

Zhu ZD, Wu Z, Liu S (1981b) Sand deserts in China. Science Press, Beijing (in Chinese)

Chapter 4
Driving Factors of Aeolian Desertification

Tao Wang, Xian Xue, Bing Liu, and Heqiang Du

Abstract This chapter reviews the natural features and driving factors of aeolian desertification in northern China. Drought, strong winds, and loose sandy surfaces are the main environmental factors that cause aeolian desertification in northern China, and these factors cause arid, semiarid, and dry subhumid lands to be sensitive to climate change and human activities. Climate change in the past 2000 years, 50 years, and the next decades is also introduced to explain the climatic driving forces of aeolian desertification. Human activities such as overgrazing, overfarming, overlogging, and inappropriate water management are analyzed as the driving forces of aeolian desertification. In northern China, there are three aeolian desertification regions. The agropastoral ecotone is located in semiarid and subhumid areas, where overfarming is the leading cause of aeolian desertification. Overgrazing and overcutting result in aeolian desertification in the pure pastoral regions located to the north of the agropastoral ecotone. In the irrigation oases of arid northwestern China, unsustainable water management and use are the fundamental causes of aeolian desertification. Different measures to address aeolian desertification should be developed based on regional characteristics.

T. Wang · X. Xue (✉) · H. Du
Northwest Institute of Eco-Environment and Resources, Chinese Academy of Sciences, Lanzhou, Gansu, China

Key Laboratory of Desert and Desertification, Chinese Academy of Sciences, Lanzhou, Gansu, China

Drylands Salinization Research Station, Northwest Institute of Eco-Environment and Resources, Chinese Academy of Sciences, Lanzhou, Gansu, China
e-mail: xianxue@lzb.ac.cn

B. Liu
Northwest Institute of Eco-Environment and Resources, Chinese Academy of Sciences, Lanzhou, Gansu, China

Key Laboratory of Desert and Desertification, Chinese Academy of Sciences, Lanzhou, Gansu, China

T. Wang et al. (eds.), *Combating Aeolian Desertification in Northeast Asia*, Ecological Research Monographs, https://doi.org/10.1007/978-981-16-9028-0_4

Keywords Aeolian desertification · Climate variability · Driving factors · Human activities

Desertification is the process of land degradation caused by climate variability and human activities. Aeolian desertification is one of the main desertification types and is characterized by wind and sand activities. The driving forces of aeolian desertification include climate variability and human activities. In recent years, climate abnormalities such as droughts caused by global warming, increasingly heavy precipitation events, and reduced average wind speeds have greatly affected the development and reversal of aeolian desertification. Increasing populations increase food demand. Improvements in land productivity and technological advancements also positively and negatively affect the desertification process. This chapter mainly takes the northern region of China as an example to analyze the driving forces of desertification development and reversal in terms of climate variability and human activities. Before analyzing the driving forces, we introduce the natural features and ecological vulnerability of northern China and explain the sensitivity of arid and semiarid regions to climate variability and human activities.

4.1 Natural Features of Aeolian Desertification

Desertification is a process of land degradation that occurs in arid, semiarid, and dry subhumid areas. The most significant feature of the arid area is infrequent and highly variable precipitation, intense evaporation, and windy conditions. Together, drought and strong winds cause sparse vegetation and poor soil quality. In particular, most arid areas consist of paleolacustrine sediments, and sandy sediments dominate the soil. These constitute the natural features of aeolian desertification in northern China.

4.1.1 Climate Characteristics

The geographical location, atmospheric circulation, and geomorphic environment of northern China determine that the region's climate characteristics are dominated by drought and strong winds. Controlled by the subtropical high-pressure airflow, the arid areas in the world are mainly concentrated at 30°S–30°N. The climate is humid in other areas north of 30°N. However, the uplift on the Qinghai-Tibetan Plateau blocks the warm and humid airflow from the Indian Ocean and cuts off the westerly circulation but exacerbates the East Asian monsoon circulation; thus, the area north of 35°N in northern China is controlled by the influence of the Mongolian high. In addition, the eastern, southern, and western sides of the arid zone in northern China are surrounded by mountain ranges of varying heights, blocking the penetration of the warm and humid summer monsoon from the Pacific Ocean. The north is the open and flat Gobi Desert and desert grassland of the People's Republic of Mongolia, and this area is conducive to the northern dry and cold air moving southward. Together, these factors drive the climate of northern China that is characterized by drought and wind (Zhao 1990; PGSC 2000).

Meteorological data show that the precipitation in northern China gradually decreases from 500 mm in the east to approximately 20 mm in the west, which is much lower than that at the same latitude level in the Northern Hemisphere. Because precipitation occurs at the end of the summer monsoon, precipitation is greatly affected by the strength of the monsoon. In some years, the summer monsoon occurs in the arid area. However, the monsoon does not occur in the arid area some years, which results in unstable interannual precipitation. The average annual change rate of precipitation in most areas is 25–40%, and in some areas, it exceeds 50%. The seasonal distribution of precipitation is also extremely uneven. Precipitation in winter accounts for less than 10% of annual precipitation, while in summer, almost 70% of annual precipitation occurs. Although the total rainfall amount in summer is large, it is often concentrated within a few days, making it difficult for plants to utilize the precipitation.

Westerly winds are blocked by the Qinghai-Tibetan Plateau, bypassing the Altai Mountains to the north and moving south at the Sino-Mongolian border, forming a northern branch of airflow. This airflow merges with the airflow generated by Mongolian high pressure, often causing windy conditions (Geng 1986). Therefore, strong winds are another major climatic feature in the arid regions of northern China. According to meteorological records, the number of wind-blown-sand days in northern China is generally 20 to 100 days, with sand and dust storms occurring on as many as 35 to 60 days and floating dust occurring on over 100 days. Wind characteristics vary in different regions. Although there are many strong winds in the western region, they mainly occur in the spring. In the eastern area, there are strong winds in winter, spring, and autumn, and the wind is weak in summer. Sand-dust storm activity is directly related to the magnitude of wind force and is also restricted by the nature of the ground. The dry, loose ground sand material in the western region is susceptible to wind blowing and forms aeolian sandy land. The total wind force in the eastern part is greater than that in the western part during the whole year, but the plant coverage is better than that in the western part. Therefore, the sand-dust storm activity is much less than that in the west.

Therefore, the climate is arid, precipitation is extremely unstable, and windy weather is the primary feature of the climate in the arid area of northern China. These are the main climatic features and dynamic conditions underlying the occurrence of aeolian desertification.

4.1.2 Hydrological Characteristics

Most of the runoff replenishment in northern China is from rainfall, so its distribution is consistent with the annual precipitation and has longitudinal and vertical zonal characteristics. However, due to the influence of natural geographical factors, runoff replenishment also has regional characteristics. According to the depth of runoff, the sand area in northern China can be divided into four subareas. They are extremely arid areas without runoff, arid areas with annual runoff depths of less than 10 mm, semiarid areas with annual runoff depths of 10–50 mm, and dry semihumid areas with runoff depths of 50–200 mm (these areas are defined based on runoff amounts that differ from the definition system based on the aridity index).

In the northwestern arid area, the rivers are called inland rivers. The inland rivers usually originate from the mountains and disappear in the desert hinterland. Along the river, some natural or artificial oases are irrigated by inland rivers. Due to the replenishment of glaciers and snowmelt water and mountain precipitation, the main characteristics of the runoff are mainly spring floods and summer floods. The flood seasons mostly occur from April to June every year. Various reservoirs have been built in the upper and middle reaches of these inland rivers, and the runoff has been mainly controlled manually (Xiao 2000).

In the semiarid grassland areas, many rivers are seasonal. Seasonal rivers are shorter in length, generally less than 400 km; the catchment area is also small; and there is no fixed channel. During flooding, seasonal rivers overflow both banks, forming many inland lakes and wetlands. Because the river is wide and shallow, it is not recharged by groundwater.

The dry subhumid area has a climatic transition, so it has many different river hydrological characteristics. Due to the instability of climate and hydrological systems, river hydrology is very sensitive to environmental changes. The overall performance is as follows: (1) the annual runoff variability is large, the flood peak variation coefficient is large, and there are many drought and flood disasters. (2) Runoff is concentrated in summer, and there is less runoff in spring. (3) The river has a large amount of sediment transport and high sediment content, and the regional differences are obvious. (4) There is a special connection between river water and groundwater.

In addition to surface water, northern China also has abundant groundwater resources. Groundwater is mainly stored in the Meso-Cenozoic tectonic basin in the form of loose sediment pore water. Groundwater characteristics also differ significantly from west to east. In the west, large thick gravel layers are deposited in the piedmont zone on the edge of the basin, an important place for storing groundwater. The Quaternary sediments in the eastern grassland area are extremely thick and rich in groundwater. Since the 1970s, due to the increase in cultivated land area and the popularization of machine well technology, the development and utilization of groundwater in northern China have been very high. Therefore, in terms of groundwater resources, consumption is greater than recharge, and the groundwater level is declining. In addition, the reuse of groundwater in arid areas has also led to increased salinity levels and decreased water quality (Xue et al. 2015).

4.1.3 Soil Characteristics

The soil types in northern China are diverse but mainly consist of Sandic Entisols distributed continuously from east to west. Relatively new sandy facies are a kind of soil with lithological characteristics of sandy sediments in new facies, also known as aeolian sandy soils. The soil is mainly distributed in inland basins and plateaus between 35°–50°N and 75°–125°E (Chen and Li 1999).

Sandic Entisol is a common product of an arid climate and aeolian sand. Aeolian sand has a wide range of material sources, including alluvial river deposits (dry deltas or alluvial fans), alluvial lake deposits, alluvial deposits, and weathered bedrock residues. Due to the arid climate and sparse vegetation, biological effects have a slow impact on the formation of sandy soil. Therefore, the formation process of sandy soil is long, and it is closely related to the process of natural vegetation growth and succession on sandy land. The evolution of aeolian sand into sandy soil can generally involve three stages: fine soil accumulation, crust formation, and organic matter accumulation. With the completion of these three stages, mobile sand or dunes gradually evolved into semifixed sand, fixed sand, and sandy grasslands (Chen et al. 1998).

Because the process of soil formation is slow, the content of organic matter is low, the water holding capacity is poor, and the soil easily blows away. Therefore, sandy soil provides the sand source for the development of aeolian desertification in northern China.

4.1.4 Vegetation Characteristics

Due to the arid climate and barren soil, vegetation in northern China is sparse. Forest cover is low, and its coverage rate is less than 5%. Forests are mainly distributed in the Daxinganling, Xiaoxinganling, and Changbai Mountains in eastern China, the Taihang Mountains in central China, and the Qilian Mountains in western China.

In most areas outside the mountains, grassland is the dominant ecosystem in northern China. Grassland type is consistent with the distribution of precipitation. At the same time, grasslands are also affected by temperature and show characteristics based on longitude and latitude. Meadow steppe, typical steppe, desert steppe, and steppe desert are distributed from east to west. From south to north, the warm temperate zone transitions to the moderate temperate zone. The Son-Nen Sandy Land is located in the moderate temperature, forest-steppe zone in the temperate steppe area. Hulun Buir, Horqin, Hunshandake, and other sandy areas and their surrounding areas are in the moderate temperature typical steppe zone.

Ulanqab-Wulate is a moderate-temperature desert steppe. The eastern half of the Hobq Sand Belt and the Mu Us Sandy Land are in the typical warm temperate steppe. The western half of the Hobq sand belt and the Hedong Sandy Land in Ning Xia are in the warm temperate desert steppe (Liao and Jia 1996).

In those places west of the Helan Mountains, Wushaoling, and East Qilian Mountains, the annual precipitation is less than 200 mm. The vegetation type is temperate desert. The Ulan Buh Desert, Tengger Desert, Badain Jaran Desert, Western Hexi Corridor, Qaidam Basin Desert, Gurbantunggut Desert, and surrounding areas are in the moderate-temperature desert zone. Eastern Xinjiang, southern Xinjiang, the Kumtag and Taklimakan Deserts, and their surrounding areas are in warm-temperature deserts. Vegetation types gradually change from xerophytes to severe xerophytes and ultra-xerophytes from east to west in the region.

Vegetation is an essential factor in protecting surfaces from wind erosion, and the influence of vegetation is especially obvious in grasslands dominated by annual species. Temperature and precipitation greatly determine the greening season of grasslands in northern China. In the years when there is less precipitation in spring, the vegetation turns green later, and the large sandy surface is directly exposed and partly exposed. At the same time, spring is also the season of frequent strong winds in northern China. Wind force acts directly on the surface of the sand, resulting in aeolian desertification (Wang 2011).

According to the above brief introduction to the climate, soil, hydrology, and vegetation, we can see that the aeolian desertified areas in northern China are characterized by drought, strong winds, loose surface materials, and sparse vegetation. These characteristics make the ecosystem extraordinarily fragile and sensitive, and its ability to resist climate fluctuations and the impact of human activities is low. When influenced by external factors, aeolian desertification occurs extremely easily.

4.2 Climatic Driving Force of Aeolian Desertification

Under vulnerable natural conditions, it is difficult for the ecosystem to withstand the impact of climate change and human activities, so aeolian desertification occurs to varying degrees. Since the Western Han Dynasty, aeolian desertification occurred in northern China, and it has aggravated or been reversed based on cyclical climate changes (Zhu and Chen 1994). Understanding the driving mechanism of climate change helps forecast aeolian desertification development in the future and develop prevention strategies in advance. This section mainly reviews climate change in northern China during the historical period and the past 50 years.

4.2.1 Historic Climate Change and Its Impact on Aeolian Desertification

Climate change over the past 2000 years in northern China has mainly occurred as changes in temperature and precipitation. By analyzing historical documents, tree rings, ice cores, lake sediments, and stalagmites, Ge et al. (2013, 2015) proposed four warm periods and four cold periods in northern China (Fig. 4.1). The four warm

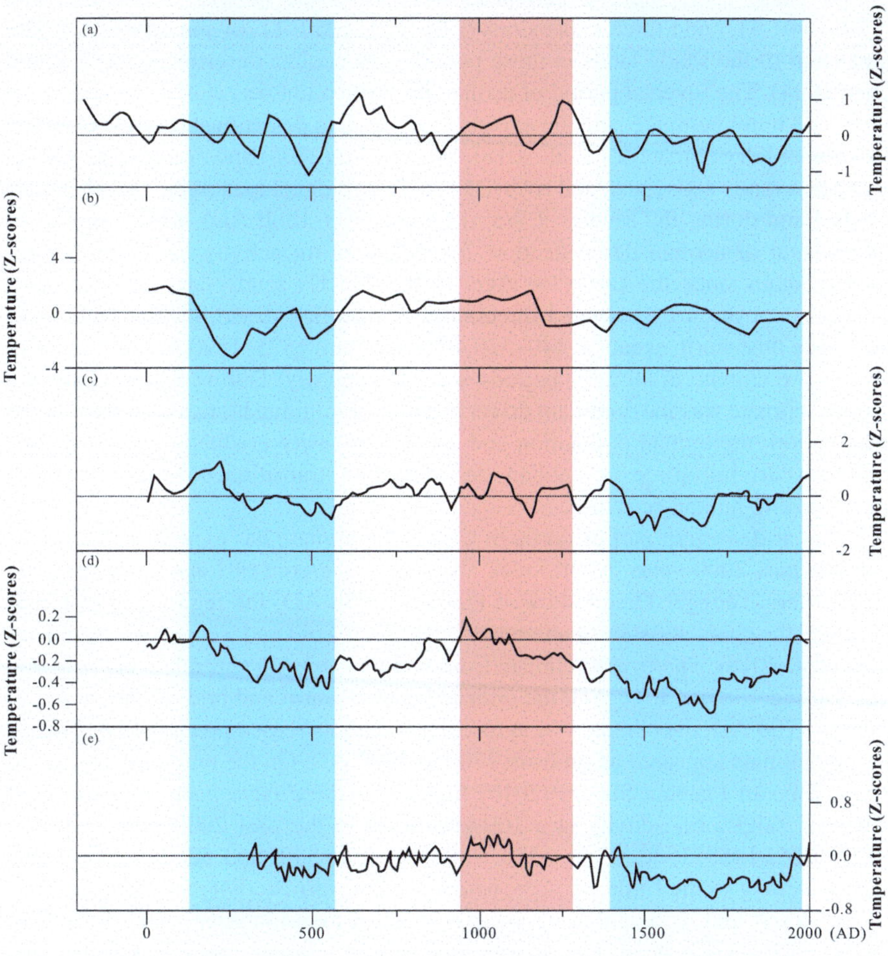

Fig. 4.1 Temperature changes in different regions in the past 2000 years (Ge et al. 2013). (**a**) Winter half-year temperature changes in the central and eastern regions of northern China; (**b**) temperature changes indicated by the sediments of Sugan Lake in Qinghai; (**c**) national temperature changes reconstructed based on multiple pieces of evidence; (**d**) Northern Hemisphere subtropical temperature changes reconstructed based on multiple pieces of evidence; and (**e**) reconstruction based on evidence of temperature changes in the Northern Hemisphere

periods were in 1–200, 550–760, 950–1300 AD, and the twentieth century. The cold periods were in 210–350, 420–530, 780–940, and 1320–1900 AD. The historic precipitation change was different between the eastern and western areas (Fig. 4.2). The boundary line of dry and wet conditions in northern China was approximately 105°E. The west to the line was dry, and the east was wet.

The sandy lands in the semiarid regions of northern China are very sensitive to climate change (Lu et al. 2005, 2013; Mason et al. 2009; Yang et al. 2013, 2015). The sandy lands in the eastern part of northern China have experienced several large fixations and activations in the past 2000 years. For example, the Hulun Buir and Horqin sandy lands have experienced three reversals (0~210, 770~1050, and 1170~1440 AD) and three expansions (210~770, 1050~1170, and 1440 AD). The expansion of the sandy lands in these periods was aeolian desertification (Zhu and Chen 1994). The reversal period of aeolian desertification was closely related to the warm and humid climate, and the expansion of aeolian desertification was related to the cold and dry climate (Li et al. 2018). Wu and Lu (2005) and Zhang et al. (2012) found that sandy soils, peat, and river-lake sediments developed in the Hunshandake Sandy Land during the Medieval Warm Period (700–1300 AD), which implicates the reversal of aeolian desertification. Sand-dust storm activity has occurred frequently again since the Little Ice Age. In the Mu Us sandy land, aeolian sand-paleosol deposition experienced three stages of aeolian desertification with high-frequency dust storm events in 440–570, 840–960, and 1525–1890 AD (Bai and Cui 2019). The climate in all three periods was cold and dry. During the previous two periods, climate was the dominant driver because of minimal human activities. In the third period, the human population and activity intensity gradually increased. The dominant driving force of aeolian desertification transitioned from climate to unsustainable human activities.

In the arid western part of northern China, the climate has experienced changes over the past 2000 years. Multi-index analysis of the core of Lake Ungertu in the southeastern Tengger Desert showed that since 988 AD, the regional climate has exhibited the characteristics of alternating changes in cold and wet and warm and dry (Cao et al. 2018). The overall climate from 988 to 1383 AD was relatively cold and wet. During 1383 to 1560 AD, the climate became warm and arid. In the period of 1560 to 1700 AD, the climate was at its coldest with increased precipitation. Then, a warmer climate appeared again from 1700 to 1900 AD. On the northeastern margin of the Tibetan Plateau, Stauch (2016) found that there have been six periods of enhanced sandstorm activity at a 100-year scale in the past 2000 years, namely, 1630–1725, 1450–1530, 1250–1350, 750–950, 390–540, and 50–225 AD. These periods belong to the Little Ice Age, when the sandstorm activity was strong.

The desert and its neighboring areas in the monsoon marginal zone of northern China have experienced two warm and humid climate periods (0–200 and 900–1300 AD) and two cold and dry climate periods (700–1000 and 1400 AD), which is consistent with the two cold and two warm climate change conditions shown by temperature integration in the Northern Hemisphere. During the warm and wet periods, aeolian desertification reversed. When the climate became colder and dry, wind and sand activities occurred, and aeolian desertification developed.

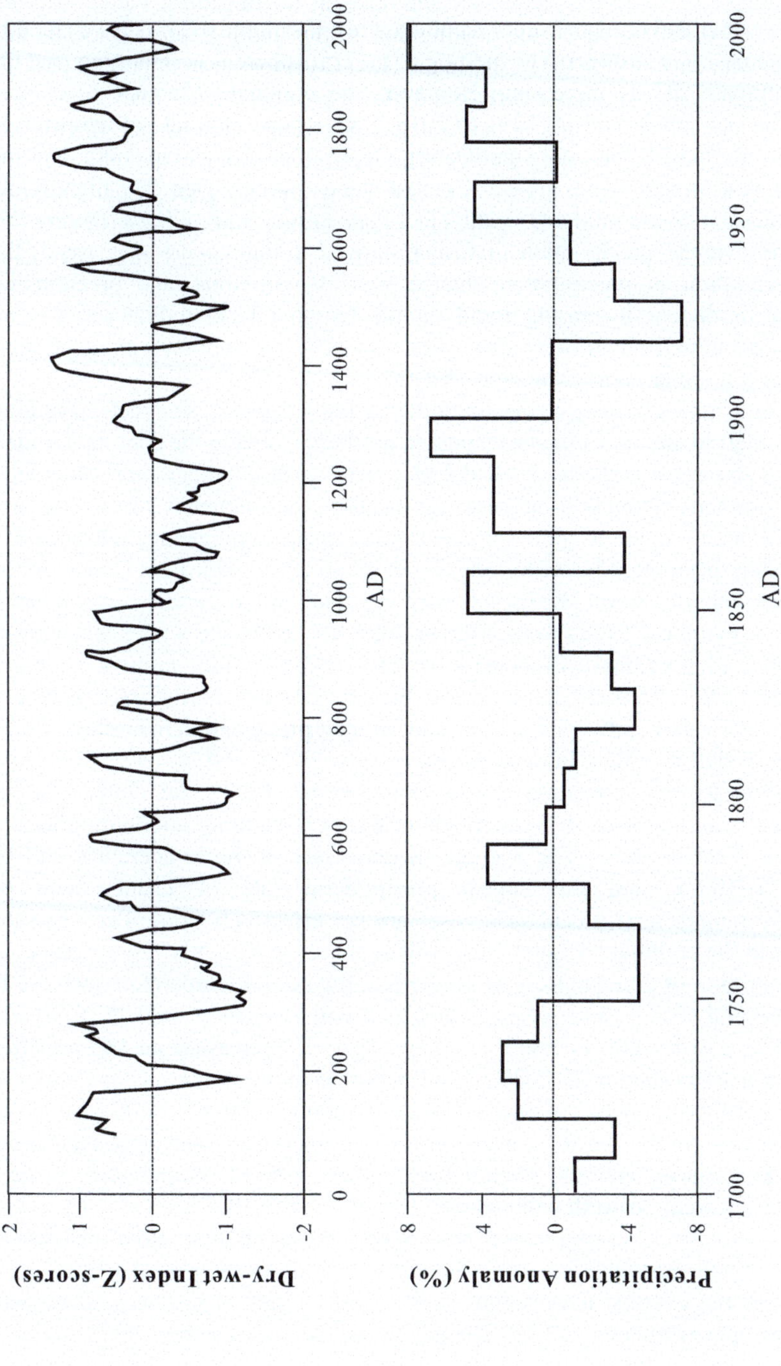

Fig. 4.2 Changes in precipitation (dry and wet) in the monsoon region of eastern China and the arid region of western China in the past 2000 years (Ge et al. 2013). **(a)** Eastern monsoon area; **(b)** western arid area

4.2.2 Recent and Future Climate Change in Northern China

Extreme weather and climate events accompanying warming in China have recently become increasingly stronger. The melting of the cryosphere is accelerating, and the level of climate risk is increasing. However, the regional difference in climate change was very obvious from 1961 to 2018. For example, the rate of temperature increase in northern China was significantly greater than that in the south, and the rate in the western region was greater than that in the eastern region. The highest rate of temperature increase on the Qinghai-Tibet Plateau was 0.37 °C/10 a. During this period, the average precipitation in China showed a slight increasing trend, the regional variation in precipitation change was also obvious, and precipitation showed a significant increasing trend on the Qinghai-Tibetan Plateau, a weak decreasing trend in the southwest area, and a substantial fluctuation with no obvious change trend in other areas (Xue et al. 2012).

Based on China's climate change over the past few decades, scientists have used climate models to simulate climate change in northern China in the coming decades. The results show that in the next few decades, near-surface temperature and precipitation in northern China will increase significantly, the frequency of extreme arid climate will increase, the scope of influence will expand significantly, and there will be significant differences between regions (Hu et al. 2015). Northern China will be the region with the most significant warming and will experience the greatest warming on mainland China from 2015 to 2025. By 2030, the winter temperature increase in northern China will increase by 2.5 °C relative to the multiyear average temperature (1961–1990). In the summer of the next 10 years, northern China will be in a clear water vapor convergence zone, and summer precipitation in northern China will increase significantly (Hu et al. 2015).

According to the simulation results (Fang et al. 2020), the climate in the agropastoral ecotone zone of northern China shows a warming and humidification trend from 2006 to 2050. The average increase rate of the temperature will be 0.2–0.5 °C/10 a, and the annual precipitation rate of change will be 1.49–15.59 mm/10 a. If greenhouse gases are not effectively controlled, then the instability in the regional climate system will increase significantly. For example, as the concentration of greenhouse gases increases, the rate of temperature increase in the agropastoral ecotone zone of northern China will increase from 0.25 °C/10 a to 0.48 °C/10 a, and the rate of change in precipitation will increase from 3.97 mm/10 a to 14.58 mm/10 a (Fang et al. 2020). The Loess Plateau in northern China will also increase significantly with obvious seasonal variation in the next 100 years. The spatial variation coefficient of the average winter temperature is the largest under the same RCP scenario, and the overall trend in the rate of temperature increase gradually from south to north will increase; under the same RCP scenario, the spatial variation coefficient of spring rainfall will be the largest, and the rate of increase in precipitation will gradually increase from north to south (Ren et al. 2019).

By 2030, the average temperature in the western part of northern China will increase by approximately 1.67 °C. The maximum increase will be 1.79 °C and will occur in the western part of Tianshan Mountain and southern Xinjiang. The minimum increase will be 1.56 °C and occur in northern Xinjiang. The eastern region (including western Inner Mongolia, most of Gansu, and central and northern Qinghai) will have a temperature increase between the previously stated values, and the temperature increasing range will be 1.65 °C. The spatial and temporal distribution of precipitation will be more complicated. Overall, precipitation in the western region will increase and relatively decrease in the northern and eastern regions. Overall, drought in northwestern China will increase significantly in the next 10 years (Feng et al. 2019).

4.3 Anthropogenic Mechanism of Aeolian Desertification

As mentioned above, aeolian desertification is mainly caused by unsustainable land use activities in the context of climate variability. Several factors mainly control unsustainable land use. First, excessive land use is often associated with an increase in population. The increase in population will inevitably lead to an increase in demand for food, water, and other necessary materials for living. When the productivity per unit area of land remains unchanged, the increased demand can only be addressed by increasing the use of land resources, such as expanding the area of cultivated land and increasing the number of animals per unit area. Land use is often directly related to state policies. The government will encourage people to change or exacerbate land use due to various needs. For example, at the end of the Qing Dynasty, to pay war compensation, the government encouraged farmers to reclaim the northern grasslands. In the 1930s, the US government promulgated various land policies to attract farmers to carry out reclamation. To build the cotton base in Central Asia, the former Soviet Union called on local people to use the Amu Darya and the Syr Darya to plant vast fields of cotton (Xue et al. 2015). Facts show that land use supported by policies is the most critical factor leading to desertification because the scale of policy is more significant than personal behavior. Changes in the original traditional production methods and improvements in levels of technology will also possibly cause aeolian desertification. When people gradually change from a nomadic civilization to a farming civilization, land reclamation occurs. The increase in drilling for water has made the use of groundwater possible, which has led to the continuous expansion of irrigated agriculture in arid areas. The unrestricted expansion of irrigated agriculture is likely to cause desertification of the land.

The human driving force behind the occurrence and development of aeolian desertification in northern China is also determined by the population size, policies, and production methods. The specific human activities are overfarming, overgrazing, overlogging, overmining, and unsustainable exploiting of water resources. According to research (Wang 2011), aeolian desertified land caused by inappropriate land use accounts for approximately 95% of the total aeolian desertification area in northern China. However, in different regions, the leading factors are different. The aeolian desertified land in northern China is mainly divided into agropastoral ecotones, pure pastoral zones, and irrigation oasis zones in inland river basins. The following sections introduce the driving mechanisms of aeolian desertification in these three regions.

4.3.1 Agropastoral Ecotone

The agropastoral ecotone in northern China mainly refers to the transition zone between the agricultural zone and the pastoral zone in northern China. In this transition zone, planting and grassland animal husbandry are spatially staggered and overlap in time. Land uses shifted with time. In different historical periods, the geographical scope of the agropastoral ecotone in northern China was not the same. Before the Qing Dynasty, the area was mainly distributed on both sides of the Great Wall, but now it is more distributed north of the Great Wall. During the historical period, factors such as climate change and ethnic migration often led to the northward or southward invasion of the agropastoral ecotone. At present, the pattern of the agropastoral ecotone is more controlled by changes in policies and land use methods.

At present, the agropastoral ecotone in northern China is mainly located to the north of the Great Wall and is a semiarid temperate zone. The ecotone consists of the Horqin region, the regions in southern Inner Mongolia and northern Hebei Province (SIM and NH), and the Ordos region (Fig. 4.3). The precipitation in this ecotone is about 200–400 mm, and the evaporation is 2000–2500 mm. The natural landscape is mainly dry steppe, including part of a sparse forest steppe and desert steppe. Almost all the regions were nomadic pastures 200 years ago. From the perspective of geographical area, the ecotone is located at the junction of semiarid and subhumid areas. From the perspective of population and cultural distribution, the ecotone is located at the junction of the Han population with a farming culture and a minority population with nomadic culture. Because this ecotone is at the edge of the monsoon, the interannual and intra-annual changes in precipitation in this area are quite significant. When the monsoon is strong, it often produces abundant precipitation, and when the monsoon is weak, the opposite occurs. These characteristics are concrete examples of the vulnerability and sensitivity of the geographical environment in the agropastoral region, and this vulnerability and sensitivity serve as the main sources of the generation and development of land desertification.

The inhabitants of the agropastoral ecotone 200 years ago were pastoral minority populations, and grazing was dominant because strong winds, drought, and a loose soil surface were not conducive to cultivation. However, since the eighteenth century, agriculture has gradually replaced animal husbandry (Wang et al. 2015). According to research (Yang and Ta 2002), farmland increased from 50,200 km^2 in 1952 to 82000 km^2 in 1998 in Inner Mongolia. The same research also showed that the farmland area increased by 1150 km^2 from 1915 to 1928 and reached 13,330 km^2 in 1949 in the Ordos region. Other studies have shown that farmland increased from 700 km^2 in 1948 to 7300 km^2 in the late 1980s and natural pasture decreased from

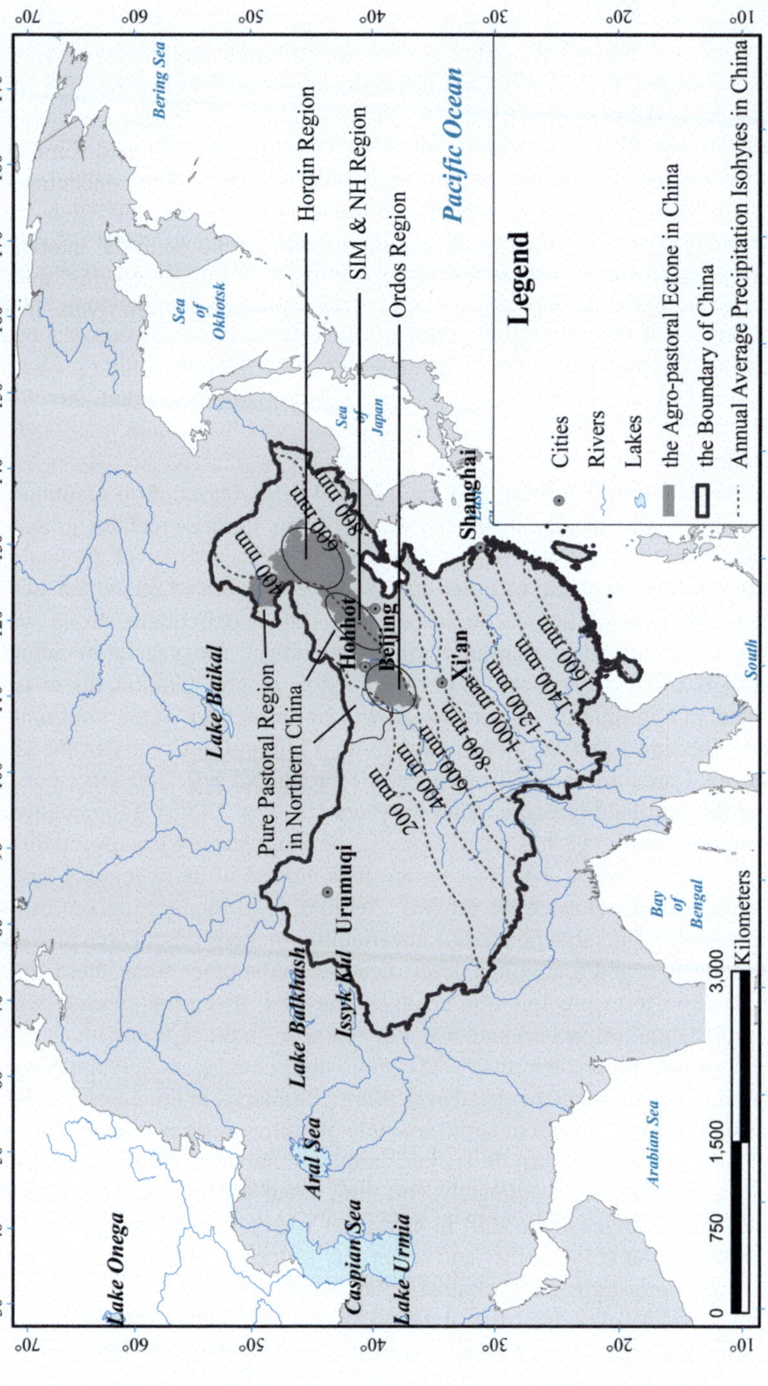

Fig. 4.3 The location of the current agropastoral ecotone and pure pastoral region in northern China (revised according to Wang et al. 2015)

8700 to 2700 km^2 in the same period in northern Hebei Province (Han and Han 2003; Sheng et al. 2003). Consequently, land use conversion extended farmlands northwestward and pushed the boundary of the historical agropastoral ecotones northward by an average of 180–220 km into the Booe typical steppe or desert steppe (Fig. 4.4) (Wang et al. 2008).

Reclamation activities have created favorable conditions for surface sandy particles to be blown, transported and accumulated under the force of the winder, which caused a significant change in the original landscape of the sandy grassland. Therefore, reclamation is the main cause of aeolian desertification in grassland ecosystems. As mentioned above, land use patterns mainly occur based on two situations: one is the change in the natural environment, and the other is the change in national policies. Changes in the natural environment often cause farmers to spontaneously change land use patterns. When the climate is dry and cold, agricultural plantations will be replaced by animal husbandry due to limited conditions. When the climate changes toward warm and humid conditions, people usually reclaim grasslands for crop plantations.

Studies have shown that there is often a period of increased crop planting area after a wet year (Xue et al. 2005a). However, when farmers reclaim grasslands spontaneously, they often overlook the geographical characteristics of the land they inhabit differently from those of other areas. These overlooked characteristics are large precipitation variability, loose surface material, and frequent strong winds. People forget that abundant rainfall is only temporary. The year after abundant precipitation is often a dry year. With a reduction in precipitation, there is not enough water to continue to support the growth of crops. Cultivated land must be abandoned (Zhu and Liu 1981). For example, in Yijinhuoluo County in the Ordos region, the total area converted from pasture to farmland was 700 km^2, but only 300 km^2 of the farmland remained after 20 years (Jia et al. 2003). When cultivated land is abandoned and grass has not yet grown, the loose surface is exposed directly to the wind. Fine particles and nutrients are thus eroded in these regions, and the sediment moves to the downwind regions. Aeolian desertification occurs in these downwind regions. The large precipitation variability in the semiarid area provides a good natural background for aeolian desertification. In the other areas, there are also long-term or short-term precipitation changes, but such precipitation changes will not have a substantial impact on land use. For example, in the region with an annual average precipitation greater than 800 mm, the change in precipitation of 100–200 mm will not change the local agriculture. Similarly, in the desert hinterland with an average annual rainfall of approximately 50 mm, a certain year or years with rainfall of 100 mm will not turn desert into farmland. However, in the agropastoral ecotone with rainfall of approximately 400 mm, small precipitation changes may lead to landscape changes and a shift in land use (Xue et al. 2005a).

Being aware of the vulnerability and sensitivity of the geographical environment and formulating long-term and sustainable land uses that adapt to environmental changes are not determined by the ability of some individuals but must rely on national policies. Large-scale army reclamation during the Qin and Han Dynasties, subsidy reclamation in the late Qing and the Republic of China, migration

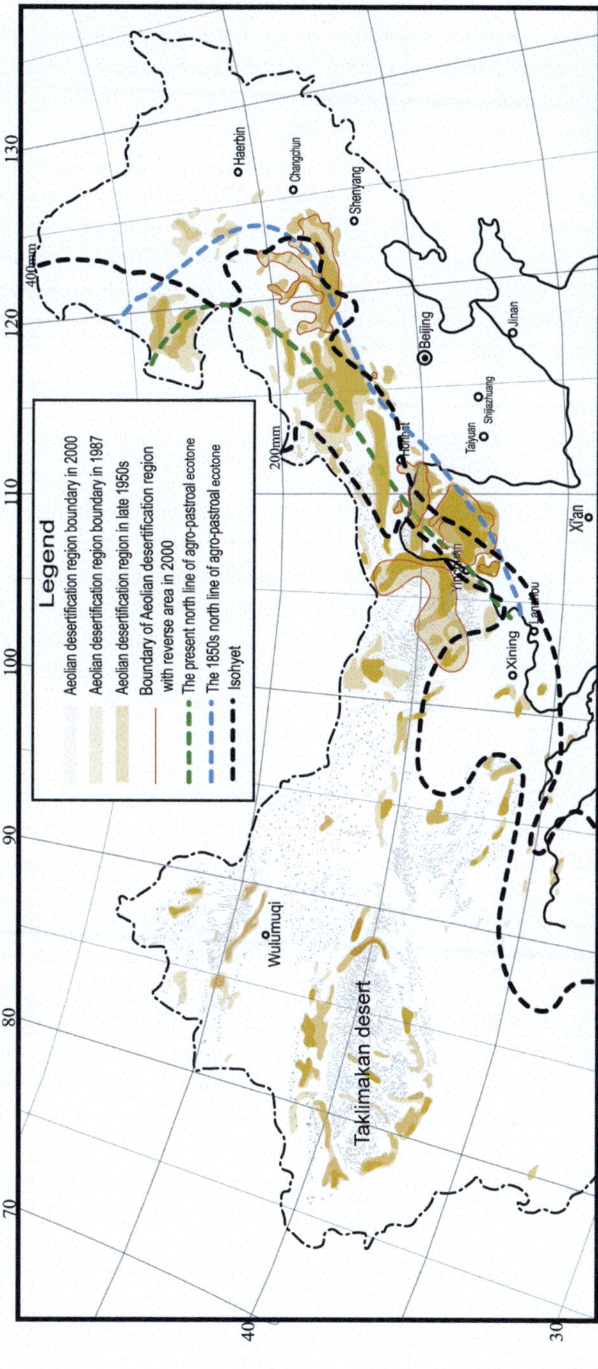

Fig. 4.4 The change in the agropastoral ecotone of northern China in the last 200 years (revised according to Wang et al. 2008)

reclamation in the 1950s and 1960s, and the conversion of farmland to forests and grasslands after the 1980s are all nationally dominated land uses. When land use changes are coordinated with the evolution of the natural environment, the human-land system is at an equilibrium, and it has a strong resistance to external interference. The reversal of aeolian desertified land in some areas of northern China in the past 20 years mainly comes from the implementation of the policy of returning farmland to forest and grassland. In contrast, when policy development only considers short-term economic benefits and ignores long-term ecological and social benefits, it leads to an unbalanced relationship between people and land. When the external conditions change slightly, various ecological problems, economic problems, and even social problems will appear in the system, such as aeolian desertification in the agropastoral ecotone before the 1980s.

4.3.2 Pure Pastoral Region

Pure pastoral areas refer to areas where grazing is the only or dominant land use activity. China's pure pastoral areas are mainly distributed in the temperate grasslands in the north and the alpine grasslands on the Qinghai-Tibetan Plateau. Aeolian desertification of the alpine grasslands on the Tibetan Plateau is largely affected by climate change (Xue et al. 2009). The pure pastoral area discussed here refers to the temperate grassland north of the agropastoral ecotone in northern China (Fig. 4.3). The precipitation in this area ranges mostly from 100 to 300 mm. Landscape types include typical steppe, dry steppe, desert steppe, and steppe desert. Overgrazing is the main cause of aeolian desertification in pure pastoral areas. In recent years, with the popularization of solar energy and wind energy, overlogging has been effectively resolved. However, before the end of the last century, of the aeolian desertification caused by the range of human activities, that caused by excessive woodcutting was the largest.

There are two reasons for overgrazing in pure pastoral areas. On the one hand, because the original grassland was reclaimed to farmland, the grassland area was greatly reduced. Even when the number of livestock remained unchanged, the livestock load per unit of grassland increased significantly. This scenario is often referred to as relative overgrazing. On the other hand, the total number of livestock has also increased significantly, which is called absolute overgrazing. Inner Mongolia is the main distribution area of pure pastoral husbandry in northern China. Research (Wang 2011) shows that the livestock amount in Inner Mongolia increased from 2447.3×10^4 sheep units in the 1950s to 6460.14×10^4 sheep units at the end of the last century. In the 1990s, the theoretical livestock carrying capacity in Inner Mongolia was 4837×10^4 sheep units, and the overload rate reached 33.56%. Overloading was even more severe in other regions. For example, overload rates in the same period were 135.70% and 138.93% in Gansu Province and Ningxia Province, respectively. In addition to the increase in livestock amount, a decrease in grassland productivity also resulted in overloading in the pure pastoral regions.

According to our investigation in Xiuzhumqin Qi (county) of Inner Mongolia, 0.54 ha of grasslands could support one sheep in 1955, but the data showed that 1.09 ha of grasslands could support one sheep in 1982 and 2.45 ha of grasslands in 2000 (unpublished data).

Overgrazing mainly occurs because an increase in the number of people and improvements in living standards increase the demand for products such as meat, milk, wool, and cashmere. In addition, improvements in herdsmen's own living standards and an increasing emphasis on education have prompted them to expand their livestock scales as much as possible to obtain increased economic benefits. These are the main reasons for the current overgrazing issue. Notably, herdsmen's instincts for pasture protection and the grazing experience will not cause severe damage to pastures due to overgrazing. However, frequent climate fluctuations in recent years have often resulted in unpredictable consequences. The abundant precipitation in the spring increased grassland biomass, and the number of livestock increased rapidly. However, when the next year's drought arrived or the rainy season was postponed, the productivity of the vegetation decreased, and the grassland was overeaten by the increased livestock, making it difficult to recover it in the short term. For example, in 2000–2001, the drought of Sunite Zhuo Qi (a county of Inner Mongolia) decreased the livestock carrying capacity of a grassland by approximately half, which made the region become one of the most severely aeolian desertified regions in northern China.

For both absolute overgrazing and relative overgrazing, the number of livestock exceeds the carrying capacity of a pasture. An excessive load on a pasture will not only decrease the vegetation coverage but also induce wind erosion. More importantly, over browsing by livestock destroys the root system, reduces the number of seeds and germination capacity, and inhibits vegetation restoration, which reduces the stability of the ecosystem. In addition, under heavy trampling of livestock, the surface crusts are broken, sands are exposed, and wind erosion increases. Remote sensor monitoring shows that the aeolian desertification occurred in 25–35% of the pure pastoral region in 2010.

The aeolian desertification process caused by overgrazing is more obvious around wells. Field investigations show that within a radius of 500 m around a water well or livestock habitat, native vegetation is destroyed and presents a sandy or gravel textured surface. Within a radius of 500–1000 m, only uneatable or even poisonous weeds grow. Forage grass gradually appears 1000 m away from wells. Therefore, the emergence of this kind of landscape with different types of grasslands centered around water wells usually represents the emergence of aeolian desertification. In addition to wells and livestock habitat sites, this aeolian desert landscape caused by overgrazing is also distributed around some small lakes.

In a grassland area, after excessive browsing and trampling damage the surface vegetation and crusts, wind erosion begins to occur. As the degree of wind erosion increases, wind erosion dens first appear on the surface. The sizes of wind erosion dens continue to increase, and sometimes large wind erosion pits are formed. Fixed dunes or semifixed dunes with higher vegetation coverage, affected by overgrazing and overlogging, will be gradually activated and eventually form mobile dunes.

Therefore, fixed sand dune activation and sandy grassland wind erosion are the two main types of land desertification that occur in pure pastoral areas. The particles eroded by wind can be transported and accumulate several meters to hundreds of meters downwind. Fine silt and clay particles easily become airborne and form dust storms in downwind areas (Liu et al. 2008). Sandy grasslands might be one of the principal dust sources for Beijing and its surrounding area. Furthermore, sand particles move downwind along the land surface and form sand patches or active dunes when they encounter barriers. After long periods of wind erosion, the land becomes coarse and loses productivity.

In the process of aeolian desertification of grasslands, overfarming and overgrazing affect each other, and they are closely related.

4.3.3 Irrigation Oasis Zone in Inland River Basins

Because irrigation oases provide habitats for animals and humans in drylands (Mainguet 1991), they are vitally essential for people living in drylands (Reynolds et al. 2007; UNCCD 2008). Irrigation oases only occupy 4–5% of arid northwestern China (Fig. 4.5) but support 90% of the local population and produce 95% of social wealth (Li et al. 2007). Water is the most fundamental resource for the sustainability of oases because it sustains the lives of local people, livestock, and agriculture, as well as environmental health and ecosystem services. However, unsustainable water use is widespread at almost all these oases are also the leading cause of aeolian desertification in these areas. Mainguet (1991) said, *where is the wrong management method, where there is wind erosion.*

Irrigation oases in northwestern China are distributed in a beaded pattern along the inland rivers from the middle reaches to the lower reaches (Fig. 4.5). The rivers originate in the mountainous areas in the upper reaches and flow through oases and deserts, which make up the inland river catchment or watershed. Environmental factors such as hydrology, climate, soil, and vegetation in the same catchment, as a subsystem, are interconnected and interdependent. By changing one or several subsystems, an entire river basin system can be changed. In addition, various natural geographic processes are consistent, and various human activities occur in a watershed system. This also means that land use in the upstream region of a basin system is closely related to environmental change in the midstream and downstream regions. A reduction in upstream water intake is sometimes controlled by natural factors, but often, a reduction is due to an excessive use of water resources from the unrestricted expansion of farmland. This conflict of water resource use in basins has occurred consistently for more than 2000 years and reached its peak in the 1980s (Xue et al. 2005b, 2015).

During the 1970s–1980s, the central and local governments called for the construction of irrigated oases in the northwest that transformed the area into a base of commercial grain and cotton. Encouraged by the policy, reservoirs were built in the upper and middle reaches to block floods and expand cultivated land. This caused

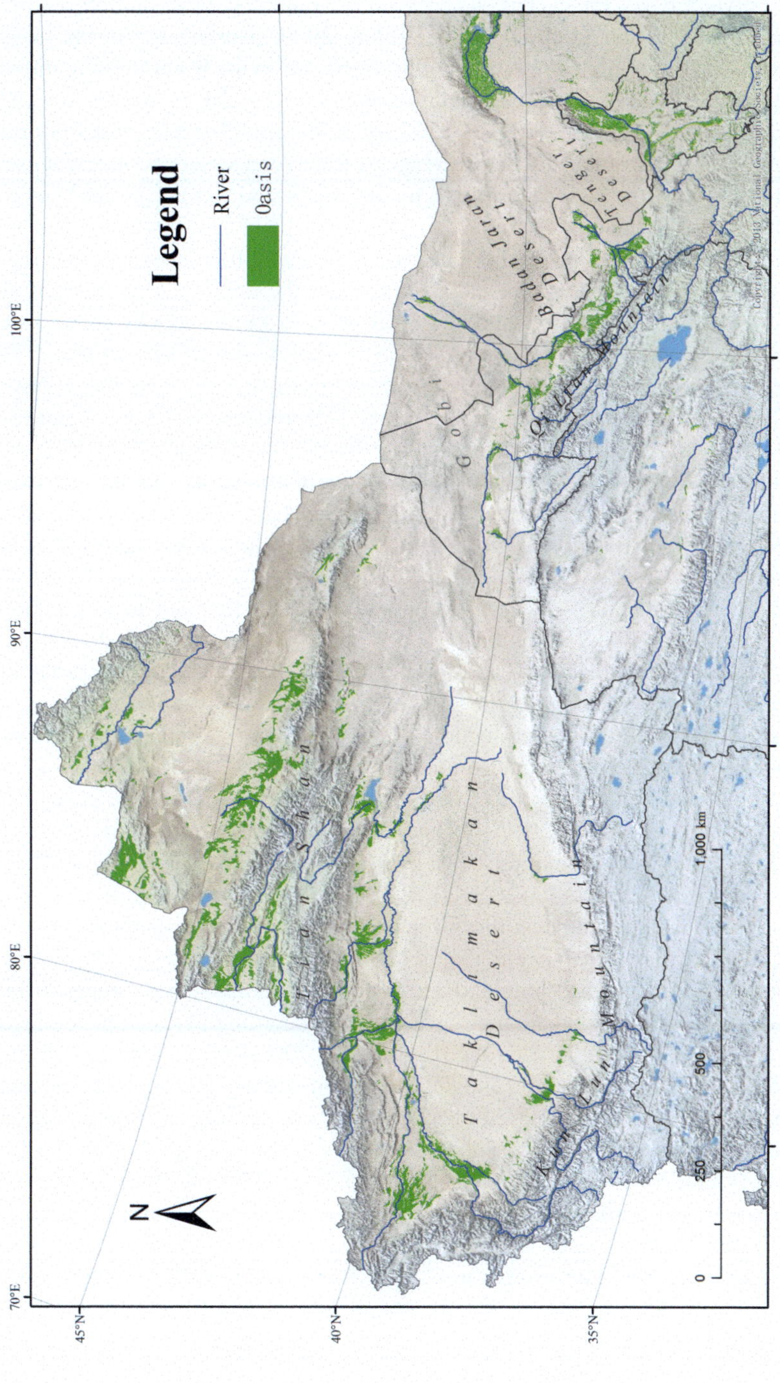

Fig. 4.5 The distribution of irrigation oases in northwestern China

the water volume in the lower reaches of the basin to continually decrease. Considering the Shiyang River Basin (Fig. 4.5), which is located on the easternmost side of the Hexi Corridor, as an example, from 1971 to 2000, overfarming in the middle reaches led to an accumulated reduction of 30×10^8 m^3 in the water flowing into the downstream oasis, with an average reduction rate of 0.83×10^8 m^3 $year^{-1}$. In 2000, the per capita water consumption was 1552 m^3 in the middle reach oases but only 537 m^3 in the downstream oasis. The shortage of fresh river water and the increase in farmland irrigation forced local farmers to dig wells and extract groundwater to irrigate their crops in the downstream oasis (Xue et al. 2015). Groundwater has been used to flood irrigated cropland over many years. Frequent flood irrigation has led to a decrease in the groundwater table and an increase in the salinity of the groundwater because of intense evaporation and high salt concentrations. Field surveys and literature analysis show that the average value of total dissolved salt content (TDS) in the downstream oasis groundwater increased from 0.7 g l^{-1} in the 1950s to 2.5 g l^{-1} in 2005. In the northern oasis region, TDS reached 6–10 g l^{-1} and even as high as 16 g l¯1 in the northern end area of the oasis (Li et al. 2006; Zhu et al. 2007). Similar patterns in water quantity and quality degradation from the overuse of water resources also exist in other arid areas (Wada et al. 2010), such as the Heihe Basin, the Sule River Basin, and the Tarim River Basin in northwestern China and the Aral region in Central Asia.

Therefore, unsustainable water use and management first led to a decline in groundwater levels and deterioration of water quality. After the groundwater table lowered, natural and artificial vegetation dried, and shelter forest systems protecting oases became degraded because the tree roots could not absorb water from the shallow soil layer. Without the protection of the shelter forest, dunes moved toward the oases and buried farmlands and settlements (Xue et al. 2015). Since the 1950s, desert areas located in the north of the Shiyang River Basin have encroached southward into the oasis by 50–70 m and destroyed 400 ha of farmland. Desert areas located in the west have moved east by 30–60 m and have destroyed 467 ha of farmland (Zhang et al. 2004; Sun et al. 2005; Dong et al. 2010). Irrigation with saline or high-sodium water resulted in the formation of alkaline soil, damaging the soil structure. The farmland areas were abandoned because of their high salinity levels and low productivity. Saline wasteland and abandoned farmland areas increased the danger of wind erosion and aeolian desertification. According to research conducted in the downstream oasis of the Shiyang River Basin, 70% of local farmland has been affected by aeolian desertification, and 85% has been affected by different degrees of salinization (Sun et al. 2007). Based on satellite remote sensing, aeolian desertification rapidly developed from the 1970s to the 1990s in the lower part of the Shiyang River Basin. Aeolian desertified land has increased by 1.13×10^4 ha at the margins of the downstream oasis (Zhang et al. 2004). Severe aeolian desertification has resulted in the lower reaches of the Shiyang River Basin becoming one of the primary sources of sand-dust storms in China (Wang et al. 2004).

4.4 Conclusion

Although climate variability and human activities are the leading causes of aeolian desertification, the natural environment of areas with aeolian desertification also plays an important role. The vulnerability and low carrying capacity of ecosystems are the main characteristics of the natural environments in arid areas. Generally, the vulnerability of ecosystems in arid areas makes these areas more prone to degradation than humid areas when short-term climate variability occurs. The lower carrying capacities of arid ecosystems result in human activities having a larger impact there than in other areas. This scenario indicates that land use and management in arid areas need more attention than those in other regions. Given global warming, extreme events can cause the arid ecosystem to become degraded rapidly. Developing sustainable land management policies to cope with future climate change is essential and urgent.

In addition, when carrying out aeolian desertification prevention and control, appropriate policies and measures should be developed according to regional characteristics. For example, in dry subhumid areas, reducing excessive farming may be the main means to control aeolian desertification, but in semiarid areas, a reasonable animal number based on the local carrying capacity and future climate change needs to be determined. Sustainable water use and management are the fundamental solution for aeolian desertification in the irrigated oases of arid areas.

References

Bai ZZ, Cui JX (2019) Desertification and its causes in mu us desert in recent 2 000 years. J Desert Res 39(2):177–185 (in Chinese with English abstract)

Cao ZH, An CB, Yin LY et al (2018) Climate change derived from Anggeertu Lake in the Tengger Desert since 988 AD. Arid Land Geography 41(6):1251–1259

Chen LH, Li FX (1999) Sandy neo-soils. In: Gong ZT et al (eds) Classification of soil system in China: theory, methods and practices. Science Press, Beijing

Chen LH, Li FX, Di XM (1998) Aeolian soils in China. Science Press, Beijing

Dong ZB, Man DQ, Luo WY (2010) Horizontal aeolian sediment flux in the Minqin area, a major source of Chinese dust storms. Geomorphology 116(1–2):58–66

Fang ZX, He CY, Liu ZF, Zhao YY, Yang YJ (2020) Climate change and future trends in the Agro-Pastoral Transitional Zone in Northern China: the comprehensive analysis with the historical observation and the model simulation. J Nat Resour35(2):358–370 (in Chinese with English abstract)

Feng SQ, Wang HE, Liu YX, Guo YF (2019) Simulation and prediction of climate change trendin the next 10 years over Northwest China. J Arid Meteorol 37(4):557–564

Ge QS, Liu HL, Zheng JY et al (2013) The climate change and social development over the last two millennia in China. Chin J Nat 35(1):9–21 (in Chinese with English abstract)

Ge QS, Hua Z, Zheng JY, et al (2015) Forcing and impacts of warm periods in the past 2000 years. Chin Sci Bull 60:1727–1734 (in Chinese with English abstract)

Geng KH (1986) Climate in desert regions in China. Science Press, Beijing

Han YW, Han JG (2003) Soil content of phosphorus and potassium in the agro-pastoral transitional zone after cropland being transformed into grassland. Acta Agrestia Sinica 11:228–233 (in Chinese with English abstract)
Hu S, Lin XG, Mo ZH (2015) Projections of spatial-temporal variation of drought in North China. Arid Land Geography 38(2):239–248
Jia BQ, Ci LJ, Gao ZG (2003) The quantitative analysis of effect factors on sandy desertification in Ordos plateau. Scientia Silvae Sinicae 39:15–20 (in Chinese with English abstract)
Li XY, Xiao DN, He XY, Chen W, Song DM (2006) Dynamics of typical agricultural landscape and its relationship with water resource in inland Shiyang River watershed, Gansu province, Northwest China. Environ Monit Assess 123:199–217
Li WD, Li ZZ, Wang JQ (2007) Evaluation of oasis ecosystem risk by reliability theory in an arid area: a case study in the Shiyang River Basin. Chin J Environ Sci 19(4):508–512
Li JC, Han LY, Zhang GM, Wu ZT (2018) Insights on historical expansions of desertification in the Hunlun Buir and Horqin Deserts of Northeast China. Ecol Indic 85:944–950
Liao GF, Jia YL (1996) Rangeland resources of China. Chinese S&T Press, Beijing
Lu HY, Miao XD, Zhou YL, Mason J, Swinehart J, Zhang JF, Zhou LP, Yi SW (2005) Late quaternary aeolian activity in the Mu Us and Otindag dunefields (North China) and lagged response to insolation forcing. Geophys Res Lett 32:365–370
Lu HY, Yi SW, Xu ZW, Zhou YL, Zeng L, Zhu FY, Feng H, Dong LN, Zhuo HX, Yu KF, Mason J, Wang XY, Chen YY, Lu Q, Wu B, Dong ZB, Qu JJ, Wang XM, Guo ZT (2013) Chinese deserts and sand fields in Last Glacial Maximum and Holocene Optimum. Chin Sci Bull 58(23): 2775–2783 (in Chinese with English abstract)
Mainguet M (1991) Desertification natural background and human mismanagement. Springer, Heidelberg
Mason JA, Lu HY, Zhou YL, Miao XD, Swinehart JB, Liu ZY (2009) Dune mobility and aridity at the desert margin of northern China at a time of peak monsoon strength. Geology 37:947–950
Project Group of Soil Classification (2000) Classification of soil system in China, 3rd edn. China Agricutlure Science and Technology Press, Beijing
Ren JY, Zhao JX, Ma HB, Peng SZ, Li BY (2019) Spatiotemporal distribution characteristics of seasonal climate change trends over loess plateau during 2015-2100. Bull Soil Water Conserv 39(5):262–271
Reynolds JF, Stafford SDM, Lambin EF et al (2007) Global desertification: building a science for dryland development. Science 316:847–851
Sheng XB, Liu YX, Sun JZ (2003) The relation between land use/cover and soil nutrients change in Bashang regions. Quater Sci 23:1
Stauch G (2016) Multi-decadal periods of enhanced aeolian activity on the North-Eastern Tibet Plateau during the last 2ka. Quat Sci Rev 149:91–101
Sun D, Dawson R, Li H, Li BG (2005) Modeling desertification change in Minqin County, China. Environ Monit Assess 108:169–188
Sun D, Dawson R, Li H, Wei R, Li BG (2007) A landscape connectivity index for assessing desertification: a case study of Minqin County. Chin Landsc Ecol 22(4):531–543
UNCCD (2008) Desertification-coping with today's global challenges in the context of the strategy of the UNCCD. UNCCD, Bunn
Wada Y, Beek LPH, Kempen CM, Reckman JWTM, Vasak S, Bierkens MFP (2010) Global depletion of groundwater resources. Geophys Res Lett 37:L20402
Wang T (2011) Desert and aeolian desertification in China. Science Press, Beijing
Wang XM, Dong ZB, Zhang JW, Liu LC (2004) Modern dust storms in China: an overview. J Arid Environ 58:559–574
Wang T, Xue X, Luo YQ et al (2008) Human causes of aeolian desertification in Northern China. Sci Cold Arid Reg 1:13
Wang T, Xue X, Zhou LH, Guo J (2015) Combating aeolian desertification in nNorthern China. Land Degrad Dev 26:118–132

Wu JW, Lu RJ (2005) Spatial pattern and landscape characteristics in Otindag Sandy Land during the medieval warm period. J Arid Land Resour Environ 19(7):110–113

Xiao HL (2000) Resource environment and Management of Water in China. Kaiming Press, Langfang, Hebei

Xue X, Wang T, Wu W, Sun QW, Zhao CY (2005a) Development and cause of desertification in agro-pastoral region of North China. J Desert Res 25(3):320–328 (in Chinese with English abstract)

Xue X., Wang T., Yao ZY, Peng GC (2005b) Sandy desertification status and region development in Shiyanghe River Basin, J Desert Res, 25(5):682–688 (in Chinese with English abstract)

Xue X, Guo J, Han BS, Sun QW, Liu LC (2009) The effect of climate warming and permafrost thaw on desertification in the Qinghai-Tibetan plateau. Geomorphology 108(3–4):182–190

Xue X, Wang T, Guo J, Yang YH, Peng F, Liu LC (2012) Impact of Global Warming on drought in China. Sci Cold Arid Reg 4(3):0201–0210

Xue X, Liao J, Hsing YT, Huang CH, Liu FM (2015) Policies, land use, and water resource management in an arid oasis ecosystem. Environ Manag 55(5):1036–1051

Yang GS, Ta WQ (2002) Sandstorm disasters and controlling ways in the agricultural and pasturing interlaced zone of Ningxia, Inner Mongolia and Shaanxi. J Desert Res 22:453–465 (in Chinese with English abstract)

Yang XP, Wang XL, Liu ZT, Li HW, Reng XZ, Zhang DG, Ma ZB, Rioual P, Jin XD, Scuderi L (2013) Initiation and variation of the dune fields in semi-arid China-with a special reference to the Hunshandake Sandy Land, Inner Mongolia. Quat Sci Rev 78:369–380

Yang XP, Scuderib LA, Wang XL, Scuderi L, Zhang DG, Li HW, Forman S, Xu QH, Wang RC, Huang WW, Yang SX (2015) Groundwater sapping as the cause of irreversible desertification of Hunshandake Sandy lands, Inner Mongolia, northern China. Proc Natl Acad Sci U S A 112: 702–706

Zhang KC, Qu JJ, Liu QH (2004) Environmental degradation in the Minqin oasis in Northwest China during recent 50 years. J Environ Syst 31(4):357–365

Zhang XY, Zhou YL, Pang JL et al (2012) The relationship between environmental changes and human activities during the medieval warm period and the little ice age in Otindag sandyland by OSL dating. Quater Sci 32(3):535–546

Zhao SQ (1990) Arid zone in China. Science Press, Beijing

Zhu ZD, Chen GT (1994) Sandy desertification in China. Science Press, Beijing

Zhu ZD, Liu S (1981) Sandy desertification processes and prevention zoning in Northern China. Chinese Forestry Press, Beijing

Zhu GF, Li ZZ, Su YH, Ma JZ, Zhang YY (2007) Hydrogeochemical and isotope evidence of groundwater evolution and recharge in Minqin Basin, Northwest China. J Hydrol 333:239–251

Chapter 5
Aeolian Desertification Processes

Tao Wang, Xueyong Zhao, and Tonghui Zhang

Abstract Aeolian desertification is a dynamic process that involves quantitative and qualitative changes, including changes in landform features, patterns, and landscape types and changes in plants, soil animals, and microbes, which are the physical processes and biological processes of aeolian desertification, respectively. The physical processes of aeolian desertification are marked by wind-sand activities and can be generally classified into four types: the sand dune (sandy land) mobilization process, the aeolian desertification of grassland shrubs process, the soil-layer wind erosion and coarsening process, and the badland formation process resulting from uneven wind dissection. The biological processes of aeolian desertification are dominated by primary vegetation being replaced by pioneer sandy vegetation and sandy vegetation, and in most cases, this process begins with a decrease in vegetation due to overuse, deforestation, overgrazing, fuel wood collection, and other uses of plant for survival and producing resources. Under favorable environmental conditions, aeolian desertification can undergo reverse succession. Therefore, it is very important to address the physical and biological processes of aeolian desertification and restoration.

T. Wang (✉)
Northwest Institute of Eco-Environment and Resources, Chinese Academy of Sciences, Lanzhou, Gansu, China

Key Laboratory of Desert and Desertification, Chinese Academy of Sciences, Lanzhou, Gansu, China

Drylands Salinization Research Station, Northwest Institute of Eco-Environment and Resources, Chinese Academy of Sciences, Lanzhou, Gansu, China
e-mail: wangtao@lzb.ac.cn

X. Zhao · T. Zhang
Northwest Institute of Eco-Environment and Resources, Chinese Academy of Sciences, Lanzhou, Gansu, China

Naiman Desertification Research Station, Northwest Institute of Eco-Environment and Resources, Chinese Academy of Sciences, Lanzhou, Gansu, China

T. Wang et al. (eds.), *Combating Aeolian Desertification in Northeast Asia*,
Ecological Research Monographs, https://doi.org/10.1007/978-981-16-9028-0_5

Keywords Physical process · Wind-sand activity · Aeolian landform · Biological process · Human activity · Biodiversity

5.1 Introduction

Aeolian desertification involves processes related to not only primary vegetation replacement by pioneer sandy vegetation and sandy vegetation but also soil coarsening and barrenness, which are considered the physical processes and biological processes of aeolian desertification, respectively. The evolution of aeolian desertification is a process that involves quantitative and qualitative changes. With the increase in the degree of aeolian desertification, the threshold wind speed of sand movement decreases, the sand transport capacity increases, and the near surface of sandy land has a relatively high wind energy environment. The aeolian desertification process is dominated by the activation of fixed sand dunes, which leads to the fragmentation of landscapes and the appearance of extensive moving sand in many areas. At the same time, the risk of soil wind erosion caused by land reclamation cannot be ignored. Under favorable environmental conditions, the process of aeolian desertification can undergo reverse succession.

5.2 Physical Processes of Aeolian Desertification

5.2.1 The Formation of Aeolian Landforms

The development of aeolian sand geomorphology is the result of the interaction between air flow and sandy surfaces and is affected by surface fluctuations, sand source supply and water content, vegetation condition, etc. (Zhu and Chen 1994). Generally, there are two dynamic processes: the wind erosion process and the wind deposition process. Wind erosion and wind deposition may occur when the wind blows across loose, dry, and exposed sand surfaces, resulting in different sand landforms.

5.2.1.1 Wind Erosion Process and Landforms

Wind erosion is the separation of surface matter from its place under the action of wind. Wind erosion can be generally divided into three processes (Fig. 5.1): deflation, namely, winds blow away sand particles; abrasion, namely, wind-ferried particles hit a highly cemented ground surface and rock surface; and attrition, namely, the collision between moving sand grains, which plays a relatively small role in the formation of wind erosion landforms compared to the first two. Deflation and abrasion can create large wind erosion landforms such as yardangs, ventifacts, and wind-scoured depressions (Livingstone and Warren 1996). Abrasions on the

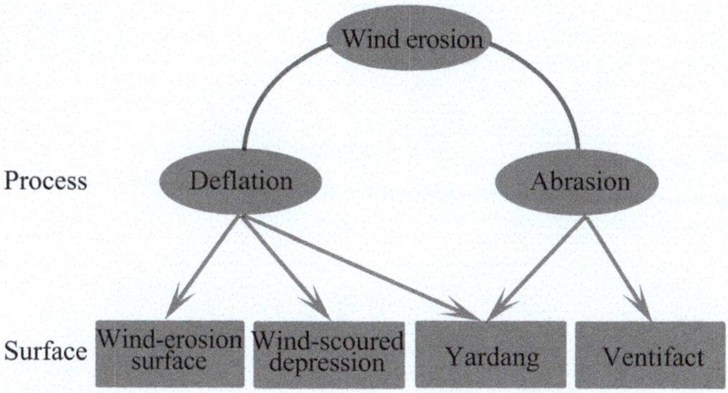

Fig. 5.1 Wind erosion processes and wind-eroded landform types

Fig. 5.2 Typical shapes of yardang landforms in arid areas (left) and semiarid areas (right) (Wang 2011)

surface of large rocks may form niches; if the objects subjected to abrasion are clasolites, then ventifacts may be formed; and when abrasion occurs in a vast region, yardang landforms may be formed.

Wind erosion geomorphology has two controlling factors, namely, erosivity and erodibility. Erosivity is a function of wind energy and time, while erodibility is the erosion degree of a land surface or object, which is related to many factors, such as the size of surface particles, vegetation coverage, soil water content, soil crust, and land terrain. Gillette et al. (1982) ranked the surface erodibility of deserts in the order of strong to weak as follows: loose soil → dune sand → deluvial and dry deposits → loose salty soil → playa beach → playa center → desert gravel. Wind erosion landform types mainly include yardangs, wind-scoured depressions, ventifacts, and stone pavement.

Yardang—"Yardang" is a Uygur word. The shapes of yardangs include deflation mushrooms, deflation columns, and deflation residual hills (Fig. 5.2). Yardang landforms are common in arid zones. At present, there are still considerable debates about their formation mechanism. It is generally believed that when abrasion occurs in a vast area, yardang landforms may be formed, and deflation is also a factor in

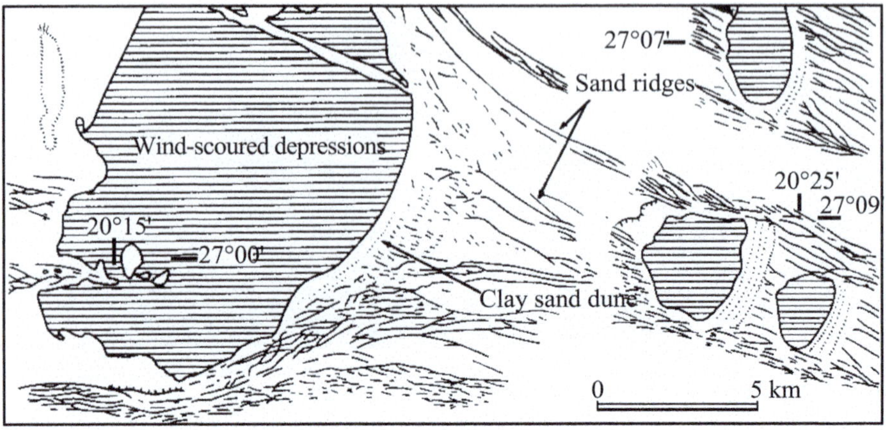

Fig. 5.3 Distribution relation among wind-scoured depressions, barchan, and other dune types in the Kalahari region (Wang 2011)

Fig. 5.4 Ventifact (left) and stone pavement (right) in the Qaidam Basin (photo courtesy of Xian Xue)

shaping yardang landforms. However, some scholars believe that abrasion does not play a leading role in the formation of yardangs.

Wind-scoured depression—Wind-scoured depressions refer to saucer-like depressions formed in arid zones. They are generally formed on erodible ground surfaces; in South Africa, they mainly occur on sand beds, and in Australia, they mainly occur on dry riverbeds or lake beds. In some cases, they coexist with yardangs. The size and depth of wind-scoured depressions are significantly different, and barchan dunes or small sand ridges, which mainly consist of clay materials, commonly appear on downwind edges (Fig. 5.3).

Ventifact—Ventifact refers to rocks scattered in a desert or gobi that have experienced long-term sand abrasion (Fig. 5.4). The prominent feature of a ventifact is that it has a polished surface or edges; there are pits and scratches on the surfaces, and the edge line is often close to the wind direction. The in situ shape of a ventifact is related to the shape of the original stone in addition to the number of stone reversals or changes in the wind direction. However, a ventifact is mostly three-sided and shaped like a small pyramid. The sizes of the smooth surfaces are also different due to the different times of wind erosion. Wind-tunnel experiments by Miotke (1982) show that tens to several hundreds of years are required to form ventifacts, and the abrasion rate varies in the range of 0.001–20 mm/a.

Stone pavement—Stone pavement refers to a stone mantle or eroded gobi in central Asia that cover the surface of sandy and clay lands and are formed through wind deflation (Fig. 5.4). Stones are often small and multiangular. It has also been suggested that deflation may not be the only cause of stone pavement, and its formation is also related to other factors, such as freezing and thawing, crystallizing, and flowing water.

5.2.1.2 Wind Deposition Process and Landforms

When the wind speed decreases, wind-sand flow is in a supersaturated state, and the sand particles carried by the wind will be deposited to form aeolian landforms. Many other factors can cause sand deposition, such as vegetation and topographic obstruction. The most important landforms formed during wind deposition processes are sand dunes.

There are many classification systems for sand dunes, but there are two main classifications: one is based on the relationship between dune shape and the wind environment or sand source, and the other is based on the genetic principle.

According to the relation between dune strike and the main wind direction, there are three major dune types: a longitudinal dune, an oblique dune, and a transverse dune (Fig. 5.5). According to the genetic classification, there are five major types of dunes, i.e., a barchan dune, linear dune, reverse dune, star dune, and parabolic dune, based on their shape and slope number. Except for the single type, sand dunes usually appear as composite and complex types. Compound dunes refer to two or more sand dunes of the same type that are linked or overlap, and complex dunes are composed of two or more different types of dunes. Approximately 46.6% of the

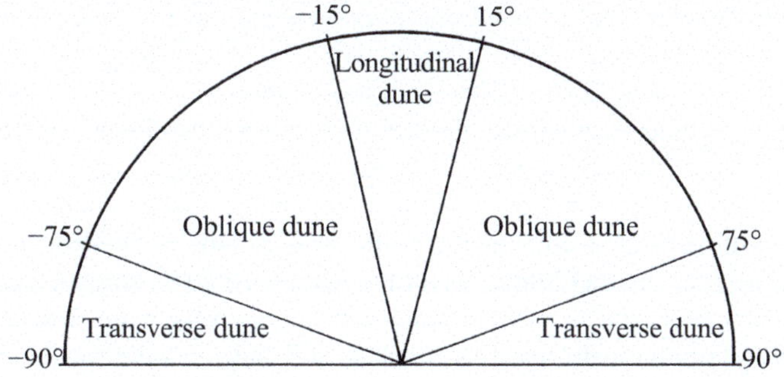

Fig. 5.5 Dune types divided according to the relation between dune strike and main wind direction

current sand sea is covered by compound or complex sand dunes according to statistical data (Fryberger and Goudie 1981).

In addition, there are also some other sand depositional landforms in the sand sea, such as sand sheets; narrow and long dunes without sandfall slopes known as zibars, phytogenetic dunes, or coppice dunes; and obstacle anchored dunes known as echo dunes, climbing dunes, and falling dunes.

5.2.2 *Physical Processes of Aeolian Desertification Development*

The physical processes of aeolian desertification development involve changes in landform features, patterns, and landscape types. Generally, the modern aeolian desertification process marked by wind-sand activities can be divided into four types: the sand dune (sandy land) mobilization process, aeolian desertification of grassland shrubs process, soil-layer wind erosion and coarsening process, and the formation of badlands process resulting from uneven wind dissection (Zhu and Chen 1994).

5.2.2.1 Sand Dune (Sandy Land) Mobilization Process

Sand dune (sandy land) mobilization refers to the process of sand being removed after the destruction of vegetation in fixed deserts formed in the historical period. However, the modern aeolian desertification process has mainly been caused by human activities. Therefore, sand dune (sandy land) mobilization occurs mainly near sandy farmland, residential areas, and livestock drinking water areas and along traffic routes.

The aeolian landforms formed in the process of sand dune (sandy land) mobilization are restricted by their original forms, but they are not the reappearance of the original forms. In addition, most have new morphological characteristics. The formation and development of parabolic dunes are the most typical representations (Fig. 5.6).

In the process of sand dune (sandy land) mobilization, the decrease in vegetation cover and plant species is obvious, and the succession of different vegetation types has the same trend: community structure evolves from complex to simple; the level gradually decreases; plants are gradually low and sparse, and the biomass decreases; and plant quality changes from superior to inferior, mainly in the decrease in edible species. In addition, the soil humus layer formed in the fixed stage is continuously lost by wind erosion, and the whole soil layer is sorted by wind, resulting in fine particles being continuously lost. As a result, soil nutrients decrease, and the soil becomes increasingly barren.

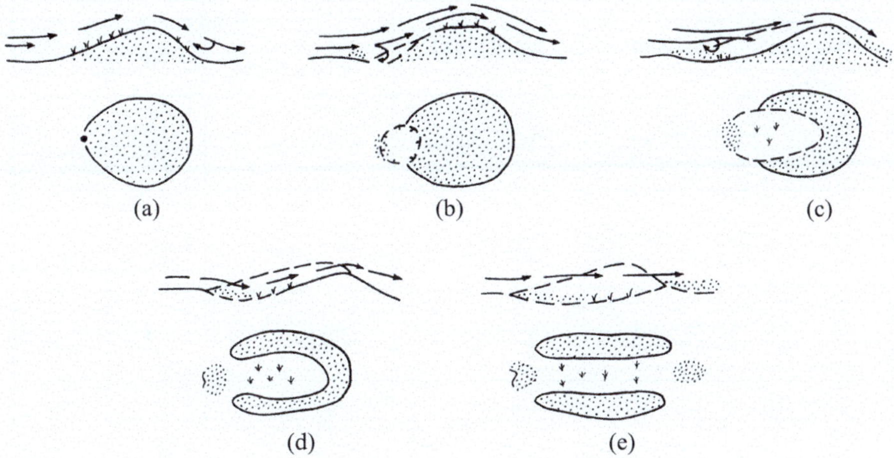

Fig. 5.6 Form and development of parabolic dunes (Wang 2011)

Fig. 5.7 Coppice dunes in southern Xinjing (photo taken by Xian Xue)

5.2.2.2 Aeolian Desertification of Grassland Shrubs

Grassland shrub aeolian desertification refers to the process of formation, growth, sand enrichment of coppice dunes on sandy grassland, and the final formation of sandy desert (Fig. 5.7). A coppice dune is a kind of aeolian landform formed by the accumulation of sand particles when wind-sand flow is obstructed by shrubs. Its formation and evolution are mainly affected by wind, sand sources, vegetation, etc. The wind must be strong enough to cause sand flow but not exceed the sandbreak capacity of shrubs, the vegetation acts as the sand trapper, and the sand source is the material forming the wind-sand flow and composing the coppice dunes.

Coppice dunes mostly appeared as shield, dome, strip, and compound shapes. Field observations and wind-tunnel experiments have shown that coppice dunes have formed through the following stages (Fig. 5.8):

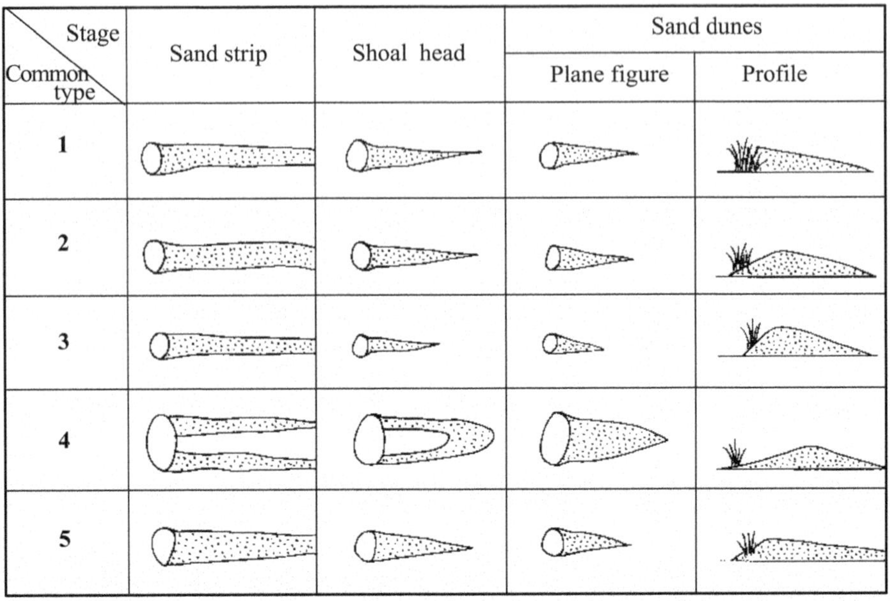

Fig. 5.8 Process of coppice dune formation in wind-tunnel experiments (Zhu and Chen 1994)

Sand-strip stage: A sand strip is mainly composed of fine grains and extends in the wind direction, with the ratio of the length to the width being dozens of times and with the height being a few centimeters. This stage is rarely seen in natural settings because of the short process time and the lack of stable unidirectional winds.

Sand-spit stage: A sand spit is a shortened sand strip, and the plane shape is an isosceles triangle with shrubs at the bottom. In arid grasslands, the height of the sand spit generally does not exceed 35 cm; its widest height is approximately 50–60 cm, and the length-width ratio ranges from 5:1 to 10:1.

Sand-mound stage: The downwind extended part of a sand spit is continuously shortened, and the height increases, finally forming an oval-shaped sand mound, with its large head facing the main wind direction. Its longitudinal profile shape is streamlined, and the length-width ratio is less than 2:1.

Sand-heap stage: The shape of a sand heap changes less compared to that of a sand mound. However, it gradually grows increasingly larger as the vegetation multiplies and entirely covers the dune due to water storage in the heap, and eventually, mature coppice dunes are formed.

The evolution of individual coppice dunes is not endless; when entering an equilibrium state, it shows a very strong zonality in relation to the habitat characteristics, such as in relation to the wind regime, sand source, water condition, and vegetation type. For example, a *Caragana sinica coppice dune in* a semiarid sandy grassland can grow to 1.5 m in height and 3 m in diameter; however, in an arid desert and gravel grassland, its height is only 20–30 cm, and its diameter is less than 3 m; a

Tamarix ramosissima coppice dune on the edge of the Taklimakan Desert generally can be as tall as 7–8 m and 10 m or more in diameter.

5.2.2.3 Soil-Layer Wind Erosion and Coarsening Process

The process of soil-wind erosion and coarsening refers to the destruction of surface soil structure by wind erosion, the loss of fine grains, the increase in coarse grains, the loss of soil fertility, the deterioration of soil performance, and the decline in land productivity, leading to the degradation of the whole ecosystem and the occurrence of aeolian sand microgeomorphology.

Erosion factors depend on natural environmental conditions such as local wind force, ground cover conditions, and water state. In his study of soil erosion on the western Great Plains of the United States, Chepil (1945) referred to sand grains < 0.84 mm as an erodible factor through a large number of observations and experiments. Chen et al. (1995) determined that the erodible factor was <0.8 mm on the Beijing Plain of China; some results in Inner Mongolia showed that the nonerosion factor of grains is larger than 1.0 mm, and in desert areas, grains with a larger than 2.0 mm nonerosion factor are present. A wind-tunnel experiment showed that sand can be moved by wind only if the water content in the sand is below 1% (He 1988).

Soil-layer coarsening can be demonstrated as aeolian desertification and gravelization in different zones. In the sandy loess area, wind erosion is the loss of the dust, and the remaining area is sand grains; here, aeolian desertification occurs. Near gobi areas, the wind force is strong, and the remaining particles are mainly larger than 2 mm in diameter. Here, the surface exhibits gravelization, such as the gravel mantle formed in the piedmont diluvial plain or undulate eluvial plain.

With the development of soil-layer wind erosion and coarsening processes, the soils become barren rapidly. For example, in the farming area of the Bashang region in Hebei Province and Ulanqab Banner in Inner Mongolia, the soil organic matter content of newly reclaimed land was between 2% and 4%, and it decreased to below 1% after 5–10 years of repeated cultivation and wind erosion; crop yields also decreased significantly, and the crop yield decreased by more than half after 9–20 years of cultivation.

5.2.2.4 Uneven Land-Wind Dissection: The Formation Process of Badland

The formation of badlands often occurs from a wind erosion blowout in the deposition area, with the soil stratum being relatively uniform and mainly consisting of silt-fine sand, subsand, or even clay. The vortex formed in the blowout produces uneven cutting on the ground, making the ground more undulated and similar to the "yardang" landforms such as niche and earth pillars.

Badlands, according to their location, can be divided into three types: (a) the depression, distributed in relatively wide intermountain depressions, is composed of

Fig. 5.9 Wind-eroded badland in northern Shangdu County of Inner Mongolia (Wang 2011)

limnetic and aeolian facies sediments, with more mud-calcium and obvious horizontal bedding in strata. (b) The shallow ravine depression mostly occurs in the middle of gently undulating plains in interdune depressions. The deposition is dominated by aeolian sediments, and vertical bedding joints similar to loess deposition are well developed (Fig. 5.9). Usually, at the junction of two branch ravines, airflows easily form eddy currents, where uneven land-wind dissection is most developed. (c) The piedmont type is distributed on windward slopes on hilly land and is composed mainly of slope deposits. The lower soil layers are mixed with gravel, and the paleosol layer parallel to the slope is obvious. Wind plays a primary role in the formation of this third type of badland. However, the formation of most badlands in the grassland region of eastern China is related to cutting by running water, especially in the wide, shallow ravine depression type; extensive human activities such as excavating sand, digging soil, and building roads in dry farming regions also trigger the development of badlands.

5.3 Biological Processes of Aeolian Desertification

The biological process of aeolian desertification refers to the processes dominated by changes in plants, soil animals, and microbes due to land desertification and/or restoration in arid, semiarid, and subhumid regions. In most cases, the aeolian desertification process starts with vegetation reductions due to overuse, deforestation, overgrazing, fuel wood collection, and other use of plants as livelihood and production resources. Therefore, it is very important to discuss the biological processes of aeolian desertification from individual plants to communities.

5.3.1 Changes in Biodiversity in the Aeolian Desertification Process

Biodiversity is an important trait of plant communities degraded to desertified land or restored from desertified land; here, in the eastern part of China, biodiversity changes mainly refer to those from restoring desertified grassland. An increase in biodiversity to some extent could promote the stability of restored ecosystems; therefore, plant coverage, diversity, productivity, and habitat traits from 60 communities were measured and analyzed. The results showed that biodiversity and biomass increased along the habitat gradient from shifting dune to semishifting dune to fixed dune to dry grassland to wet grassland to riverbank wash land (grassland on the flood plain along rivers). Plant diversity showed a significant and positive relation to restoration (Fig. 5.10) (Zuo et al. 2012), and the habitat factors landform and soil traits significantly influenced the community composition and productivity. Habitat factors first influenced community composition, and then, the composition influenced the synergic change in diversity and productivity (Fig. 5.11) (Zuo et al. 2012).

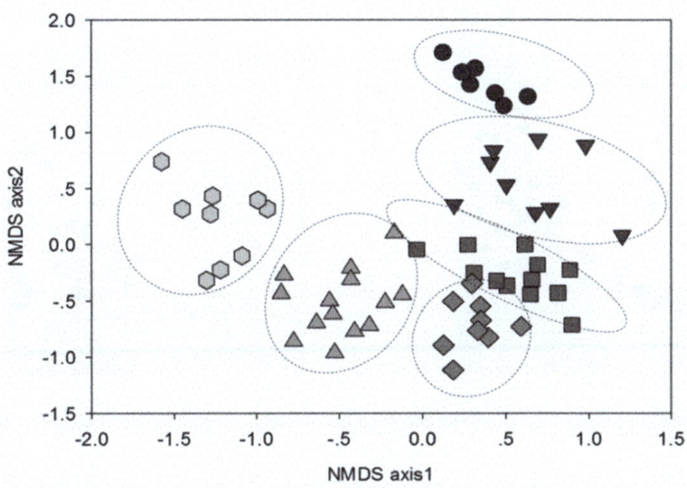

Fig. 5.10 Sixty species were divided into 6 communities (minimum stress values first axis/dimension = 49.13, R^2 = 0.28, P = 0.004; second axis/dimension = 31.66, R^2 = 0.42, P = 0.004).,shifting dune;▲, semishifting dune; ◆, fixed dune; ■, dry meadow; ▼, wet meadow; ●, riverbank grassland (Zuo et al. 2012)

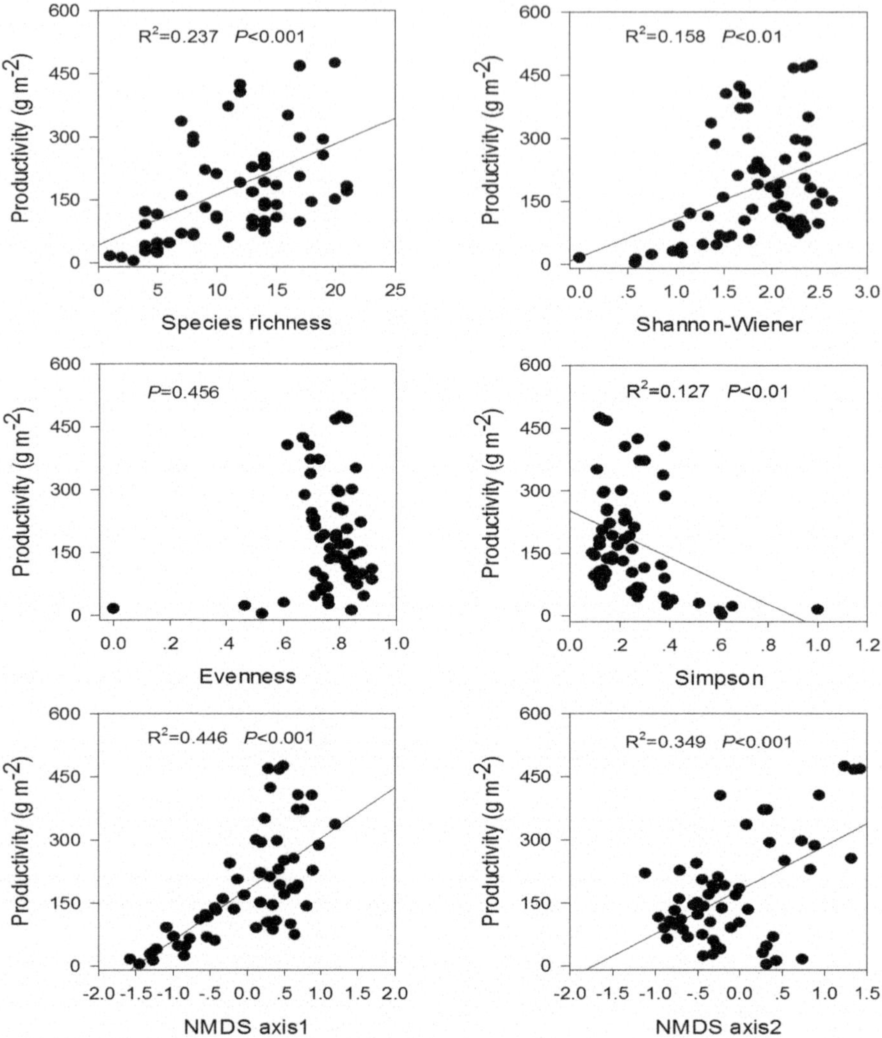

Fig. 5.11 Sandy grassland biodiversity, composition, and productivity ($n = 60$) (Zuo et al. 2012)

5.3.2 Soil Function and Microorganism Diversity Change with Restoration

Vegetation restoration is the primary stage of ecosystem restoration, and soil composition and structure restoration are a crucial process for the sustainable restoration of ecosystems in desertified regions (Raven and Johnson 1986). Therefore, when adopting biological measures for aeolian desertified land restoration, it is very important and highly necessary to take measures to enhance soil restoration in

Table 5.1 Effects of fertilizer application (g m^{-2}) on root biomass, shoot biomass, and soil moisture

Treatment	Fertilization	Root biomass (g m^{-2}) 0 cm–10 cm	Root biomass (g m^{-2}) 10 cm–20 cm	Shoot biomass (g m^{-2})	Soil moisture (%)
CK (Control)	74.9 (23.5)a	35.9 (16.9)a	87.2 (22.8)a	6.826 (0.349)a	5.660 (0.234)ab
N (N addition)	112.7 (26.3)a	25.1 (18.9)a	98.3 (25.5)a	6.868 (0.290)a	6.097 (0.239)a
P (P addition)	79.8 (21.4)a	24.0 (15.4)a	113.8 (20.8)a	7.052 (0.295)a	5.277 (0.234)b
NP (N and P addition)	80.5 (21.4)a	28.2 (15.4)a	115.9 (20.8)a	6.940 (0.290)a	5.551 (0.245)ab

Table 5.2 Effects of water addition on root biomass, shoot biomass, and soil moisture

Treatment	Water addition	Root biomass (g m^{-2}) 0 cm–10 cm	Root biomass (g m^{-2}) 10 cm–20 cm	Shoot biomass (g m^{-2})	Soil moisture (%)
CK	79.5 (11.6)a	28.3 (8.3)a	103.8 (11.3) a	6.294 (0.258)b	5.083 (0.204)b
Winter	85.7 (11.0)a	45.6 (7.9)a	103.6 (10.6) a	6.496 (0.258)b	5.450 (0.208)b
Summer	91.4 (11.2)a	42.3 (8.0)a	127.2 (10.9) a	7.935 (0.281)a	6.403 (0.206)a

Data represent the means, and the values in the brackets are standard errors. Means with different lowercase letters in the same column are significantly different at $P < 0.05$

desertified areas. Therefore, an experiment on the restoration of a degraded grassland with water and fertilizer additions was carried out in the Horqin Sandy Land. This experiment occurred in a degraded grassland, mainly due to overgrazing, with the goal of determining how the degraded grassland was restored with the water and nutrient additions. The results showed that N and water additions increased both the aboveground and underground biomass of the aeolian desertified land during restoration (Tables 5.1 and 5.2). That is, there is a possibility of promoting the restoration of degraded soil by fertilization and water addition, but the prioritized amounts of added N and water need more in-depth research, particularly when economic costs and benefits are also taken into consideration.

Different water addition amounts and times (winter and summer) led to differences in soil respiration. The experiment showed that a water addition in summer could lead to a significant increase in the soil respiration rate in all N-addition plots and the soil respiration rate increased by 20% in 2010 and by 29.2% in 2011. Water addition through snow in winter could lead to an increase in the soil respiration rate, but this increase was only significant in 2011 by 11.8% (Fig. 5.12).

Integrative research has found that there are different relationships between soil respiration and soil temperature and soil moisture. The soil thermal sensitivity (Q10)

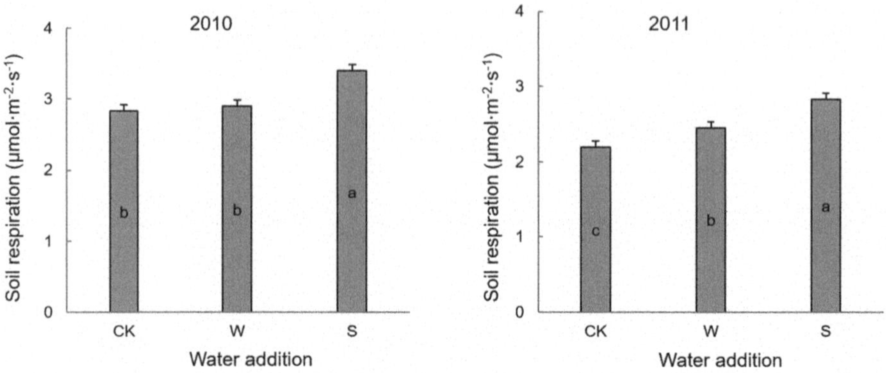

Fig. 5.12 Soil respiration in response to different water additions in 2010 and 2011. Bars with different letters are significantly different ($P < 0.05$). Error bars indicate one SE of the mean

was in the order of summer water addition (W) > winter water addition (snow manipulation = S) > control. In comparison to soil temperature, soil moisture contributed more to soil respiration. This result is in accordance with those of other studies in this area (Table 5.3).

5.3.3 Changes in Soil Structure with Restoration

Habitat change includes near-ground climatic changes, soil changes, and other related plant or plant community changes. Here, the changes in soil and soil moisture are defined as the principle change in desertification and are discussed. The near-ground microclimatic changes on dunes will be discussed in other chapters.

Near-ground climatic traits changed along the desertification restoration gradient such that soil surface temperature was decreased by soil and plant moderation and soil moisture also decreased due to vegetation restoration leading to water consumption. The air temperature near the ground at less than 50 cm also decreased during vegetation restoration. It is clear that restoration of aeolian desertification in Horqin Sandy Land can lead to an improvement in the ambient environment of dunes, including temperature, moisture, and wind velocity (Harazono and Li 1999).

Organic matter is not only an important constituent of soils but also a structural component of water and nutrient supply. Soil C accumulation in the process of sandy land restoration plays an important role in maintaining soil carrying capacity, including nutrient and water holding capacity, together with soil texture. Measurements of the texture and soil organic matter content showed that in an enclosure after 8 years, the coarse sand (2–0.1 mm) content at a depth of 0–20 cm decreased from 73% to 51%, fine sand (0.1–0.05) increased from 20% to 36%, and silt and clay (<0.05 mm) increased from 7% to 13% (Fig. 5.13).

Table 5.3 Relationship of soil respiration with temperature and soil moisture across all treatments

Year	Fertilizer	Water	Equation R = Aexp(BT)					Equation $R = A$ + BW + CW^2				
			A	B	R^2	P	Q_{10}	A	B	C	R^2	P
2010	CK	CK	0.470	0.060	0.225	0.000	1.82	2.182	−0.299	0.061	0.861	0.000
		W	0.811	0.045	0.128	0.000	1.57	3.792	−0.526	0.060	0.422	0.001
		S	0.761	0.045	0.130	0.000	1.57	3.764	−0.681	0.078	0.735	0.000
	N	CK	0.477	0.062	0.169	0.000	1.86	4.102	−0.554	0.060	0.480	0.000
		W	0.714	0.050	0.115	0.001	1.65	5.092	−0.921	0.093	0.332	0.004
		S	0.611	0.068	0.109	0.002	1.97	7.133	−0.808	0.064	0.233	0.001
	P	CK	0.569	0.049	0.189	0.000	1.63	3.232	−0.226	0.025	0.422	0.001
		W	0.574	0.051	0.114	0.001	1.67	−2.678	1.505	−0.077	0.513	0.000
		S	0.548	0.063	0.146	0.000	1.88	−4.242	2.053	−0.119	0.236	0.020
	NP	CK	0.903	0.042	0.116	0.001	1.52	2.887	−0.196	0.025	0.384	0.000
		W	0.817	0.036	0.098	0.002	1.43	−2.058	1.160	−0.052	0.365	0.001
		S	0.674	0.052	0.187	0.000	1.68	2.448	0.057	0.011	0.287	0.010
2011	CK	CK	0.229	0.073	0.290	0.000	2.08	1.300	−0.028	0.052	0.469	0.000
		W	0.161	0.088	0.369	0.000	2.41	2.162	−0.502	0.102	0.458	0.000
		S	0.186	0.088	0.410	0.000	2.41	7.090	−2.448	0.273	0.687	0.000
	N	CK	0.184	0.080	0.259	0.000	2.23	0.792	0.177	0.027	0.579	0.000
		W	0.160	0.087	0.291	0.000	2.39	0.666	0.102	0.035	0.468	0.000
		S	0.191	0.089	0.278	0.000	2.44	8.367	−2.752	0.277	0.644	0.000
	P	CK	0.181	0.078	0.295	0.000	2.18	2.389	−0.873	0.151	0.810	0.000
		W	0.213	0.078	0.289	0.000	2.18	0.386	0.394	0.022	0.561	0.000
		S	0.142	0.097	0.483	0.000	2.64	2.593	−0.334	0.069	0.227	0.019
	NP	CK	0.219	0.070	0.258	0.000	2.01	0.548	0.105	0.046	0.543	0.000
		W	0.185	0.078	0.306	0.000	2.18	2.664	−0.807	0.101	0.788	0.000
		S	0.184	0.082	0.372	0.000	2.27	2.802	−0.785	0.111	0.579	0.000

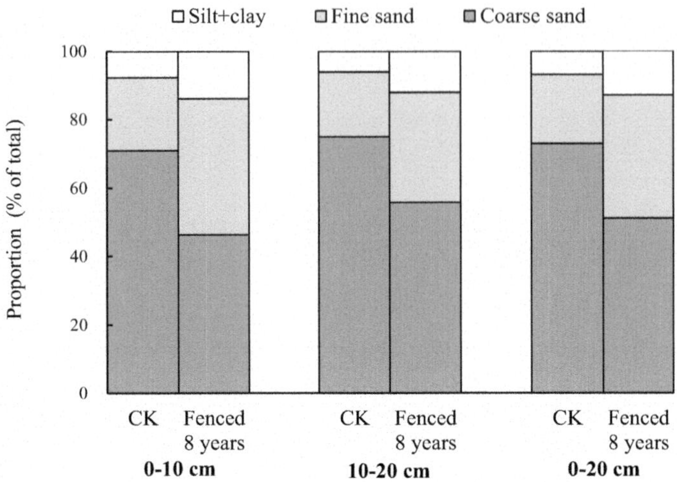

Fig. 5.13 Soil texture changes before and after 8 years of enclosure

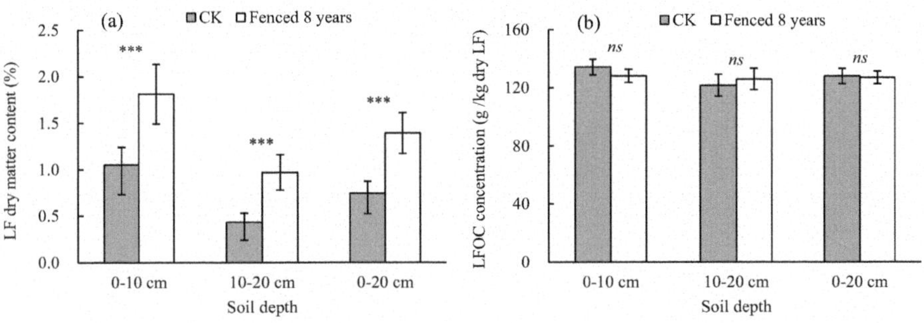

Fig. 5.14 Light fraction dry matter content (%); (**a**) light organic C (LFOC); (**b**) change. ***, significant at 0.001; ns, significant at 0.05

The light soil organic C, easily decomposed, increased from 0.7% to 1.39% (Fig. 5.14a) but without a significant change in the averaged C content before and after enclosure (Fig. 5.14b), while the total soil C content at a depth of 0–20 cm increased from 2.87 g kg^{-1} to 4.45 g kg^{-1}. In terms of the C reserve, the light soil organic C increased from 221.0 gC m^{-2} (increased by 84%), and the soil total organic C reserve increased by 435 gC m^{-2} (increased by 55%).

There were linear relationships between the content of soil total organic C, light dry matter content, fine sand, and silt + clay (Fig. 5.15).

Analysis of soil moisture in relation to precipitation is crucial to providing theoretical support for plant selection for vegetation restoration and land management.

Continued observations over 4 years found that the soil moisture increased with increasing precipitation; in particular, the shifting dune soil moisture increased

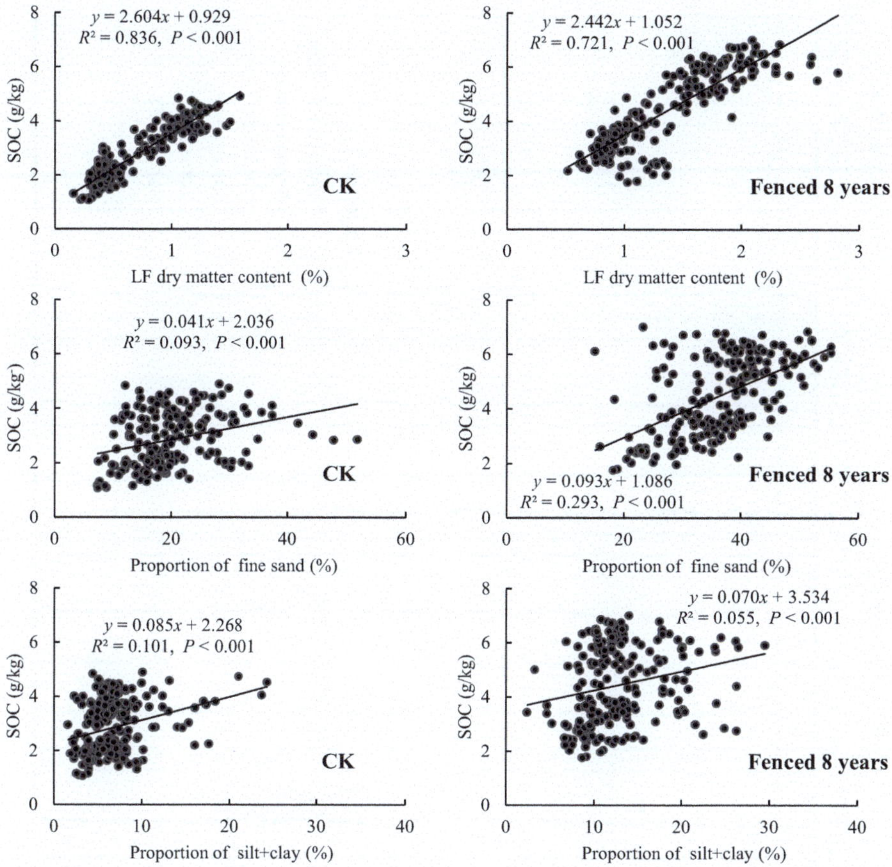

Fig. 5.15 Correlation between total soil organic C content and LF matter, fine sand content, and silt + clay content

significantly following large amounts of precipitation. When the precipitation amount was 34.4 mm, 35.1 mm, 47.1 mm, 56.3 mm, and 47.4 mm, the soil moisture increased by 126.9%, 81.7%, 122.4%, 90.7%, and 45.1%, respectively, after rainfall events at depths of 0–20 cm. The increasing pattern at depths of 20–160 cm was similar to that at depths of 0–20 cm, but the magnitude of the increase decreased with increasing depth. When the rainfall event was greater than 40 mm or the accumulated precipitation in the following few days was greater than 40 mm, the soil moisture at depths of 140–160 cm presented an increasing trend (Fig. 5.16). Correlation analysis showed that soil moisture at a depth of 0–100 cm was significantly related to precipitation ($R = 0.239 \sim 0.438$, $P < 0.05$) and the soil moisture at a depth of 0–40 cm was significantly related to precipitation intensity ($R = 0.284$, 0.244, $P < 0.05$). It was found that the soil moisture at a depth of 0–20 cm was negatively related to the precipitation intervals ($R = -0.293$, $P < 0.05$). The above analysis deductively showed that when precipitation was greater than 10 mm, precipitation

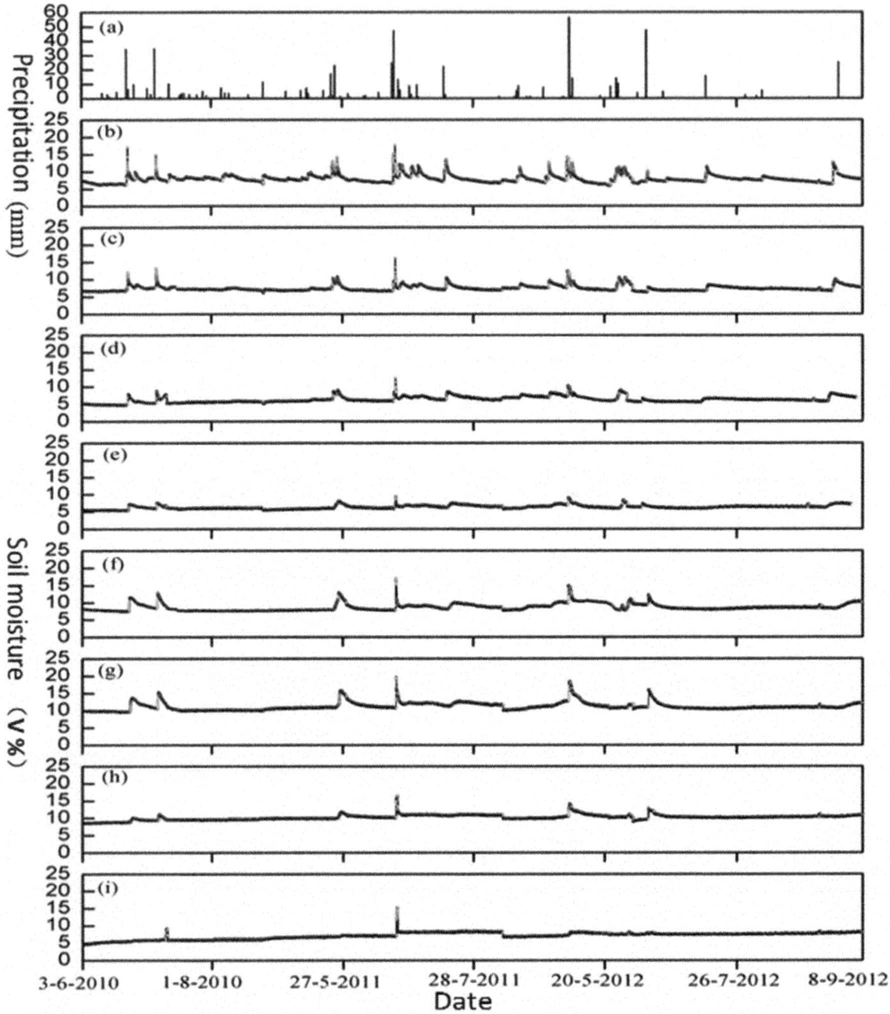

Fig. 5.16 Precipitation (mm) and soil moisture over 4 years (Liu et al. 2015, 2016)

could be effective for plants after sufficiently wetting the soil and precipitation amounts greater than 40 mm could enough provide water to reach the groundwater. That is, precipitation could be recharged into groundwater and used relatively stably by plants with root systems down to depth (Liu et al. 2015, 2016).

Continued observations over 4 years found that when rainfall was less than 20 mm, the potential water replenishment from rainfall to underground water bodies was consistent, and when rainfall was greater than 30 mm, the replenishment significantly increased. The potential underground water replenishment was significantly related to rainfall magnitude ($R = 0.907$, $P < 0.001$) and intensity ($R = 0.659$, $P < 0.001$) but not to the interval of rainfall events.

5.3.4 *Impacts of Land Use/Cover Change on Soil C and Texture*

Land use change is the primary driving force for land degradation and aeolian desertification. Restoration through modifications of land use could, to a large extent, enhance aeolian desertification restoration, by improving soil texture and increasing soil C and N.

Compared to that in nonaeolian desertified land, the soil C content of a restored woodland over 30 years, which was originally a severely desertified grassland, increased by 74% at a depth of 0–5 cm in the Horqin Sandy Land, while the soil C content of a cropland and grazed land increased by 16.9% and 39.2% over approximately 30 years, respectively. The soil C content of the grazed grassland increased by 44.6%. There was a significant relationship between land use and soil texture (Fig. 5.17). The soil C reserve decreased by 121 g m^{-2} from the grassland to the cropland and increased by 261 g m^{-2} from the grassland to the woodland within 30 years. After 16 years of being enclosed, the soil C reserve increased by 111 g m^{-2} when the severely aeolian desertified grassland (Li et al. 2012).

5.3.5 *Physiological Adaptation of Plants to Different Aeolian Desertified Lands*

Revegetation is an important means of restoring degraded land or ecosystems, and the proper selection of plants and adaptive patterns of plant layouts are critically

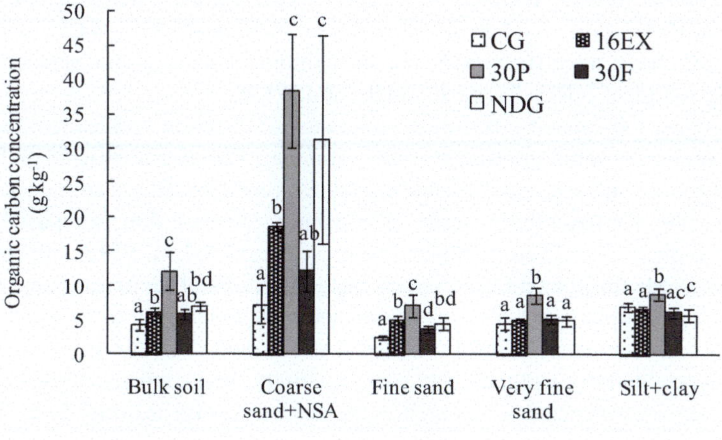

Fig. 5.17 Impacts of land use change on the soil texture and C (CG, continued grazing; 16EX, desertified grassland enclosed for 16 years; 30P degraded grassland afforested for 30 years; 30F grassland reclaimed for 30 years; NSA, nonwater-stable aggregate; NDG, nondesertified grassland; and a, b, c, and d refer to different significance levels ($P < 0.05$)

important for sustainable desertification reversal. To properly select plants and design adaptive plant patterns for restoration, a series of physiological studies have been carried out, and the research showed that different landforms due to aeolian desertification, such as a shifting dune (SD), semishifting dune (SSD), semifixed dune (SFD), fixed dune (FD), and interdune grassland (IDG), have different water, N, and pH values and sustain different vegetation types. N concentrations in green leaves (N_g) and senesced leaves (N_s) and the gas exchange characteristics of 35 species of shrubs and herbs (some species existed in several habitats) were measured across the above five habitats. To further analyze the physiological responses to soil drought and rewatering and the influence of fertilization, 2 annual herbs, *Setaria viridis* (L.) Beauv. and *Digitaria ciliaris*; 2 perennial herbs, *Pennisetum centrasiaticum* Tzvel. and *Calamagrostis pseudophragmites* (Hall. f.) Koel.; and 2 shrubs, *Caragana microphylla* Lam. and *Hedysarum fruticosum* Pall. var. *lignosum* (Trautv.) Kitagawa, were planted in 30 pots for each plant species with different irrigation and N fertilization treatments.

The N_g, N_s, net photosynthetic rate (P_n), water use efficiency (WUE), and intrinsic water use efficiency (WUE_i) decreased; leaf nitrogen use efficiency (NUE) and photosynthetic nitrogen use efficiency (PNUE) differed significantly; and nitrogen resorption efficiency (NRE), transpiration rate (E), and stomatal conductance (g_s) did not differ across the five habitats with decreasing N content. The N_g, N_s, P_n, WUE, and WUE_i decreased, and leaf NUE and PNUE increased significantly with decreasing soil N status across the above habitats. There were low N_g, N_s, and P_n values and high NUE and PNUE values for the SD and SSD, while there were high N_g, N_s, and P_n values and low NUE and PNUE values for the IDG, FD, and SFD.

For all species in the five habitats, there were significantly correlated relationships among different leaf nitrogen and water use traits. N_g correlated positively with N_s but correlated negatively with leaf NUE and PNUE, and leaf NRE was not affected by N_g. Leaf PNUE correlated positively with leaf NRE, leaf NUE, WUE, and WUE_i, while it correlated negatively with N_s. In addition, leaf mass per area correlated negatively with leaf NRE, P_n, g_s, E, WUE, and WUE_i.

To compare the capacity of drought tolerance of different species from arid and semiarid regions, annual grasses, *P. centrasiaticum*, *Tribulus terrestris*, and *Bassia dasyphylla*, from shifting and semishifting dunes were adopted as sample species for measuring the photosynthetic traits. The results showed that compared to the unfertilized plants, fertilized plants had increased leaf FV/FM, ET0/CS0, and PICS in well-watered and light drought conditions, and fertilization exacerbated fluctuations in the chlorophyll, a fluorescence characteristic during the drought period.

Biomass and biomass allocation between shoots and roots are key indicators of vegetation productivity and mass and energy activity. Irrigation and fertilization can promote degraded vegetation restoration, such as restoring shifting dunes to fixed dunes.

The total biomass, belowground biomass, and root/shoot ratio were significantly affected by nutrient and irrigation regimes, as well as their interactive effects

Table 5.4 Biomass (aboveground, belowground, and total) and the root/shoot ratio for *P. centrasiaticum*. Values are means ±SE ($n = 3$) (N and water addition or cross treatment; also refers to Table 5.3)

	Total biomass	Aboveground biomass	Belowground biomass	Root/shoot ratio
N0W0 (No N and Water added)	5.65 ± 0.50	1.56 ± 0.20	4.10 ± 0.31	2.67 ± 0.16
N0W1 (No N added but Watered)	4.46 ± 0.20	0.80 ± 0.04	3.66 ± 0.17	4.57 ± 0.11
N1W0 (N added but no water)	57.58 ± 5.11	21.86 ± 3.07	35.72 ± 2.54	1.69 ± 0.21
N1W1 (N and Water added)	36.19 ± 2.08	19.45 ± 3.28	16.74 ± 1.24	0.93 ± 0.19

(Tables 5.4 and 5.5). Aboveground biomass was highly significantly affected by nutrient regimes (Chen et al. 2019a).

Approximately 1.03% to 5.88% of the variation in the total biomass, belowground biomass, and root/shoot ratio was explained by irrigation regimes. Nutrient regimes explained 42.16%, 31.18%, 36.93%, and 16.76% of the variation in the aboveground biomass, belowground biomass, total biomass, and root/shoot ratio, respectively. Irrigation regimes×nutrient regimes explained 5.36%, 2.15%, and 5.56% of the variation in the belowground biomass, total biomass, and root/shoot ratio, respectively. Both irrigation regimes and irrigation regimes×nutrient regimes had no significant effect on the aboveground biomass (Table 5.5). Typically, research shows that biomass decreases along gradients from interarea > fixed dune > shifting dune. Irrigation and fertilization could change the variation in plants but not the general trend in the changes during the slow vegetation restoration from dune to fixed dune when disturbance is moderated.

The relative water content (RWC) of the leaves decreased with time, and the RWC of *T. terrestris* was higher than that of *B. dasyphylla*. The RWC of *T. terrestris* leaves under different treatments decreased by 18.26–30.28% (Fig. 5.18a, b). At the same time, the RWC of *B. dasyphylla* decreased by 26.22–47.64%. The RWC was reduced slightly in the first phase of 1–4 D, decreased rapidly from 4 D to 7 D, and then decreased in the last phase of 7–10 D. The RWC differences in 1 D between the control, P addition, and P reduction were not significant ($P > 0.05$). In comparison with the control, when P was reduced by 30% and 60%, the RWC decreased and was significantly different between 7 D and 10 D after the treatment, and when P was added at 30% and 60%, the content was higher than that in the control. When P was reduced by 60% on 10 D, *B. dasyphylla* had the lowest RWC at 43.95%, whereas that of *T. terrestris* was 62.20%. The RWC differences in the control, P addition, and P reduction were significant ($P < 0.05$).

MDA is one of the key enzymes used by plants in response to drought stress, and its content in *T. terrestris* increased continuously over the whole drought-stress period. When precipitation was reduced by 60%, the content was 2.927 mmol. g^{-1} FW on the seventh day of the experiment, and it was higher in *T. terrestris* than in *B. dasyphylla*, with the latter occurring in most cases from shifting dunes. This result showed that in comparison to *T. terrestris*, *B. dasyphylla* is more tolerant to drought.

Table 5.5 Summary of GLM examining the effects of irrigation regimes and nitrogen regimes on biomass and biomass allocation between shoots and roots, ANOVA

	df	Aboveground biomass		Belowground biomass		Total biomass		Root/shoot ratio	
		F	SS(%)	F	SS(%)	*F*	SS(%)	*F*	SS(%)
Irrigation regimes	1	0.50	0.27 ns	46.53	5.88**	16.58	2.69**	11.10	1.03**
Nutrient regimes	1	74.93	42.16**	246.73	31.18**	227.47	36.93**	181.29	16.76**
Irrigation regimes× nutrient regimes	1	0.14	0.08 ns	42.41	5.36**	13.25	2.15**	60.14	5.56***
Total			42.16		42.42		41.77		23.35

Notes: The intercept was included in the model. *df* = degrees of freedom. SS (%): proportion of sum of squares to the total sum of squares (Type I). Asterisks denote significance at $P < 0.05$ (*) and $P < 0.01$ (**)

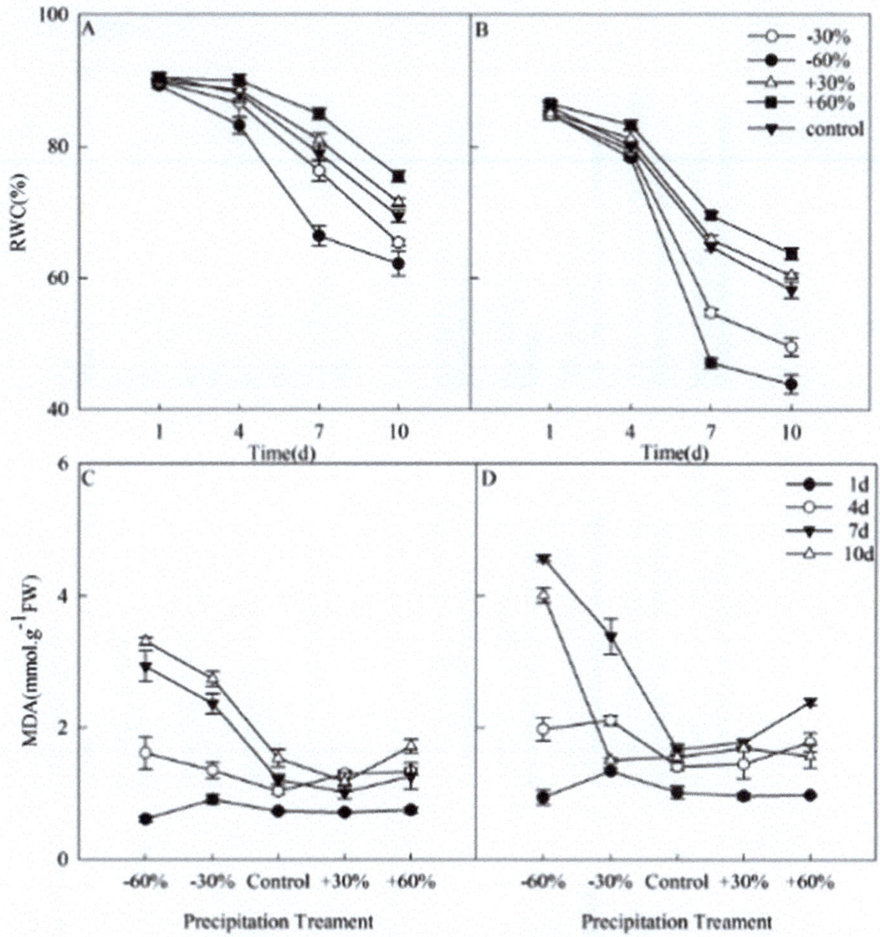

Fig. 5.18 Influence of different rainfall treatments on the RWC of *T. terrestris* (**a**) and *B. dasyphylla* (**b**) and the MDA of *T. terrestris* (**c**) and *B. dasyphylla* (**d**). Values were assigned as the mean ± SE (one-way ANOVA and Fisher's LSD test at $P < 0.05$)

The POD activities of *T. terrestris* and *B. dasyphylla* were weak, and the rate was less than $1H_2O_2.g^{-1}$ FW.min^{-1} (Fig. 5.19a, b), but the POD activity of *T. terrestris* was higher than that of *B. dasyphylla*. The POD activity of *T. terrestris* increased in the first phase (1–4 D) and then decreased in the second phase (4–7 D) and increased in the last phase (7–10 D) except for when P was added at 60%. When P was reduced by 60%, the POD activity was maintained at its highest rate (0.668 and 0.709) on 4 D and 10 D, respectively, and had a significant difference from that of the control and P addition ($P<0.05$). The POD activity of *B. dasyphylla* first decreased (1–4 D) and then increased (4–7 D); in the last phase, the POD activity decreased (7–10 D), except for P addition at 60%. When P was reduced by 60% on the 7 D and 10 D, the

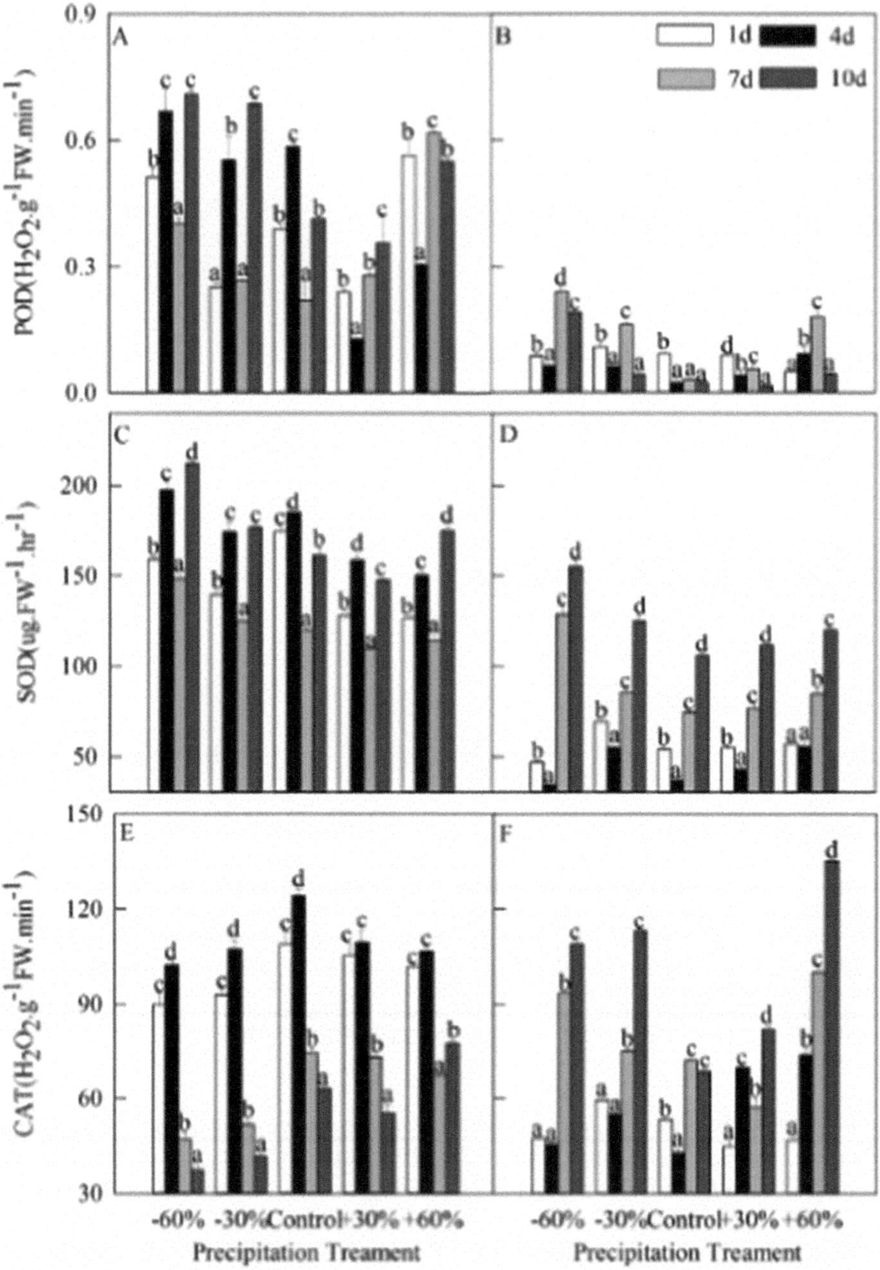

Fig. 5.19 Influence of different rainfall treatments on antioxidant enzymes. (**a**) POD of *Tribulus terrestris*; (**b**) POD of *Bassia dasyphylla*; (**c**) SOD of *Tribulus terrestris*; (**d**) SOD of *Bassia dasyphylla*; (**e**) CAT of *Tribulus terrestris*; and (**f**) CAT of *Bassia dasyphylla*. Values were assigned as the mean ± SE (one-way ANOVA and Fisher's LSD test at $P < 0.05$)

POD activity was the highest (0.709); was 6.70 and 7.09 times that of the control, respectively; and had a significant difference from that of the control and *P* addition ($P < 0.05$). The difference in the POD activity between the control and P addition at 60% was not significant ($P > 0.05$). The SOD activity of *T. terrestris* was significantly higher than that of *B. dasyphylla*.

Plants can successfully adaptively mitigate changes in osmotic regulatory substances in stressed habitats. To assess the adaptability of sandy land plants, an experiment on changes in osmotic regulatory substances of *T. terrestris and B. dasyphylla* was carried out, and the results showed that the soluble protein of *T. terrestris* was significantly higher than that of *B. dasyphylla* ($P < 0.05$) (Fig. 5.20). During the study period, the soluble protein of *T. terrestris* was 23.212–40.733 mg/g FW, and the soluble protein of *B. dasyphylla* was 9.541–23.171 mg/g FW. The soluble protein content of *T. terrestris* first increased (0–7 D) and then decreased (7–10 D). On 7 D, when P was reduced by 60%, the soluble protein (40.733 mg/g FW) was significantly higher than in the other treatments ($P < 0.05$). When P was reduced by 30% or 60%, the soluble protein of *B. dasyphylla* increased. From 4 D to 10 D after rain, the soluble protein content of the P reduction treatment was significantly higher than that of the P addition treatment ($P<0.05$). When P was reduced by 60% on 10 D, the soluble protein (23.171 mg/g FW) was the highest. From 0 D to 7 D after rain, the soluble protein of the control ranged from 13.998 to 14.132 mg/g FW, and the differences were not significant ($P > 0.05$). When P was added at 60%, the soluble protein continuously increased from 11.406 to 12.262 mg/g FW, and the differences were not significant ($P > 0.05$).

On 4 D, in comparison to the P addition and P reduction treatments, the control resulted in higher levels of soluble sugars in *T. terrestris* and *B. dasyphylla* (Fig. 5.20c, d). The soluble sugar content of *T. terrestris* continuously increased. Between 4 and 10 D after rain, the soluble sugar contents of the control increased from 25.556 to 27.986 μg/g FW, and the differences were not significant ($P > 0.05$); however, the differences were significantly higher than those 1 D. On 10 D, the soluble sugar of the P addition treatment was significantly lower than that of the P reduction treatment and higher than that of the control ($P < 0.05$). When P was reduced by 60%, the soluble sugar (48.145 μg/g FW) was the highest and had significant differences with that in the other treatments ($P < 0.05$). When P was added at 30% and in the control, the soluble sugar content of *B. dasyphylla* increased continuously. When P was added at 60% and P was reduced by 30% or 60%, the soluble sugar content first increased (0–7 D) and then decreased (7–10 D). The soluble sugar content in the control was lower than that in the P reduction treatment and higher than that in the P addition treatment. On 7 D, when P was reduced by 60%, the soluble sugar content was 53.945 μg/g FW and was significantly different from that of the other treatments ($P < 0.05$).

The free proline content of *T. terrestris* was significantly higher than that of *B. dasyphylla* (Fig. 5.20e, f). When P was reduced by 60%, the free proline content of both plants increased, but the content of *B. dasyphylla* was significantly lower than that of *T. terrestris*. On 7 D, when P was reduced by 60%, the free proline

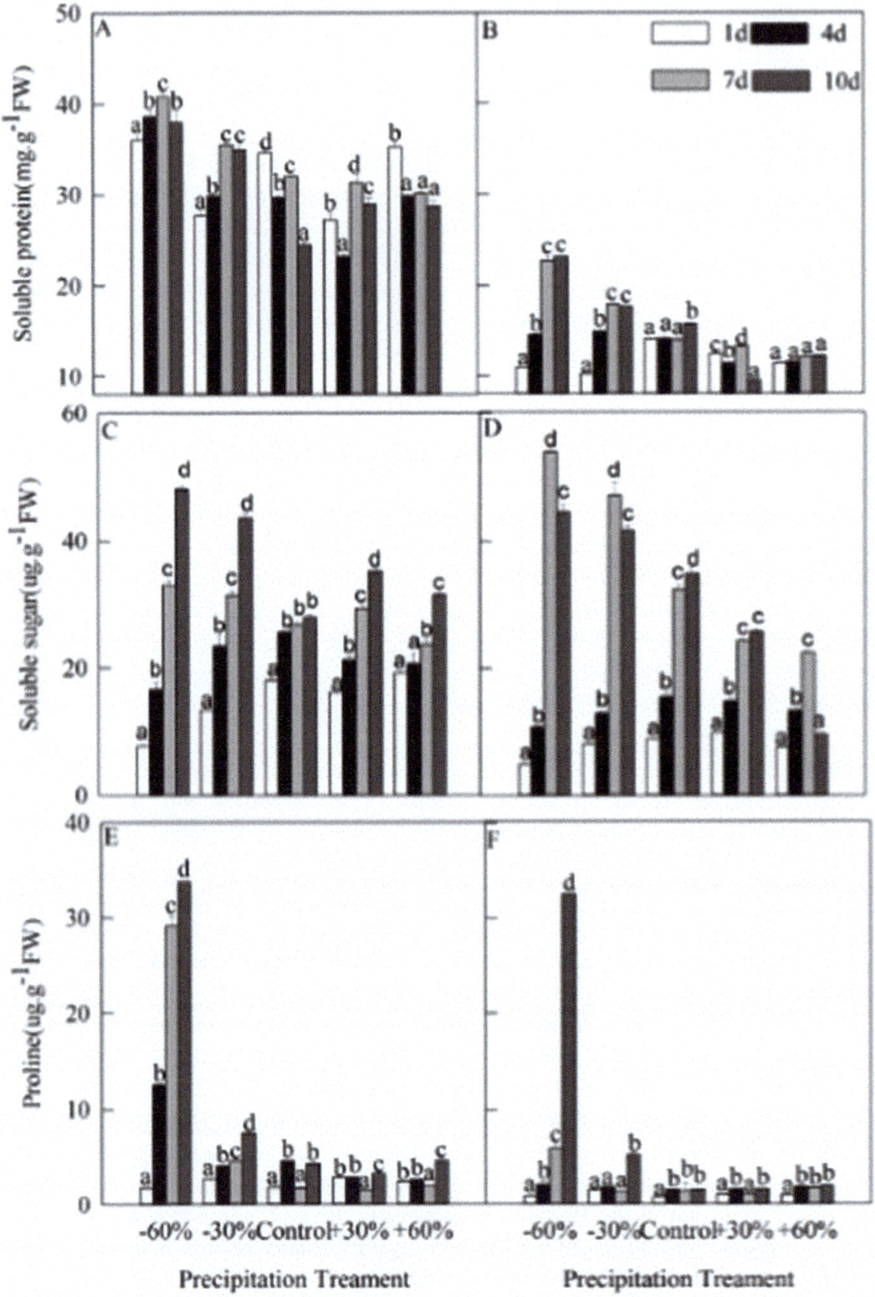

Fig. 5.20 Influence of different rainfall treatments on osmotic adjustment substance. (**a**) Soluble protein of *Tribulus terrestris*; (**b**) soluble protein of *Bassia dasyphylla*; (**c**) soluble sugar of *Tribulus terrestris*; (**d**) soluble sugar of *Bassia dasyphylla*; (**e**) free proline of *Tribulus terrestris*; and (**f**) free proline of *Bassia dasyphylla*. Values were assigned as the mean ± SE (one-way ANOVA and Fisher's LSD test at $P < 0.05$)

content was 6.35–18.76 times that of the other treatments. On 10 D, when P was reduced by 60%, the free proline content was 4.50–10.63 times that of the other treatments. From 4 D to 10 D after rain, when P was added at 60% and in the control, the free proline content was not significantly different from that of the control ($P > 0.05$). On the tenth day, when P was reduced by 60%, the free proline content was 5.54 times that on 7 D and was 6.27–20.99 times that of the other treatments.

From the above analyses, it is clear that plants, either on dunes or fixed dunes, become more adaptive or resistant to stresses by mitigating physiological process and utilizing their functional traits if proper management measures, such as fencing and revegetating in combination with moderating interferences, are adopted (Chen et al. 2019b).

5.4 Human Activity Impacts on the Development and Reversal of Aeolian Desertification

According to related research, desertification processes in the last 100 years were mainly caused by land use changes, particularly in eastern China and Mongolia. In approximately 1872, cropland was cultivated around the lakes of the Horqin Sandy Land and along rivers in a general pattern of patches from the southeastern and southern part of China, and some Korean and Japanese farmers managed paddy rice lands using this pattern. Later, the expansion of cropland continued and caused an increase in population and livestock, leading to the overuse of plants, water, and land.

Increasing populations and croplands lead to water shortages and compromise the basic equilibrium point between ecological improvement and sustainable development. Thus, in recent years, the central and local governments as well as researchers have called for industry development to occur according to regional water capacity.

Therefore, the mechanisms for combating desertification should focus great importance on moderating land use pressure. The key to realizing the goal of desertification reversal or neutrality is at least to neutralize or reduce land resource use to moderate development.

The revegetation of aeolian desertified land is becoming increasingly influential in affecting land cover change in China, and its success depends heavily on the plant species and density and mosaic pattern of different plants according to the trade-off between water supply and plant water consumption (Li 2016).

The response of plants to habitat changes is a multiscale process. At the community level, the water consumption of individuals in a community plays an important role in revegetation. Based on several years of experiments and observations, the Naiman Desertification Research Station of the Chinese Academy of Sciences provided a quantitative assessment of the water consumption of 13 main plant species for revegetation in drylands in China (Fig. 5.21).

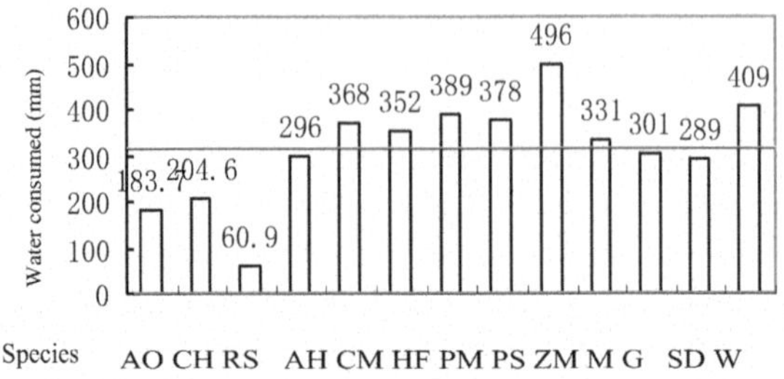

Fig. 5.21 Water consumption of the main plants in drylands in China (Zhao and Zhang 2001). The red bar refers to the average annual precipitation of approximately 350 mm. The red bar in the figure refers to the average annual precipitation of 330–350 mm. *AO Artemisia ordosica*, *CH Caragana korshinskii* Kom., *RS Reaumuria songarica* (Pall.) Maxim, *AH Artemisia halodendron*, *CM Caragana microphylla*, *HF Hedysarum fruticosum*, *PM Pinus sylvestris* var. *mongolica*, *PS Populus simonii*, *ZM Zea mays* L., *M* millet (*Setaria italica*), *G* grassland vegetation, *SD* shifting dune vegetation, *W* wheat (*Triticum aestivum* L.). Red line refers to the long-term averaged annual precipitation 350 mm

It is clear from Fig. 5.21 that *C. microphylla*, *H. fruticosum*, *P. sylvestris* var. *mongolica*, *P. simonii*, and *A. halodendron* need an annual water supply of approximately 300 mm, which is close to the annual precipitation in Horqin Sandy Land in eastern China and Hexi Corridor in western China. *Z. mays* consumes approximately 500 mm of water, which was higher than the average precipitation of 350 mm in the Horqin Sandy Land. Approximately 200–300 mm of irrigation water is needed from groundwater. *Artemisia ordosica*, *Reaumuria songarica* (Pall.) Maxim., and *Caragana korshinskii* Kom. are mainly distributed in western China and need much less water.

P. sylvestris var. *mongolica and P. simonii* and *A. halodendron* have been widely used as primary species for revegetation in the eastern part of China and in the western part of China, where there is an annual precipitation less than 200 mm, mostly in the lower areas with a large groundwater water supply. Together with local plant species, such as *H. ammodendron* and *Hedysarum scoparium*, these plants were planted as demonstration or example species to establish various shelterbelts, sand-fixation vegetation, and street trees around villages and along streets and transportation systems. When *P. sylvestris* var. *mongolica* and *P. simonii* were planted at a density of 15–25 or 25–35 plants/mu, natural restoration of degraded vegetation could be successful, even without irrigation and local grass natural invasion (Zhao and Zhang 2001).

From the above analyses, it is concluded that native plant species play a key role in revegetation in combating desertification because these plants can successfully respond to drought, high temperatures, and salinity to some extent through diversified mechanisms. These mechanisms include resolvable proteins, enzymes, photosynthetic processes, and even natural thinning of dense vegetation. Based on research and monitoring findings, researchers have determined the threshold value of the water requirement for minimal survival and growth. According to these findings, the stakeholders engaged in combating aeolian desertification can choose the right plant species and plant trees or bushes at a more adaptive density and with a species arrangement that obtain the expected revegetation goal.

Land use change is one of the driving forces leading to aeolian desertification, and land use change is driven by the needs of the region of concern for survival and development. Therefore, to combat aeolian desertification, climate change, land use change, marketing, land use policy, and, more importantly, the expectation of the locals for survival and development need to be considered systematically. When we take measures to promote ecological conservation and development, we can combat aeolian desertification even in the regions in underdeveloped areas.

References

Chen GT, Zhang JX, Guan YZ et al (1995) Blown sand control in Beijing region. Articles collection of the Institute of Desert, Chinese Academy of Sciences, vol 4. Science Press, Beijing (in Chinese)

Chen JL, Zhao XY, Liu XP et al (2019a) Growth and physiological responses of two psammophytes to precipitation manipulation in Horqin sandy land, eastern China. Plants Basel 8(7):244

Chen JL, Zhao XY, Zhang YQ et al (2019b) Effects of drought and rehydration on the chlorophyll fluorescence parameters and physiological responses of *Artemisia halodendron*. Water 11(4): 793

Chepil WS (1945) Dynamics of wind erosion: nature of movement of soil by wind. Soil Sci 60:305–320

Fryberger SG, Goudie AS (1981) Arid geomorphology. Prog Phys Geogr 5:420–428

Gillette DA, Adams J, Muhs D et al (1982) Threshold friction velocities and rupture moduli for crusted desert soils for input of soil particles to the air. J Geophys Res 87C:9003–9005

Harazono Y, Li SG (1999) Micrometeorology of dune and vegetation at the semi-arid area at Naiman in Inner Mongolia. Bull Nat Institute Agro-environ Sci 24

He DL (1988) The wind velocity of threshold sand affected by rainfall. J Desert Res 8(4):18–26 (in Chinese with English abstract)

Li XR (2016) Fundamental eco-hydrology of ecological restoration and recovery in sandy desert regions of China. Scientific Press, Beijing (in Chinese)

Li YQ, Zhou XH, Brandle JR et al (2012) Temporal progress in improving carbon and nitrogen storage by grazing enclosure practice in a degraded land area of China's Horqin sandy grassland. Agric Ecosyst Environ 159:55–61

Liu XP, He YH, Zhang TH et al (2015) The response of infiltration depth, evaporation, and soil water replenishment to rainfall in mobile dunes in the Horqin sandy land, northern China. Environ Earth Sci 73:8699–8708

Liu XP, He YH, Zhao XY et al (2016) The response of soil water and deep percolation under *Caragana microphylla* to rainfall in the Horqin sand land, northern China. Catena 139:82–91

Livingstone I, Warren A (1996) Aeolian geomorphology. Addison Wesley Longman Limited, Singapore

Miotke F (1982) Formation and rate of formation of ventifacts in Victoria land, Antarctica. Polar Geogr Geol 6:97–113

Raven P, Johnson G (1986) Biology. Times Mirror/Mosby, St. Louis

Wang T (2011) Deserts and aeolian desertification in China. Elsevier in Amsterdam and Science Press in Beijing

Zhao XY, Zhang TH (2001) Management of land resources and land degradation in Horqin sandy land. In: Integrated land management in dry areas. UNU Desertification Series No.4, United Nations University

Zhu ZD, Chen GT (1994) Sandy desertification in China. Science Press, Beijing (in Chinese)

Zuo XA, Knops JMH, Zhao XY et al (2012) Indirect drivers of plant diversity-productivity relationship in semiarid sandy grasslands. Biogeosciences 9:1277–1289

Chapter 6
Impacts of Aeolian Desertification and Dust Storms on Ecosystems, Economic Development, and Human Health

Yasunori Kurosaki, Toshihiko Kinugasa, Mandakh Nyamtseren, Shulin Liu, and Shinji Otani

Abstract Aeolian desertification has impacts on ecosystems, agriculture, industry, transportation, and human health, and these impacts are caused by wind erosion, sediment deposition, sand-dust storms (SDSs), and fine airborne dust particles. Section 6.1 will explain the mechanism of SDS occurrence based on erosivity and erodibility. This section also describes the seasonality of SDSs, the increase in their frequency from the 1990s to 2000s, and the reasons for this increase. Section 6.2 describes the impacts of aeolian desertification on ecosystems. In terms of wind erosion, windblown sand particles physically injure plants. In addition, wind erosion removes soil and thereby exposes roots to the air. Sediment deposition enhances grass mortality and expands the area of erodible land surfaces; as a result, desertification accelerates. Section 6.3 provides a variety of concrete examples of aeolian desertification in China. Examples show the progress of aeolian desertification and its effects on agriculture, industry, and transportation. Section 6.4 explains the impacts of aeolian desertification on human health in dust source regions and downwind regions. SDSs, which can be disasters, occur only in dust source regions and lead to deaths and injuries. Fine airborne particles cause cardiovascular and respiratory diseases in both regions.

Y. Kurosaki (✉)
Arid Land Research Center, Tottori University, Tottori, Japan
e-mail: kuro@tottori-u.ac.jp

T. Kinugasa
Faculty of Agriculture, Tottori University, Tottori, Japan

M. Nyamtseren
The Institute of Geography and Geoecology, Mongolian Academy of Sciences, Ulaanbaatar, Mongolia

S. Liu
Key Laboratory of Desert and Desertification, Northwest Institute of Eco-Environment and Resources, Chinese Academy of Sciences, Lanzhou, Gansu, China

S. Otani
International Platform for Dryland Research and Education, Tottori University, Tottori, Japan

T. Wang et al. (eds.), *Combating Aeolian Desertification in Northeast Asia*, Ecological Research Monographs, https://doi.org/10.1007/978-981-16-9028-0_6

Keywords Sand-dust storm · Wind erosion · Sediment deposition · Sand burial · Death · Disease

6.1 Aeolian Desertification and Sand-Dust Storms (SDSs)

6.1.1 Airborne Soil Particles in Source and Downwind Regions

Strong surface winds (i.e., storms) lift soil particles into the free atmosphere over drylands. Such storms are known as sand and dust storms (SDSs). The raised soil particles are advected with diffusion by prevailing wind. Finally, the particles return to the Earth's surface in a wide downwind area through gravitational settling, turbulence, and precipitation. Thus, the transport of soil particles has three phases: emission, advection, and deposition. In the case of Northeast Asia, major regions of SDSs are drylands around the Gobi Desert and the Taklimakan Desert. The particles travel eastward by the prevailing westerlies, and they reach the Korean Peninsula, Japan, and further remote regions.

According to the Udden-Wentworth grade scale (Udden 1914; Wentworth 1922), soil particles are divided into sand ($63 < d \leq 2000$ μm), silt ($4 < d \leq 63$ μm), and clay ($d < 4$ μm); generally, silt and clay are called dust. The motion of soil particles near the surface depends on the sizes of the particles, and this motion can be classified into three modes: creep ($d >$ approximately 500 μm), saltation (approximately $70 < d < 500$ μm), and suspension ($d <$ approximately 70 μm) (Bagnold 1941). Based on these definitions, both sand and dust particles are contained in source regions, but most soil particles reaching downwind areas are composed of dust particles.

The concentration and size of dust particles increase toward dust sources. Kanai et al. (2005) conducted observations in Beijing in China and in Fukuoka in Japan in the spring from 2001 to 2003. The average concentration and the mode of the sizes were 307.8 μg m^{-3} and >11 μm in Beijing and 34.7 μg m^{-3} and 3.3–4.7 μm in Fukuoka, respectively. In addition to high concentrations and large dust particles, strong winds during dust events are also nonnegligible characteristics in upwind regions.

Dust has a considerable effect on human activity. However, its type and degree vary and depend on the distance from its source due to the differences in dust concentrations, particle sizes, and wind speeds at different sources. As SDS is a type of storm, it is a disaster in dust source regions. It kills humans, livestock, and wild animals. A SDS also has a substantial economic impact as it destroys infrastructure and damages transportation systems due to low visibility and agriculture due to wind erosion. The extents of the disaster decrease at greater distances from dust sources. However, finer dust particles have effects in downwind regions. They cause health problems related to respiratory diseases. In addition, when finer dust particles mix with microorganisms and anthropogenic air pollutant, different types of health problems can occur, and ecosystems can be impacted. There are reports that dust particles originating from Northeast Asia reach very distant downwind regions such as Greenland (Bory et al. 2003) and North America (Husar et al. 2001) and travel around the Northern Hemisphere (Uno et al. 2009). Suspended dust particles change shortwave and longwave radiation, and they work as condensation nuclei, thereby affecting the climate on a global scale.

6.1.2 Factors That Cause SDSs

There are a large number of factors that affect SDS, such as wind, soil particle size, soil moisture, soil crust, stone, snow cover, salt accumulation, and vegetation. The factors can be divided into two broad categories: aeolian erosivity and erodibility (hereafter, erosivity, and erodibility). Erosivity is the ability of wind to move soil particles, and it is typically expressed by wind friction velocity (e.g., Bagnold 1941; Kurosaki et al. 2011). For simplification, wind velocity is often employed as a substitute (e.g., Kurosaki and Mikami 2003), and hereafter, we will use wind velocity in this section. Erodibility is characterized as the susceptibility of soil and land surfaces to the movement of soil particles. While erosivity is composed of wind alone and can be measured by wind velocity, erodibility is influenced by a multitude of factors (Fig. 6.1). Monitoring all erodibility factors with accuracy is impossible with the current technology. In addition, differences in units of the factors affecting erodibility make it difficult to measure.

The threshold wind velocity (hereafter, threshold velocity), which is the minimum wind velocity required for soil particles to move, can be recognized as a unified index of a variety of erodibility factors, and its unit is conveniently the same as that of wind velocity. The many factors controlling erodibility complicate discussion on the causes of a SDS. However, by employing the threshold velocity as an index of

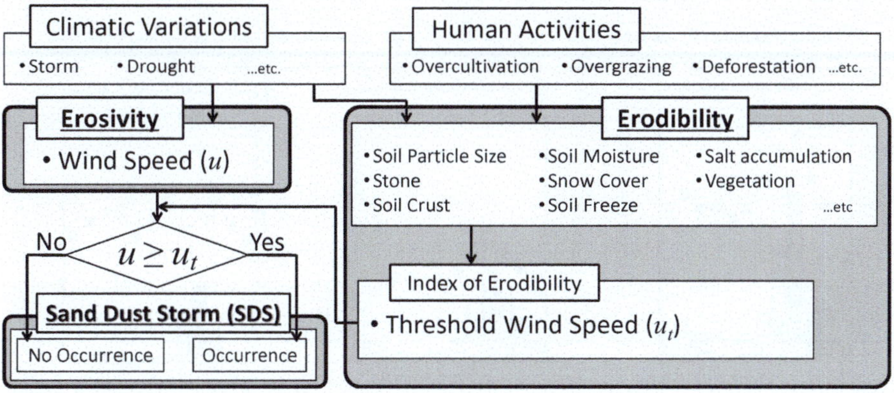

Fig. 6.1 Relationship among the erosivity factor, erodibility factors, and the occurrence of an SDS (modified from Kurosaki et al. 2011)

erodibility, the discussion can be simplified: when wind velocity exceeds threshold velocity, a SDS occurs.

Climatic variations and human activities influence erosivity and erodibility (Fig. 6.1). Regarding climatic variation, storms not accompanied by precipitation are connected to only the erosivity factor (i.e., surface wind velocity), storms accompanied by precipitation are connected to both erosivity and erodibility factors, and drought is connected to erodibility factors. Human activities such as overcultivation, overgrazing, and deforestation are connected to erodibility factors. Various desertification factors, including climatic variations and human activities (UNCCD 1994), are connected to erodibility factors. Figure 6.1 shows that desertification generally accelerates the intensity and frequency of SDSs by increasing erodibility, although some desertification types, such as salt accumulation, reduce erodibility.

6.1.3 Seasonality of SDSs

Meteorological elements such as air temperature, precipitation, and wind speed have been observed every 3 or 6 synoptic hours (i.e., 00, 03, . . ., 21 UTC) for long periods at meteorological observatories. According to WMO (2019), the elements include "present weather," which is the weather existing at the time of observation, or under certain conditions, during the hour preceding the time of observation. Its weather conditions are divided into 100 types (ww = 00 to 99, where "ww" is the code of present weather), as shown in Code Table 4677 in the WMO (2019). Kurosaki and Mikami (2003) categorized ww = 06 as suspended dust, ww = 07 and 08 as dust occurrences but not SDSs, and ww = 09, 30–35, and 98 as SDSs. The research also defined a strong wind as a wind whose velocity exceeds 6.5 m s^{-1}.

Figures 6.2 and 6.3 show that dust events, including SDSs, occur most frequently in spring (i.e., March to May) in Northeast Asia. Figure 6.2 shows that erosivity (i.e.,

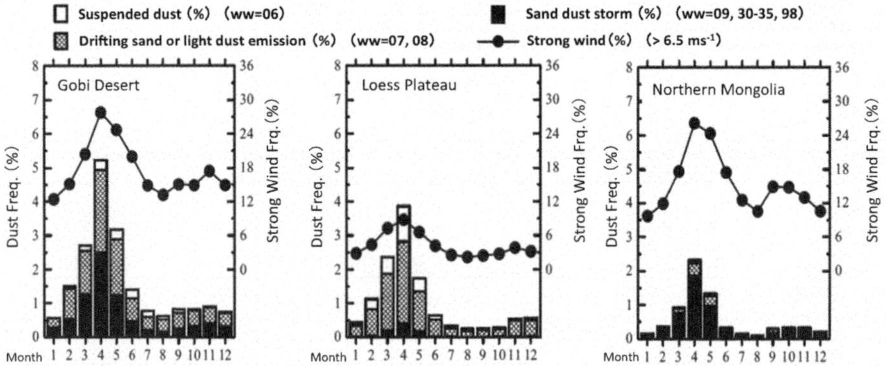

Fig. 6.2 Seasonal variations in the frequency of strong winds and dust event. (**a**) Gobi Des. (**b**) Loess Plateau. (**c**) Northern Mongolia (modified from Kurosaki and Mikami 2005)

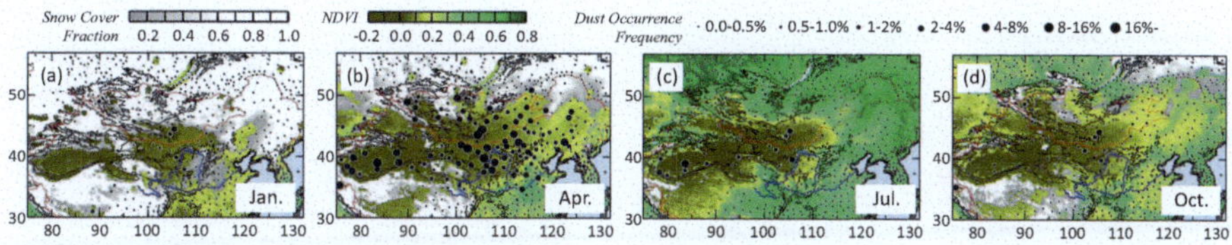

Fig. 6.3 Seasonal variations in the frequency of snow cover fraction, NDVI and dust occurrence (modified from Kurosaki and Mikami 2007)

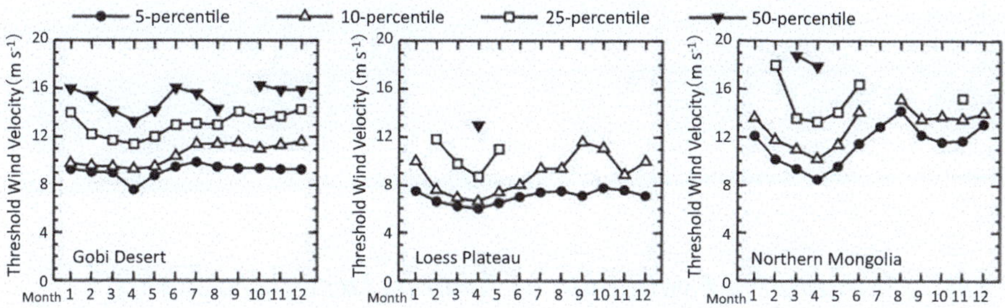

Fig. 6.4 Seasonal variations in the 5th, 10th, 25th and 50th percentiles of threshold velocity for dust occurrence (ww = 07–09, 30–35 and 98) (modified from Kurosaki and Mikami 2007)

strong wind frequency) is the highest in spring. Figure 6.3 shows that the snow cover fraction and normalized difference vegetation index (NDVI), which can be an index of erodibility factors, are low in spring. We cannot survey all erodibility factors, but we can determine the seasonal variation in erodibility by statistically estimating threshold velocity. Figure 6.4 shows the 5th, 10th, 25th, and 50th percentiles of threshold velocity (Kurosaki and Mikami 2007). The statistical values cannot be estimated when dust occurrences (ww = 07–09, 30–35, and 98) are rarely observed. The 5th percentile shows very small seasonal variations in the Gobi Desert and the Loess Plateau, but other percentiles show high erodibility in spring and low erodibility in summer and winter. This suggests that seasonal variations in soil and land surface conditions occur in normal years even in very dry regions, but almost no seasonal variations occur in years of very high erodibility (e.g., drought years). In northern Mongolia, whose land cover type is grassland, the 5th percentile shows very wide seasonal variation. We can expect this seasonal variation to be based on seasonal erodibility factors such as snow and vegetation, as shown in Fig. 6.3.

We recognize that since both erosivity and erodibility are highest in spring, SDSs occur most frequently in spring in Northeast Asia.

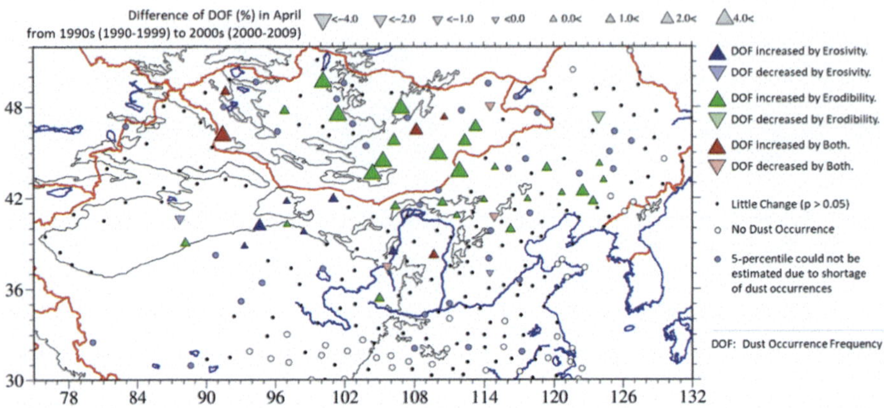

Fig. 6.5 Differences in DOF in April from the 1990s to the 2000s and the causes of these differences in terms of erosivity and erodibility (drawn based on results of from Kurosaki et al. 2011)

6.1.4 Quantitative Evaluation of Aeolian Desertification by an Erodibility Index

When SDSs become active, we often find reports that indicate desertification as the cause of the SDSs, especially in the media. However, as discussed in Sect. 6.1.2, we need to examine the causes from aspects of not only erodibility, which can be recognized as a condition of desertification with some exceptions, but also erosivity. If we have data on both wind velocity and threshold velocity, then we can easily examine them, but there are no meteorological observatories that are monitoring threshold velocity, although wind speed is monitored at all observatories. However, as shown in Sect. 6.1.3, we can estimate percentiles of threshold velocity.

Figure 6.5 shows a difference in dust occurrence frequency (hereafter, DOF) in April from the 1990s to the 2000s. By employing strong wind frequency as an erosivity index and the 5th percentile of threshold velocity as an erodibility index, we can examine the reasons for the decadal change in DOF in terms of both erosivity and erodibility. The figure shows that DOF increased at many observatories in Mongolia, eastern Inner Mongolia, and northeastern China. It also shows that the 5th percentile threshold velocity dropped (i.e., erodibility increased) at most observatories in such regions. This result suggests that desertification was the reason for the increase in dust occurrence. However, we still have two major questions that should be resolved in the future: one is which factors increase erodibility, and the other is what types of desertification are connected to erodibility factors.

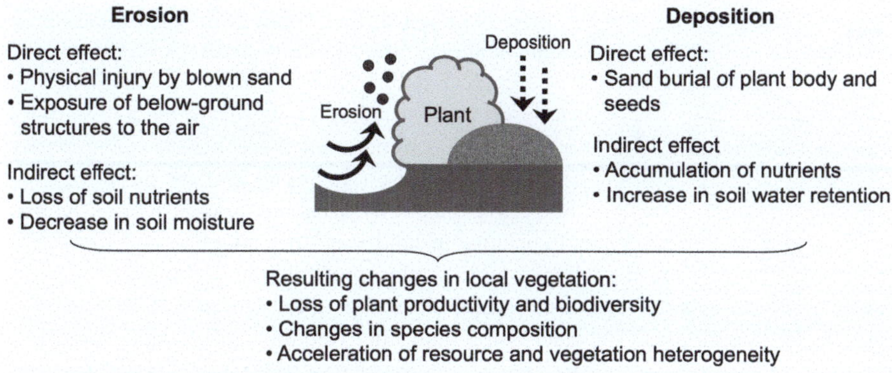

Fig. 6.6 Outline of the impact of aeolian desertification on plants

Fig. 6.7 Roots of (**a**) *Populus simonii* and (**b**) *Salix psammophila* exposed to the air by soil excavation due to wind erosion in Kubuqi desert (**a**) and Mu Us sandy land (**b**) in Inner Mongolia, China (photo by N. Yamanaka)

6.2 Impacts of Aeolian Desertification on Local Ecosystems

6.2.1 Effect of Aeolian Desertification on Terrestrial Ecosystems: An Overview

Aeolian desertification affects ecosystems through two main physical processes, wind erosion, which consists of detachment and transport of soil particles, and the subsequent deposition of transported soil particles (Khalaf 1989, Ravi et al. 2010, Wang 2013). These processes affect plant growth both in direct and indirect ways and consequently local vegetation (Fig. 6.6). Wind erosion directly affects plant growth and survival by physically damaging plants by windblown sand particles (Fryrear et al. 1973; Precheur et al. 1978). In addition to physical injury, wind erosion removes soil and thereby exposes roots to the air (Fig. 6.7). Root removal causes water loss and elevated temperatures in root systems and lowers plant

Fig. 6.8 (**a**) Adventitious roots of *Populus simonii* emerged from buried stems and branches in the Kubuqi Desert in Inner Mongolia, China, and (**b**) a *Tamarix* coppice dune (or nebkha) in Xinjiang, China (photo by F. Yamamoto (**a**) and by N. Yamanaka (**b**))

efficiency in terms of water and nutrient uptake. As a consequence, photosynthetic abilities decrease, and plant growth is impaired. Wind erosion affects plant growth indirectly by modifying soil water and nutrient conditions. Wind removes fine soil that is rich in nutrients and contributes to the ability of soil to hold water, consequently causing the loss of land that sustains vegetation (Ravi et al. 2010).

Wind-eroded sediment redeposits in areas near or distant from the source are spatially heterogeneous due to topography and roughness of vegetation. Plant burial is a direct consequence of sediment deposition on plants and can have both positive and negative effects on plant growth (Maun 1998; Kent et al. 2001). When the burial is shallow and relatively short term, it benefits plants by increasing water and nutrient availability and lowering the temperature around root systems. Sand burial sometimes induces the production of adventitious roots, which can contribute to the acquisition of more nutrients and water (Fig. 6.8a). However, when burial is deeper and long lasting, soil burial has detrimental effects on plant growth as it reduces light availability for photosynthesis and limits the oxygen available to roots. In addition to direct effects, sediment deposition affects plants indirectly through the accumulation of nutrients and improvements in soil moisture retention. Generally, windblown sediment is fine and more fertile than surface soil particles, and therefore, its deposition results in nutrient accumulation and enhanced soil water retention (Ravi et al. 2010). Sediment deposition and soil erosion processes cause a redistribution of resources in an ecosystem, and topography and vegetation roughness promote resource distribution spatial heterogeneity by affecting the aeolian soil particle transport processes (Ravi et al. 2011).

The direct and indirect effects of soil erosion and sediment deposition generally cause a decrease in plant productivity and a loss of biodiversity. As plant responses to wind erosion and deposition are different among species, the species composition of vegetation changes with the progression aeolian desertification. A typical example of a shift in species composition that occurs with aeolian desertification is the development of shrubland and the formation of coppice dunes (or nebkha, Fig. 6.8b), which occur in many arid ecosystems (Ravi et al. 2010, 2011). Wind erosion and sediment deposition in shrubby grasslands physically injure grass species and bury these species in sand, which enhance grass mortality and expand the area of erodible land surface (Li et al. 2009). Fine and fertile sediments are trapped by shrub canopies and deposited around them, forming fertile islands that enhance shrub growth and cause further sand trapping. This process enhances resource heterogeneity and accelerates the vegetation shift toward shrubland (Schlesinger and Pilmanis 1998). Thus, positive feedback between vegetation heterogeneity and resource localization mediated through aeolian processes can accelerate the impact of aeolian desertification.

6.2.2 *Effect of Wind Erosion on Plants in Northeast Asian Ecosystems*

Root exposure to air due to wind erosion affects plant growth, and this issue has been studied in arid regions of China, especially Inner Mongolia. Most of these studies are on shrub or semishrub species, including *Artemisia ordosica* (Li et al. 2010a, b), *Artemisia wudanica* (Liu et al. 2014a), *Calligonum arborescens* (Luo and Zhao 2015), and *Calligonum mongolicum* (Fan et al. 2018), and some are on rhizomatous perennial grasses, such as *Psammochloa villosa* (Yu et al. 2008) and *Phragmites communis* (Liu et al. 2014b). These studies revealed that removal of belowground structures by wind erosion reduces plant survival, growth, and reproduction depending on the severity of soil erosion. In addition, some of these studies emphasized the contribution of rhizome connections to plant tolerance against wind erosion conditions (Yu et al. 2008, Luo and Zhao 2015, Fan et al. 2018). Wind erosion induces changes in plant biomass allocation, although its ecological significance is not fully understood; plants that experience wind erosion allocate more biomass to vegetative growth than to reproduction (Li et al. 2010b), more biomass to belowground structures than to aboveground structures (Li et al. 2010b; Liu et al. 2014a), and more biomass to the structures of vertically oriented growth than to those of horizontally oriented growth (Liu et al. 2014b). In the sand dune shrub *Artemisia wudanica*, the negative effects of wind erosion on plant growth have been shown to be compensated by subsequent sand burial, probably due to the induction of the production of adventitious roots that favor the acquisition of more nutrients (Liu et al. 2014a).

6.2.3 *Effect of Deposition on Plants in Northeast Asian Ecosystems*

There have been an abundant number of studies on the effects of sand burial on plant germination, growth, and reproduction in Northeast Asia in recent years. Most of these studies were performed in northern China, especially Inner Mongolia. In our literature survey, we found studies on the effects of sand burial on plant germination for 40 species, including 23 shrub/semishrub species, 4 perennial grasses, and 11 annual forbs (Table 6.1). In these studies, in general, shallow sand burial of most approximately 1 cm did not reduce and rather increased germination, although severe sand burial decreased germination and subsequent seedling emergence. The response of germination to sand burial differed among species (Zhu et al. 2009; Liu et al. 2013; Wang et al. 2019), and these differences can cause a shift in species composition with aeolian desertification.

The effects of sand burial on plant growth and reproduction have also been studied intensively in northern China, although the number of studies is limited compared to that on germination. We found studies on 17 species, including 1 tree, 13 shrubs/semishrubs, and 3 herbaceous species (Table 6.2). Among these studies, partial burial usually did not have a negative effect on plant growth and survivorship and enhanced plant survival in many cases. When burial depth increased and plants suffered complete burial, plant mortality increased sharply. In some studies, the negative effects of sand burial on growth and survivorship were shown to be alleviated by clonal integration through stolons or rhizomes (Yu et al. 2004, Yu et al. 2002, Liu et al. 2007, Luo and Zhao 2015). Sand burial either increased or decreased the root-to-shoot ratio (Zhao et al. 2007, Liu et al. 2008, Tang et al. 2016) and sometimes induced the formation of adventitious roots (Fig. 6.8a). When experimental sand burial was performed on steppe plant communities, the plant response to sand burial was found to differ among species (Ye et al. 2017, 2019), implying that sand burial contributes to species coexistence and results in biodiversity in regions with intense aeolian sand transport.

6.2.4 Effect of Long-Range Transported Sediments on Downwind Asian Ecosystems

As described in Sect. 6.1, part of the wind-eroded sediment released from dry regions of Northeast Asia travels eastwardly by the prevailing westerlies. Wind-transported sediment is then deposited in downwind marine and terrestrial ecosystems. Sediment deposition is recognized as an important source of nutrients in those ecosystems because sediment contains nutrients essential for the growth of plants and phytoplankton, such as nitrogen, phosphorous, and iron (Harrison et al. 2001; Goudie and Middleton 2006; Maher et al. 2010).

Iron is essential for chlorophyll formation but is less available in marine ecosystems; therefore, such ecosystems sometimes respond strongly to iron input that occurs through sediment deposition (Martin 1990; Moore et al. 2001; Meskhidze et al. 2005). Satellite and ground observations of dust and chlorophyll-a have revealed that heavy Asian dust events could enhance phytoplankton blooms in the South China Sea, probably due to iron input (Wang et al. 2012, Du et al. 2021). A

Table 6.1 List of studies on the effect of sand burial on seed germination in Northeast Asian ecosystems

Species	Family	Life form	Literature
Agriophyllum squarrosum	Chenopodiaceae	AF	Zheng et al. (2005), Liu et al. (2013)
Anabasis aphylla	Chenopodiaceae	S	Wang et al. (2017)
Artemisia halodendron	Asteraceae	S	Li et al. (2012)
Artemisia ordosica	Asteraceae	S	Zheng et al. (2005), Zheng et al. (2009), Liu et al. (2013)
Artemisia scoparia	Asteraceae	AF	Li et al. (2012)
Artemisia sieversiana	Asteraceae	AF	Li et al. (2012)
Artemisia sphaerocephala	Asteraceae	S	Huang and Gutterman 1999, Zheng et al. 2005, Zheng et al. 2009
Artemisia wudanica	Asteraceae	S	Li et al. (2012)
Bassia dasyphylla	Chenopodiaceae	AF	Liu et al. (2013)
Bassia hyssopifolia	Chenopodiaceae	AF	Liu et al. (2013)
Calligonum alaschanicum	Polygonaceae	S	Ren et al. (2002)
Calligonum arborescens	Polygonaceae	S	Ren et al. (2002)
Calligonum caput-medusae	Polygonaceae	S	Ren et al. (2002)
Calligonum chinese	Polygonaceae	S	Ren et al. (2002)
Calligonum densum	Polygonaceae	S	Ren et al. (2002)
Calligonum junceum	Polygonaceae	S	Ren et al. (2002)
Calligonum leucocladum	Polygonaceae	S	Ren et al. (2002)
Calligonum mongolicum	Polygonaceae	S	Wang et al. (2019)
Calligonum mongolicum	Polygonaceae	S	Ren et al. (2002)
Calligonum potaninii	Polygonaceae	S	Ren et al. (2002)
Calligonum rubicundum	Polygonaceae	S	Ren et al. (2002)
Caragana korshinskii	Fabaceae	S	Zheng et al. (2005), Zheng et al. (2009)
Caragana microphylla	Fabaceae	S	Zhu et al. (2004)
Ceratocarpus arenarius	Chenopodiaceae	AF	Liu et al. (2013)
Chenopodium aristatum	Chenopodiaceae	AF	Liu et al. (2013)

(continued)

Table 6.1 (continued)

Species	Family	Life form	Literature
Corispermum lehmannianum	Chenopodiaceae	AF	Liu et al. (2013)
Cousinia affinis	Asteraceae	PF	Liu et al. (2013)
Eremosparton songoricum	Fabaceae	S	Liu et al. (2011)
Haloxylon ammodendron	Chenopodiaceae	S	Wang et al. (2019)
Hedysarum fruticosum	Fabaceae	S	Zheng et al. (2005)
Hedysarum laeve	Fabaceae	S	Zhu et al. (2004), Zheng et al. (2009)
Horaninovia ulicina	Chenopodiaceae	AF	Liu et al. (2013)
Leymus racemosus	Poaceae	PG	Huang et al. (2004b), Liu et al. (2013)
Leymus secalinus	Poaceae	PG	Zhu et al. (2009), Zhu et al. (2014)
Medicago sativa	Fabaceae	PF	Zheng et al. (2005)
Minuartia regeliana	Caryophyllaceae	AF	Liu et al. (2013)
Nitraria sphaerocarpa	Zygophyllaceae	S	Wang et al. (2019)
Petrosimonia sibirica	Chenopodiaceae	AF	Liu et al. (2013)
Psammochloa villosa	Poaceae	PG	Huang et al. (2004a), Zhu et al. (2009)
Stipagrostis pennata	Poaceae	PG	Liu et al. (2013)

The life forms S, PF, PG, and AF represent shrub or semishrub, perennial forb, perennial grass, and annual forb, respectively

similar contribution of dust deposition to phytoplankton blooms was also found in the Yellow Sea (Tan and Shi 2012; Tan and Wang 2014), Japan Sea (Jo et al. 2007), and North Pacific Ocean (Bishop et al. 2002; Han et al. 2011; Wan et al. 2020). Enhanced phytoplankton growth can stimulate the productivity of marine ecosystems. For example, iron fertilization of phytoplankton by Asian dust is suggested to contribute to sardine productivity in the North Pacific Ocean (Qiu 2015).

Compared to that on marine ecosystems, the effect of nutrient input through dust deposition on terrestrial ecosystems has been minimally investigated in downwind Asian regions, probably because it is difficult to detect that effect. However, deposition of nutrients, such as nitrogen and phosphorus, together with aeolian sediment has been reported in those regions (Choi et al. 2001; Mori et al. 2003; Yoshioka et al. 2009; Chiwa 2010), implying that long-range transported dust affects downwind terrestrial ecosystems to a certain degree.

Table 6.2 List of studies on the effects of sand burial on plant growth in Northeast Asian ecosystems

Species	Family	Life form	Literature
Agriophyllum squarrosum	Chenopodiaceae	AF	Li et al. (2015)
Artemisia frigida	Asteraceae	S	Liu et al. (2008)
Artemisia gmelinii	Asteraceae	S	Liu et al. (2008)
Artemisia halodendron	Asteraceae	S	Liu et al. (2008), Qu et al. (2017)
Artemisia ordosica	Asteraceae	S	Li et al. (2010a), Li et al. (2010b), Zheng et al. (2012)
Artemisia sphaerocephala	Asteraceae	S	Zheng et al. (2012)
Artemisia wudanica	Asteraceae	S	Liu et al. (2008), Liu et al. (2014a)
Calligonum arborescens	Polygonaceae	S	Luo and Zhao (2015)
Caragana intermedia	Fabaceae	S	Xu et al. (2013a)
Cynanchum komarovii	Asclepiadaceae	S	Xu et al. (2013b)
Hedysarum laeve	Fabaceae	S	Liu et al. (2007)
Lespedeza daurica	Fabaceae	S	Qu et al. (2017)
Nitraria sphaerocarpa	Zygophyllaceae	S	Zhao et al. (2007)
Potentilla anserina	Rosaceae	PF	Yu et al. (2004)
Psammochloa villosa	Poaceae	PG	Yu et al. (2002)
Salix cheilophila	Salicaceae	S	Teraminami et al. (2013)
Ulmus pumila	Ulmaceae	T	Shi et al. (2004), Tang et al. (2016)

The life forms T, S, PF, PG, and AF represent tree, shrub or semishrub, perennial forb, perennial grass, and annual forb, respectively

6.3 Impacts of Aeolian Desertification on Agriculture, Industry, and Transportation

The impacts of aeolian desertification on agriculture, industry, and transportation are introduced based on data from statistical yearbooks or published papers.

Aeolian desertification usually results in a substantial loss of available land resources. Aeolian desertification can not only reduce land productivity and land area but also further accelerate soil wind erosion and cause various types of blown sand damage, such as ruined croplands (resulting in the need to replant crops), economic losses, people and livestock casualties, destroyed buildings and houses, and interrupted traffic and communication. Aeolian desertification severely affects local agriculture, industry, and transportation.

6.3.1 Impacts of Aeolian Desertification on Agriculture

Windblown sand hazards are caused by a series of sand activities that occur during the process of aeolian desertification on farmland, such as sand grain deflation, transportation, and deposition and mobile sand dune forward movement. The types of blown sand hazards in different regions are not the same due to different sand sources. In summary, the types of blown sand hazards in the croplands of semiarid regions of East China are dominated by mobile sand dune forward movement and cultivated soil deflation, but those in the croplands in irrigated oasis regions of West China are mobile sand dune forward movement and sand-driven winds; for croplands in semihumid desert regions, wind erosion and sand beaten seedlings are the most hazardous (Wang et al. 2011).

Aeolian desertification farmland is distributed extensively in China, with an area of 3.56×10^6 hm^2 (Table 6.3). In the farming-pastoral zone of the Inner Mongolia Autonomous Region, such as in the rain-fed farmland region of Houshan, aeolian desertification is very severe (Zhu et al. 1989), and the area of aeolian desertification already accounted for 32.4% of that of local farmland area in Shangdu County at the end of the 1980s.

Aeolian desertification is particularly harmful to agriculture. The spring sowing season is from April to May each year; in desertified areas, the seeds and fertilizers are often blown away, the seedlings are uprooted, the soil moisture disperses, and the seedlings dry or are buried. Some places must be replanted repeatedly, even late in the agricultural season. Once a surface experiences wind erosion, the soil fertility of that farmland or pasture will degrade, and its yields will decrease, with organic content and nutrients blowing away by wind.

Aeolian desertification has become a concern and affects local agriculture. Seven Qis of the Houshan area in Ulanqab Meng of the Inner Mongolia Autonomous

Table 6.3 Area of aeolian desertification of dry farmland in the western region of China in 1999 (hm^2)

Province	Total area of dryland in wind erosion regions	Area of aeolian desertification	Percentage (%)
Ningxia	205,554	108,281	52.68
Shanxi	242,616	5745	2.37
Qinghai	168,779	425	0.39
Inner Mongolia	2,877,380	2,694,789	93.65
Tibet	26,159	22,245	85.04
Xinjiang	1,399,763	354,248	25.31
Gansu	1,022,671	364,328	35.63
Sichuan	7883	7869	99.82
Total	5,890,806	3,557,929	60.40

Data from the General Office of National Environmental Protection (2001): "Investigation report of the present state of the ecological environment in the Western Region of China"

Region experienced severe aeolian desertification, and there is 3.2×10^5 hm^2 dry farmland where the 1 cm thick surface soil layer has been blown away by wind every year, including 840.48 tons organic content, 5.41×10^4 tons nitrogen, and 8.30×10^6 tons physical clay particles. With continuous wind erosion, the original chestnut soil mold cover with a deep soil layer has been blown away. An investigation at the beginning of the 1990s showed that parts of the cultivated layer and grass-growing layer were blown away, the calcareous crust and sand-gravel layer under them were exposed, and the land production capacity was almost lost (Zhu and Chen 1994). In the northern region of Shangdu County of the Inner Mongolia Autonomous Region, the organic content in soil was approximately 4.129% at the beginning stage of reclamation and decreased to 1.189% in the period of moderate aeolian desertification and to only 0.757% in the period of severe aeolian desertification and abandoned cultivation. Using the Houshan area of Ulanqab Meng as an example, the loss of chemical fertilizer from aeolian desertified land in China is approximately 1.70×10^8 tons every year, with a total value of 10.575 billion yuan RMB (Zhu and Chen 1994). In terms of the loss of organic content, the amount of nitrogen and phosphorus in soil lost to wind erosion is as high as 5.95×10^7tons, which is equal to 2.68×10^8 tons of chemical fertilizer and 17 billion yuan RMB. The deterioration of soil quality will lead to a gradual decrease in single grain production. For example, in rain-fed farmland regions, the single production of grain totaled 1335 kg/hm^2 in the 1960s, 1275 kg/hm^2 in the 1970s, 900 kg/hm^2 in the 1980s, 450 kg/hm^2 in the 1990s, and approximately 150 kg/hm^2 in drought years (Wang et al. 2011).

Different degrees of aeolian desertification covered 50,200 km^2 of different types of land by 2000 in the Horqin region. The loss of biomass that occurs during the process of aeolian desertification results in ecosystem degradation, which occurs in conjunction with the destruction of ecosystem structure and function, vegetation degeneration, and soil sandification. The result of this process is that desert ecosystems are formed with lower productivity and biomass. In 2000, the losses of biomass on aeolian desertified rain-fed farmland and grassland were 6.86×10^5 and 1.01×10^7 tons, respectively, for a total of 1.08×10^7 tons in the region (Table 6.4, Wang et al. 2005).

The depths of the surface soil layer in in the eastern region of the Xilin Gol grassland are approximately 30–50 cm, with 50–60% fine sand grain content in the soil layer that is less than 0.5 mm. The surface soil layer was blown away within 2 years after reclamation, and its movement depth was approximately 20–25 cm, totaling 2000–2500 m^2 surface soil loss from one hectare of land. However, on the bit-ridged windward sections, even subsoil and parent sand material were exposed on the surface, and the fine sand grain content of the 0.05 mm diameter increased to 60–90% (Zhu et al. 1989).

The sand cutting or sand burying from sand-driving wind limits the normal growth of crop and forage grasses. If seeds of plants have been entrained or buried in sand, then people must replant them. For example, in Zhangbei County, Hebei Province, every year, an area of 2.07×10^4 hm^2 needs to be replanted on 8.0×10^4 hm^2 of farmland that is threatened by aeolian desertification, and in 1984, the amount of seeds needed for replanting was as much as 4.85×10^5 kg

Table 6.4 Total loss of aboveground biomass on different types of desertification lands in the Horqin region in 2000

Land types	Slight aeolian desertification		Moderate aeolian desertification		Severe aeolian desertification		Very severe aeolian desertification	
	Area (km^2)	Aboveground biomass loss (ton)	Area (km^2)	Aboveground biomass loss (ton)	Area (km^2)	Aboveground biomass loss (ton)	Area (km^2)	Aboveground biomass loss (ton)
Rain-fed farmland	14,780.2	532,087.2	2111.5	154,139.5	0.0	0.0	0.0	0.0
Grazing grassland	10,557.3	2,905,369.0	6334.4	2,974,001. 0	4222.9	1,982,652.0	4222.9	2,219,556.0

Quoted from Wang et al. (2005)

(Yang 1985). The precipitation in 13 villages in the middle of Naiman Qi of the Horqin region was approximately 30 mm from March to July in 1980. With long-term drought, blown sand is very strong, and the number of days with strong winds in spring and summer is approximately 15, which causes a seed-planting period delay of 7–15 days. The area of damaged plants is approximately 0.91 hm^2, and 4.1×10^3 hm^2 of buckwheat and grain that cannot be harvested occurs in the entire Qi. In only 20 years, 33.3 hm^2 of fertile lands were buried by shifting sand dunes in Yaoledianzi Village, with average reduction of 1.67 hm^2 every year. Due to the decrease in farmland, people must reclaim barren land, which leads to aeolian desertification, and the agricultural ecological environment has further deteriorated (Liu and Zhao 1993). In June 1985, seedlings and leaves of corn and other crops withered and died due to windblown sand in Yaoledianzi Village under a strong wind lasting 5–6 days, with 30% of the area damaged; this event resulted in corn ripening approximately 10 days late and broomcorn ripening half a month late (Liu and Zhao 1993).

In addition, due to the deposition of windblown sand, farmlands and pastures will experience sand accumulation. Moreover, sand accumulation can form sand dunes and directly cause soil resource loss. In 1973, in the experimental field of Jiangqiao village, Tailai County, Heilongjiang Province, with intensively reclaimed fixed dunes, farmlands within 100 m of the leeward side of the dune were buried by sand in merely 1 year, with a 4.2 cm thick sand cover layer occurring on average. A piece of farmland that covered 9 hm^2 in Hamutai Village had an area of 9000 m^3 of sand soil within 3 years, with a 10 cm thick sand cover layer on average (Ma 1985).

In the 1950s, the area of land experiencing blown sand hazards was approximately 280 hm^2 in Tongyu County, Jilin Province, and this area increased to 14.4 thousand hm^2 in the 1970s. The total area experiencing blown-out seeds, seedlings beaten by windblown sand, and damaged plantings was approximately 2.16×10^5 hm^2 from 1949 to 1980. The average size of damaged planting areas was approximately 280 hm^2 in a year, and this size reached 1.87×10^4 hm^2 in 1972. There were 700 hm^2 of farmlands that had no harvest due to wind erosion and sand burial in 1982, and the reduction was approximately 4.5×10^5 tons. The area of farmland experiencing aeolian desertification in Maxiagen Village of Xinhua Countryside was 20 hm^2 in the 1950s, which increased to 60 hm^2 in the 1960s, 136.7 hm^2 in the 1970s, and 250 hm^2 in the 1980s. There is a sand ridge from southeast to northwest in Houmaxiagen Village. With extensive reclamation and pastures, the boundary of the shifting sand dune that extended 500 m east had already entered the western regions of the village. Forty residents had to be removed because windblown sand encroached on their houses, stalls, and wells (Meng et al. 1991).

Shifting sand dune belts in the Pingchuan regions north of the Heihe River are approximately 40 km long, with an area of 12 km^2. Blown sand severely damaged the oasis around the periphery of the dune, especially the deflation of windblown sand on the soil. The average deflation depth of soil in cropland is approximately 3–5 cm, and the organic content loses are 400–700 kg/(hm^2 a) due to conspicuous loss of fine grain and organic content. Because of these factors, the amount of land experiencing aeolian desertification land in the Pingchuan Oasis to the north of the

Heihe River was 8% in the 1940s, and this amount increased to 20% in the beginning of the 1960s and even increased up to 30% in some severe sections, with coverage less than 3%. Some croplands had to be abandoned due to the encroachment of shifting sand dunes, and farmland moved southward by 200–500 m. Crescent dunes, barchan chains, coppice mounds, and land depressions developed extensively on abandoned farmland (Wang et al. 2011). Mobile sand dunes moved forward 8–10 m in the Qingtuhu District of Minqin County, and sand dune fields occurred in eastern Wuwei, Yumen, Jinta, etc., with 1.27×10^5 hm^2 of cropland in the oasis occurring before shifting dunes were severely abandoned due to sand burial and other wind-blown sand hazards. There were 679 villages that were buried in sand along windblown sand roads (Zhu and Chen 1994).

Burial by sand is a common hazard for crops, forage grasses, and fruit trees in oases surrounded by sandy deserts and the Gobi Desert, and sand burial often causes serious economic loss. For example, the strong sand storm of May 5, 1993 destroyed 66,400 hm^2 of croplands in Jinchang. In Wuwei, Gulang, Jingtai, and Zhongwei, approximately 10 cm of topsoil was eroded, with the maximum depth reaching 50 cm; on average, approximately 5 m^3 of soil was blown away per hm^2. At the same time, some croplands were covered by sand, with the maximum mum thickness reaching 1.5 m, and sand accumulation per hectare cropland totaled an average of 9 m^3. In addition, 55,000 m of canals were buried by sand, 750 poles were blown down, 22,500 m of power line were blown down, 90,000 trees were broken, and 32,000 sheep and almost 10,000 other livestock were killed or missing (Yang 1996).

In many places, aeolian desertification leads to land degradation, destruction of soil structure, and loss of soil nutrients. The natural restoration of soil fertility takes decades, centuries, and even thousands of years. If man-made measures are used to restore the fertility of the soil, then the amount of input required is difficult to calculate. Grassland degradation caused by aeolian desertification has resulted in the dominant grass species suitable for livestock feed becoming gradually reduced or even completely lost. Forage grass has become low and sparse, yields have decreased obviously, and the carrying capacities of pastures have decreased greatly.

According to the statistics of the National Statistics Bureau, the total output value of the farming industry in 1999 was 381.66 billion yuan. The average annual loss of crop production due to desertification is 18.68 billion yuan. As a result of land desertification, the loss of grassland per unit area is 0.76 thousand yuan km^2, which is equivalent to the 1999 price. The results show that the loss of animal husbandry caused by aeolian desertification is 8.02 billion yuan per year (Liu 2006).

6.3.2 Impacts of Aeolian Desertification on Industry and Transportation

With the development of aeolian desertification, windblown sand activities frequently occur. Aeolian desertification directly or indirectly blocks rivers, reservoirs,

Fig. 6.9 A post vehicle on the S101 Road affected by severe grassland aeolian desertification (photo taken on June 26, 2018) (taken by Shulin Liu)

and canals; damages mine bases, factories, railway subgrades, bridges, and culverts in some areas; buries or erodes railways, roadbeds of highways, and road surfaces; forces highway traffic to stop; and even causes highways to be abandoned. Sand accumulation on roadbeds may cause trains to jump their tracks; therefore, the mechanical removal of sand accumulated on roadbeds during strong sand storms is essential. This scenario not only affects the normal operation of railways but also greatly increases train maintenance costs. Strong sand-dust storms can also cause side slope deflation and sand accumulation on the surface of the highway. This can increase oil consumption, shorten the useful life of a car, and cut off traffic (Wang et al. 2011) (Fig. 6.9).

China's areas experiencing aeolian desertification have the largest reserves of oil, natural gas, exposed coal mines, and salt and alkali lake mines in the country. Every year, there are substantial losses due to work being stopped, sand entering machines, and wear damage caused by aeolian desertification. For example, most of the eight exploited salt lakes were buried by windblown sands. The area of salt produced in the early 1950s in Chahannaoer Salt Lake was 15 km^2, which was reduced to 4 km^2 in the 1990s. The Yabulai, Hetunchi, Dayanhaizi, and North Dachi salt lakes have been seriously endangered. According to a 1993 survey, Jirantai Salt Lake was damaged by windblown sand. Within the range of the 37.19 km^2 salt mine, there were 16.88 km^2 salt mines buried by windblown sand with a thickness of 30 cm. The removal of sand before mining increases has led to an additional economic burden, and the shifting sand invasion will lead to too quick abandonment of the salt lake (Zhou et al. 1998).

The Yellow River transports an average of 1.6 billion tons of sand a year, 25% of which originates in aeolian desertified areas. Approximately 50,000 kilometers of irrigation canals are damaged by wind and sand every year (Wang et al. 2011). According to previous research reports and related research results, the total length of the irrigation channels in the aeolian desertification area of northern China is 126 thousand km, of which 51 thousand km is often damaged by wind and sand (Liu 2006). Damage to water conservancy is mostly manifested in silting hydraulic engineering and rivers. A typical example is that of the Longyangxia Reservoir of

Qinghai Province; the blown sands entering the reservoir are approximately 1.41×10^6 m^3 every year (Dong et al. 1989).

For example, a strong sand-dust storm occurred on April 28, 1983, which resulted in 14 people dying, 37 becoming wounded, 44,837 livestock dying, 101 civilian houses being damaged, 378 stalls being damaged, 195 wells being buried, and 9 km of communication routes being damaged in Ih Ju Meng. During the period when a strong sand-dust storm occurred on May 5, 1993, the traffic on the Wuda-Jilantai Railway was cut off for 4 days, and the Lanzhou-Xinjiang Railway was interrupted for 31 hours. The strong sand-dust storm of May 5, 1993 destroyed 35 kV and 6 kV power transmission lines and cut off the water supply in Jinchang city; as a result, industrial production was partly stopped and caused a direct economic loss of 83 million yuan (Yang 1996).

Damage to communication establishments can be seen through the collapse of electrical poles and broken wires caused by wind and blown sand electrification around wires. According to observations in the field, the voltage of blown sand electrification around wires reaches up to 2700 V, which strongly disturbs communication and electricity transmission and even endangers humans. Sand dust not only makes the atmosphere hazy and impedes people working normally but also disrupts civil aviation and causes economic loss.

The damage to air transportation from aeolian desertification mainly refers to the damage caused by sand-dust storms. During periods of strong sand-dust storms, the visibility is very poor, and in extreme cases, the storm may block transportation. This kind of storm has two main impacts: first, its affects the normal takeoff and landing of an airplane; second, it affects the airplane itself and its mechanical movement and easily causes mechanical accidents. In recent years, due to the impact of sand-dust storms, air transport has been impacted. For example, when the "Black Storm" occurred in the Hexi region on May 5, 1993, sand was blowing in the sky over Lanzhou, visibility at Lanzhou Zhongchuan International Airport was extremely low, and flights had to be halted and turned back with no ability to land (Yang 1996). According to the China Youth Daily, more than 1200 flights across the country were delayed by dust and sandstorms that began on March 26, 2004, causing major economic and financial losses for many airlines. Since March 2004, there have been six sand-dust events in northern China, including one strong sand-dust storm, two sand-dust storms, and three sand-blowing events. Due to high winds, dust, and weather, 130 flights were delayed at Beijing Capital International Airport, and 27 were forced to land at airports around the capital, most of which were delayed for more than an hour. "Each second an airplane stays in the sky is equivalent to dropping a 10-yuan bill from the sky" said the Beijing Office of China Eastern Airlines.

As a result, aeolian desertification deteriorates the ecological environment, destroys living conditions, aggravates the occurrence of natural disasters, restricts economic development, deepens the poverty level, seriously affects social stability, and greatly harms the national economy and social development of China (Zhang et al. 1994). In 2005, the direct economic loss caused by aeolian desertification in ten provinces of North China was as high as 48.79 billion yuan (Table 6.5, Ma et al.

Table 6.5 Economic losses from aeolian desertification in the ten provinces of northern China in 2005 (billion yuan)

Region	Grassland carrying capacity loss	Nutrient loss of farmland	Loss of land abandonment	Pollution loss	Direct loss	Total loss
Inner Mongolia	2.68	5.86	9.98	5.69	1.99	26.19
Qinghai	0.27	0.14	0.80	0.00	0.31	1.52
Gansu	0.12	1.09	1.78	0.7	0.19	3.88
Xinjiang	0.32	1.78	0.26	0.78	0.86	4.00
Ningxia	0.06	0.17	0.41	0.05	0.05	0.74
Heilongjiang	0.26	0.08	6.68	0.00	0.03	7.06
Shaanxi	0.59	0.26	0.33	0.12	0.12	1.42
Shanxi	0.14	0.59	0.00	0.19	0.00	0.91
Jilin	0.94	0.00	0.17	0.00	0.00	0.21
Hebei	0.94	0.00	1.75	0.19	0.00	2.87
The total	5.38	9.99	22.16	7.71	3.56	48.79

Note: Quote from Ma et al. (2008)

2008). This economic loss was mainly due to the land abandonment and lost grasslands that occurred in response to aeolian desertification (Ma et al. 2008). It is even estimated that the direct economic loss caused by land desertification in China reaches 128.14 billion yuan each year (Liu 2006).

6.4 Impacts of Aeolian Desertification on Human Health in Local and Downwind Regions

6.4.1 Overview

Aeolian desertification and dust storms can affect human health through complex pathways: (1) direct effects such as deaths and injuries caused by dust storm events and acute and/or chronic diseases caused by airborne particles and human aspiration toxicity hazards; (2) indirect effects resulting from the drying of water source and food production decreases leading to malnutrition and starvation; and (3) combinations of these two types of effects. Recently, agriculture, mining, and rapid development have contributed to dust generation and community exposure in Inner Asian drylands. In particular, the physical, chemical, and biological properties of airborne dust-related desertification have a direct impact on human health, including respiratory diseases and worsening allergy-related diseases (Shepherd 2016). Health consequences are worldwide, particularly affecting populations in desertified areas.

Populations far from the source regions are exposed to airborne particles or "Asian dust," when long-range atmospheric transport carries dust, for example,

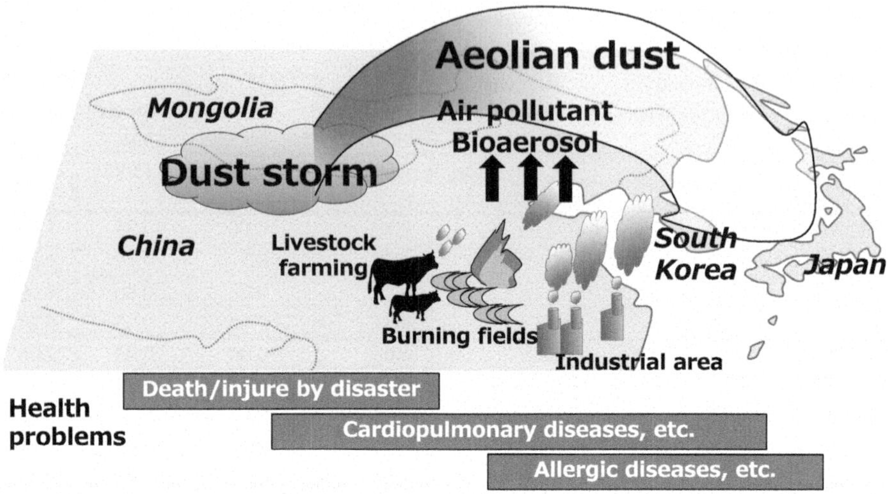

Fig. 6.10 Positional diagram of aeolian dust in Northeast Asia

from China and Mongolia to Japan and Korea (Fig. 6.10). Airborne particles can adsorb anthropogenic atmospheric pollutants during transport, including ammonium, sulfate, and nitrate ions and heavy metal compounds not believed to originate from the soil (Onishi et al. 2012). Thus, exposure to windblown dust is increasingly linked to a range of health problems. In this section, we focus on both Inner Asian drylands, one region experiencing desertification progression and the East Asia countries located in downwind regions.

6.4.2 Impacts of Aeolian Desertification in Local Regions

Severe dust events related to aeolian desertification constitute major disasters in source regions. The following are examples of cases in Mongolia: during April 17–20, 1980 and May 5–6, 1993, between 9 and 16 people and 100,000–675,000 head of livestock died because of severe snow and dust storms (Dulam 2005); 52 people lost their lives; and 320,000 head of livestock were killed because of an intense dust storm that occurred across a broad area of Mongolia on May 26–27, 2008 (Otani et al. 2017). Such a major disaster may also lead to posttraumatic stress disorder; Mu et al. (2013) noted that the quality of life of the victims may decline over time due to indirect impacts such as livestock losses leading to economic damage after extreme dust storms.

There are currently no reports of outbreaks of dust-related infectious diseases in inland Asia; however, coccidioidomycosis (also known as valley fever), which is a debilitating fungal disease caused by inhalation of arthroconidia, breaks out during and after dust storms in the southwestern United States and northern Mexico. A portion of patients with coccidioidomycosis develop severe pulmonary and/or chronic disease, and approximately 1% develop a life-threatening disseminated disease, such as meningitis (Gade et al. 2020). A causal microorganism of infectious disease should be monitored even in Asia.

Chronic exposure to aeolian dust is associated with premature death due to cardiovascular disease, respiratory disease, and lung cancer. Inhalation of dusty air exposes individuals to not only hazardous fine mineral particulates but also harmful combinations of pollutants, spores, bacteria, fungi, and potential allergens carried with mineral dusts (Shepherd 2016). As an epidemiologic example, the outpatient morbidity associated with respiratory disease in areas with a high frequency of occurrence of dust events in Mongolia (Kurosaki and Mikami 2005) has remained stably high. The morbidity rate of respiratory diseases in Omnogovi Province, located in the Gobi Desert, was 2157.6 per 10,000 inhabitants in 2011, approximately twice the Mongolian national average (1048.1 per 10,000 inhabitants) (State Implementing Agency of Health, Government of Mongolia 2012, Otani et al. 2017).

Recently, some areas in Inner Asia have been concerned with large-scale mining's potential dust risk to the health of the local population. Generally, mining has negative effects on vegetation and leads to aeolian desertification (Fang et al. 2019). Sternberg and Edwards (2017) noted that atmospheric dust from multiple sources may enhance human particulate exposure. Greater awareness of dust in Inner Asia reflects community concerns about related health implications. Future human well-being in the region will require more thorough information on dust emissions in the changing environment.

6.4.3 *Impacts of Aeolian Desertification in Downwind Regions*

In downwind regions of Asian dust events such as Japan, South Korea, and Taiwan, mortality, ambulance transportation, hospitalization, and changes in symptomatic, functional, and examination findings have been reported as effects of aeolian desertification on human health. Recent epidemiological studies on mortality have shown the following: adverse health effects on all-cause and cerebrovascular disease mortality were observed in South Korea and Japan (Kashima et al. 2016); mortality from all causes and circulatory diseases for the elderly population were significantly associated with winter Asian dust event days compared with normal days (Wang and Lin 2015); elderly, economically inactive, and nonmarried people had higher risks of all-cause mortality and cardiorespiratory mortality during days with extreme dust events in Hong Kong (Ho et al. 2018). In relation to cardiovascular diseases, it has been suggested that Asian dust is a potential trigger of acute myocardial infarction (Matsukawa et al. 2014).

Asian dust events have coincided with increases in hospital admissions and clinical visits for allergy-related diseases such as asthma, allergic rhinitis, and conjunctivitis (Shepherd 2016; Goudie 2014). Children have been found to be particularly vulnerable, and exposure to dust particles transported by Asian dust is associated with increased hospital admissions for childhood asthma (Kanatani et al. 2010). Nakamura et al. (2016) suggested that Asian dust exposure increases emergency department visits by children with bronchial asthma and respiratory diseases. Worsening asthma symptoms caused by Asian dust might be attributed to the combination of particulate matter from soil and anthropogenic pollutants (Watanabe et al. 2011).

Even in healthy people, Asian dust can induce symptoms such as itchy eyes and skin, nasal congestion, and sore throats (Otani et al. 2011). Such symptoms during Asian dust events might reflect allergic reactions to Asian dust particle-bound metals, air pollutants, and microbial allergens (Otani et al. 2012, 2014). Analysis of Asian dust particles in Japan has shown the presence of ammonium ions, sulfate ions, nitrate ions, and heavy metal compounds that are considered not to originate from soil. Recent studies have shown that microbes such as bacteria and fungi can migrate vast distances during Asian dust events by attaching themselves to dust particles (Maki et al. 2014). Asian dust events might trigger some kind of hypersensitivity to metals and microbial allergens.

Aeolian desertification may lead to many types of health problems for people inhabiting both local and downwind regions through different mechanisms. The importance of combating desertification has become increasingly important from environmental and health aspects, in not only drylands in Inner Asia but also downwind areas in East Asia.

6.5 Conclusions

The frequency of sand-dust storms increased from the 1990s to 2000s in China and Mongolia. When we divided the causes of these sand-dust storms into erosivity and erodibility, the main cause of the storms was the increase in erodibility, especially in Mongolia, eastern Inner Mongolia, and northeastern China. This result suggests that aeolian desertification progressed in such regions during the period. The condition of long-term desertification is not well studied for wide areas in East Asia. However, national statistics and research reports show that aeolian desertification has been worsening in many different places in China for a long time.

Aeolian desertification affects ecosystems and agriculture in two ways: wind erosion and sediment deposition. Wind erosion decreases surface soil and fertility, windblown sand particles cause physical injury, and this type of erosion removes soil and thereby exposes roots to the air. A significant amount of sand burial (i.e., sediment deposition) enhances grass mortality and expands the area of erodible land surface; as a result, desertification accelerates. There are cases where windblown sand from a shifting sand dune encroaches on houses, stalls, wells, and canals, and croplands must be abandoned. Aeolian desertification also damages mine bases, factories, railways, bridges, culverts, highways, communication establishments, electricity transmissions, and air transportation.

Aeolian desertification has impacts on human health in not only dust source regions but also their downwind regions. Severe dust events are disasters that lead to deaths and injuries. Chronic exposure to aeolian dust is associated with cardiovascular and respiratory diseases. There is a report that the morbidity rate associated with respiratory disease in Omnogovi Province, Mongolia, where dust emissions frequently occur, is two times that of the Mongolian national average. Additionally, in downwind regions, there are reports of cardiovascular and respiratory diseases related to dust. A characteristic of dust in downwind regions is that dust particles adsorb anthropogenic pollutants during transport. The effect of microbes in the dust is also a concern, even though their effect is not well clarified.

References

Bagnold RA (1941) The physics of blown sand and desert dunes. Methuen, London

Bishop JKB, Davis RE, Sherman JT (2002) Robotic observations of dust storm enhancement of carbon biomass in the North Pacific. Science 298(5594):817–821

Bory AJM, Biscaye PE, Grousset FE (2003) Two distinct seasonal Asian source regions for mineral dust deposited in Greenland (NorthGRIP). Geophys Res Lett 30(4):1167. https://doi.org/10.1029/2002gl016446

Chiwa M (2010) Characteristics of atmospheric nitrogen and sulfur containing compounds in an inland suburban-forested site in northern Kyushu, western Japan. Atmos Res 96(4):531–543

Choi JC, Lee M, Chun Y et al (2001) Chemical composition and source signature of spring aerosol in Seoul, Korea. J Geophys Res Atmospher 106(D16):18067–18074

Dong G-G, Shen J-Y, Jin J (1989) The distribution of the land desertification in China and its harmfulness (in Chinese with English abstract). J Arid Land Resour Environ 3(4):33–42

Du S, Ariful Islam GM, Xiang R et al (2021) The dust deposition process and biogeochemical impacts in the Northern South China Sea. Asia-Pac J Atmos Sci 57(1):77–87

Dulam J (2005) Discriminate analysis for dust storm prediction in the Gobi and steppe regions in Mongolia. Water Air Soil Pollut Focus 5(3):37–49

Fan B, Zhao C, Zhang X et al (2018) Impacts of sand burial and wind erosion on regeneration and growth of a desert clonal shrub. Front Plant Sci 9:1696. https://doi.org/10.3389/fpls.2018.01696

Fang A, Dong J, Cao Z et al (2019) Tempo-spatial variation of vegetation coverage and influencing factors of large-scale mining areas in Eastern Inner Mongolia, China. Int J Environ Res Public Health 17(1):47. https://doi.org/10.3390/ijerph17010047

Fryrear DW, Stubbendieck J, McCully WG (1973) Grass seedling response to wind and windblown sand1. Crop Sci 13(6):622–625

Gade L, McCotter OZ, Bowers JR et al (2020) The detection of Coccidioides from ambient air in Phoenix, Arizona: evidence of uneven distribution and seasonality. Med Mycol 58(4):552–559

General Office of National Environmental Protection (2001) Investigation report of the present statue of ecological environment in the Western Region of China

Goudie AS (2014) Desert dust and human health disorders. Environ Int 63:101–113

Goudie AS, Middleton NJ (2006) Desert dust in the global system. Springer, Berlin, Heidelberg

Han Y, Zhao T, Song L et al (2011) A linkage between Asian dust, dissolved iron and marine export production in the deep ocean. Atmos Environ 45(25):4291–4298

Harrison SP, Kohfeld KE, Roelandt C et al (2001) The role of dust in climate changes today, at the last glacial maximum and in the future. Earth Sci Rev 54(1):43–80

Ho HC, Wong MS, Yang L et al (2018) Influences of socioeconomic vulnerability and intra-urban air pollution exposure on short-term mortality during extreme dust events. Environ Pollut 235: 155–162

Huang Z, Gutterman Y (1999) Germination of *Artemisia sphaerocephala* (Asteraceae), occurring in the sandy desert areas of Northwest China. S Afr J Bot 65(3):187–196

Huang Z, Dong M, Gutterman Y (2004a) Factors influencing seed dormancy and germination in sand, and seedling survival under desiccation, of *Psammochloa villosa* (Poaceae), inhabiting the moving sand dunes of Ordos, China. Plant Soil 259(1):231–241

Huang Z, Dong M, Gutterman Y (2004b) Caryopsis dormancy, germination and seedling emergence in sand, of *Leymus racemosus* (Poaceae), a perennial sand-dune grass inhabiting the Junggar Basin of Xinjiang, China. Aust J Bot 52(4):519–528

Husar RB, Tratt DM, Schichtel BA et al (2001) Asian dust events of April 1998. J Geophys Res Atmos 106(D16):18317–18330

Jo CO, Lee J-Y, Park K-A et al (2007) Asian dust initiated early spring bloom in the northern East/Japan Sea. Geophys Res Lett 34(5):L05602. https://doi.org/10.1029/2006gl027395

Kanai Y, Ohta A, Kamioka H et al (2005) Characterization of aeolian dust in east China and Japan from 2001 to 2003. J Meteorol Soc Japan Ser II 83A:73–106

Kanatani KT, Ito I, Al-Delaimy WK et al (2010) Desert dust exposure is associated with increased risk of asthma hospitalization in children. Am J Respir Crit Care Med 182(12):1475–1481

Kashima S, Yorifuji T, Bae S et al (2016) Asian dust effect on cause-specific mortality in five cities across South Korea and Japan. Atmos Environ 128:20–27

Kent M, Owen NW, Dale P et al (2001) Studies of vegetation burial: a focus for biogeography and biogeomorphology? Progr Phys Geograph Earth Environ 25(4):455–482

Khalaf FI (1989) Desertification and aeolian processes in the Kuwait Desert. J Arid Environ 16(2): 125–145

Kurosaki Y, Mikami M (2003) Recent frequent dust events and their relation to surface wind in East Asia. Geophys Res Lett 30(14):1736. https://doi.org/10.1029/2003GL017261

Kurosaki Y, Mikami M (2005) Regional difference in the characteristic of dust event in East Asia: relationship among dust outbreak, surface wind, and land surface condition. J Meteorol Soc Japan Ser II 83A:1–18

Kurosaki Y, Mikami M (2007) Threshold wind speed for dust emission in east Asia and its seasonal variations. J Geophys Res Atmos 112(D17):D17202. https://doi.org/10.1029/2006JD007988

Kurosaki Y, Shinoda M, Mikami M (2011) What caused a recent increase in dust outbreaks over East Asia? Geophys Res Lett 38(11):L11702. https://doi.org/10.1029/2011GL047494

Li J, Okin GS, Alvarez LJ et al (2009) Sediment deposition and soil nutrient heterogeneity in two desert grassland ecosystems, southern New Mexico. Plant Soil 319(1–2):67–84

Li S-L, Werger MJA, Zuidema PA et al (2010a) Seedlings of the semi-shrub *Artemisia ordosica* are resistant to moderate wind denudation and sand burial in Mu Us sandland, China. Trees 24(3): 515–521

Li S-L, Zuidema PA, Yu F-H et al (2010b) Effects of denudation and burial on growth and reproduction of *Artemisia ordosica* in Mu Us sandland. Ecol Res 25(3):655–661

Li X, Jiang D, Alamusa et al (2012) Comparison of seed germination of four *Artemisia* species (Asteraceae) in northeastern Inner Mongolia, China. J Arid Land 4(1):36–42

Li J, Qu H, Zhao H et al (2015) Growth and physiological responses of *Agriophyllum squarrosum* to sand burial stress. J Arid Land 7(1):94–100

Liu T (2006) Desertification economic loss assessment in China (in Chinese with English abstract). J Desert Res 26(1):40–46

Liu X-M, Zhao H-L (1993) Study on integrated controlling of eco-environment in Horqin Sands (in Chinese with English abstract). Gansu Science and Technology Press, Lanzhou

Liu F, Ye X, Yu F et al (2007) Responses of *Hedysarum Laeve*, a guerrilla clonal semi-shrub in the Mu Us sandland, to local sand burial. Front Biol China 2(4):431–436

Liu B, Liu Z, Guan D (2008) Seedling growth variation in response to sand burial in four *Artemisia* species from different habitats in the semi-arid dune field. Trees 22(1):41–47

Liu H-L, Shi X, Wang J-C et al (2011) Effects of sand burial, soil water content and distribution pattern of seeds in sand on seed germination and seedling survival of *Eremosparton songoricum* (Fabaceae), a rare species inhabiting the moving sand dunes of the Gurbantunggut Desert of China. Plant Soil 345(1–2):69–87

Liu H-L, Zhang L-W, Yin L-K et al (2013) Effects of temperature, dry storage, and burial on dormancy and germination of seeds of 13 desert plant species from sand dunes in the Gurbantunggut Desert, Northwest China. Arid Land Res Manag 27(1):65–78

Liu B, Liu Z, Lü X et al (2014a) Sand burial compensates for the negative effects of erosion on the dune-building shrub *Artemisia wudanica*. Plant Soil 374(1–2):263–273

Liu B, Liu Z, Wang L et al (2014b) Responses of rhizomatous grass *Phragmites communis* to wind erosion: effects on biomass allocation. Plant Soil 380(1–2):389–398

Luo W, Zhao W (2015) Effects of wind erosion and sand burial on growth and reproduction of a clonal shrub. Flora Morphol Distribut Funct Ecol Plants 217:164–169

Ma W-L (1985) Status and potentials of desertification on the Songuajiang Nenjiang Plain (in Chinese with English abstract). J Desert Res 5(3):45–53

Ma G-X, Shi M-J, Zhao X-T et al (2008) Monetary accounting of economic loss of sandy desertification in North China (in Chinese with English abstract). J Desert Res 28(4):627–633

Maher BA, Prospero JM, Mackie D et al (2010) Global connections between aeolian dust, climate and ocean biogeochemistry at the present day and at the last glacial maximum. Earth Sci Rev 99(1–2):61–97

Maki T, Puspitasari F, Hara K et al (2014) Variations in the structure of airborne bacterial communities in a downwind area during an Asian dust (Kosa) event. Sci Total Environ 488-489:75–84

Martin JH (1990) Glacial-interglacial CO_2 change: the iron hypothesis. Paleoceanography 5(1): 1–13

Matsukawa R, Michikawa T, Ueda K et al (2014) Desert dust is a risk factor for the incidence of acute myocardial infarction in Western Japan. Circ Cardiovasc Qual Outcomes 7(5):743–748

Maun MA (1998) Adaptations of plants to burial in coastal sand dunes. Can J Bot 76(5):713–738

Meng X-X, Lu X-G, Wang Q et al (1991) Land desertification and controlling measures in southwest Songnen Plain (in Chinese with English abstract). Sci Geogr Sin 11(4):353–360

Meskhidze N, Chameides WL, Nenes A (2005) Dust and pollution: a recipe for enhanced ocean fertilization? J Geophys Res 110(D3):D03301. https://doi.org/10.1029/2004jd005082

Moore JK, Doney SC, Glover DM et al (2001) Iron cycling and nutrient-limitation patterns in surface waters of the World Ocean. Deep-Sea Res II Top Stud Oceanogr 49(1):463–507

Mori I, Nishikawa M, Tanimura T et al (2003) Change in size distribution and chemical composition of kosa (Asian dust) aerosol during long-range transport. Atmos Environ 37(30): 4253–4263

Mu H, Otani S, Shinoda M et al (2013) Long-term effects of livestock loss caused by dust storm on Mongolian inhabitants: a survey 1 year after the dust storm. Yonago Acta Med 56:39–42

Nakamura T, Hashizume M, Ueda K et al (2016) Asian dust and pediatric emergency department visits due to bronchial asthma and respiratory diseases in Nagasaki, Japan. J Epidemiol 26(11): 593–601

Onishi K, Kurosaki Y, Otani S et al (2012) Atmospheric transport route determines components of Asian dust and health effects in Japan. Atmos Environ 49:94–102

Otani S, Onishi K, Mu H et al (2011) The effect of Asian dust events on the daily symptoms in Yonago, Japan: a pilot study on healthy subjects. Arch Environ Occup Health 66(1):43–46

Otani S, Onishi K, Mu H et al (2012) The relationship between skin symptoms and allergic reactions to Asian dust. Int J Environ Res Public Health 9(12):4606–4614

Otani S, Onishi K, Mu H et al (2014) Associations between subjective symptoms and serum immunoglobulin E levels during Asian dust events. Int J Environ Res Public Health 11(8): 7636–7641

Otani S, Kurosaki Y, Kurozawa Y et al (2017) Dust storms from degraded drylands of Asia: dynamics and health impacts. Land 6(4):83. https://doi.org/10.3390/land6040083

Precheur R, Greig JK, Armbrust DV (1978) The effects of wind and wind-plus-sand on tomato plants. J Am Soc Hortic Sci 103:351–355
Qiu Y (2015) Iron fertilisation by Asian dust influences North Pacific sardine regime shifts. Prog Oceanogr 134:370–378
Qu H, Zhao HL, Zhao XY et al (2017) Effects of sand burial on the survival and growth of two shrubs dominant in different habitats of northern China. Environ Monit Assess 189(4):149. https://doi.org/10.1007/s10661-017-5866-x
Ravi S, Breshears DD, Huxman TE et al (2010) Land degradation in drylands: interactions among hydrologic–aeolian erosion and vegetation dynamics. Geomorphology 116(3–4):236–245
Ravi S, D'Odorico P, Breshears DD et al (2011) Aeolian processes and the biosphere. Rev Geophys 49(3):RG3001. https://doi.org/10.1029/2010rg000328
Ren J, Tao L, Liu X-M (2002) Effect of sand burial depth on seed germination and seedling emergence of *Calligonum* L. species. J Arid Environ 51(4):603–611
Schlesinger WH, Pilmanis AM (1998) Plant-soil interactions in deserts. Biogeochemistry 42(1): 169–187
Shepherd G (ed) (2016) Global assessment of sand and dust storms. United Nations Environ Program, Nairobi
Shi L, Zhang ZJ, Zhang CY et al (2004) Effects of sand burial on survival, growth, gas exchange and biomass allocation of *Ulmus pumila* seedlings in the Hunshandak Sandland, China. Ann Bot 94(4):553–560
State Implementing Agency of Health, Government of Mongolia (2012) Health indicators 2011, Uaanbaatar
Sternberg T, Edwards M (2017) Desert dust and health: a Central Asian review and steppe case study. Int J Environ Res Public Health 14(11):1342. https://doi.org/10.3390/ijerph14111342
Tan S-C, Shi G-Y (2012) Correlation of dust storms in China with chlorophyll a concentration in the Yellow Sea between 1997–2007. Atmosph Ocean Sci Lett 5(2):140–144
Tan S-C, Wang H (2014) The transport and deposition of dust and its impact on phytoplankton growth in the Yellow Sea. Atmos Environ 99:491–499
Tang J, Busso CA, Jiang D et al (2016) Experimental sand burial affects seedling survivorship, morphological traits, and biomass allocation of *Ulmus pumila* var. *sabulosa* in the Horqin Sandy Land, China. Solid Earth 7(4):1085–1094
Teraminami T, Nakashima A, Ominami M et al (2013) Effects of sand burial depth on the root system of *Salix cheilophila* seedlings in Mu Us Sandy Land, Inner Mongolia, China. Lands Ecol Eng 9(2):249–257
Udden JA (1914) Mechanical composition of clastic sediments. Geol Soc Am Bull 25(1):655–744
UNCCD (1994) Convention to combat desertification. Adopted 17 June 1994, entry into force 26 December 1996. A/AC.241/27
Uno I, Eguchi K, Yumimoto K et al (2009) Asian dust transported one full circuit around the globe. Nat Geosci 2(8):557–560
Wan S, Sun Y, Nagashima K (2020) Asian dust from land to sea: processes, history and effect from modern observation to geological records. Geol Mag 157(5):701–706
Wang X (2013) Sandy desertification: borne on the wind. Chin Sci Bull 58(20):2395–2403
Wang Y-C, Lin Y-K (2015) Mortality associated with particulate concentration and Asian dust storms in Metropolitan Taipei. Atmos Environ 117:32–40
Wang T, Wu W, Zhao H et al (2005) Evaluation on biomass loss through process of aeolian desertification Take the Horqin Region, Inner Mongolia as example (in Chinese with English abstract). J Desert Res 25(4):453–456
Wang T et al (2011) Deserts and aeolian desertification in China. Science Press, Beijing
Wang S-H, Hsu NC, Tsay S-C et al (2012) Can Asian dust trigger phytoplankton blooms in the oligotrophic northern South China Sea? Geophys Res Lett 39(5):L05811. https://doi.org/10.1029/2011gl050415
Wang TT, Chu GM, Jiang P et al (2017) Effects of sand burial and seed size on seed germination, seedling emergence and seedling biomass of *Anabasis aphylla*. Pak J Bot 49(2):391–396

Wang G, Yu K, Gou Q (2019) Effects of sand burial disturbance on establishment of three desert shrub species in the margin of oasis in northwestern China. Ecol Res 34(1):127–135
Watanabe M, Yamasaki A, Burioka N et al (2011) Correlation between Asian dust storms and worsening asthma in Western Japan. Allergol Int 60(3):267–275
Wentworth CK (1922) A scale of grade and class terms for clastic sediments. J Geol 30(5):377–392
WMO (2019) Manual on codes, international codes Vol. I.1, Annex II to the WMO technical regulations): part A - Alphanumeric Codes, WMO Publ 306
Xu L, Huber H, During HJ et al (2013a) Intraspecific variation of a desert shrub species in phenotypic plasticity in response to sand burial. New Phytol 199(4):991–1000
Xu L, Yu FH, Werger M et al (2013b) Interactive effects of mechanical stress, sand burial and defoliation on growth and mechanical properties in *Cynanchum komarovii*. Plant Biol 15(1): 126–134
Yang T-Y (1985) Land desertification on the fringe area of Inner Mongolia Plateau adjacent Hebei Province and its rehabilitation (in Chinese with English abstract). J Desert Res 5(4):25–35
Yang G-S (1996) Black wind storms and agricultural hazard reduction measures in northwest China (in Chinese with English abstract). J Desert Res 16(2):97–104
Ye X, Liu Z, Gao S et al (2017) Differential plant species responses to interactions of sand burial, precipitation enhancement and climatic variation promote co-existence in Chinese steppe vegetation. J Veg Sci 28(1):139–148
Ye X, Liu Z, Zhang S et al (2019) Experimental sand burial and precipitation enhancement alter plant and soil carbon allocation in a semi-arid steppe in north China. Sci Total Environ 651: 3099–3106
Yoshioka K, Kamiya H, Kano Y et al (2009) The relationship between seasonal variations of total-nitrogen and total-phosphorus in rainfall and air mass advection paths in Matsue, Japan. Atmosph Environ 43(22):3496–3501
Yu F, Chen Y, Dong M (2002) Clonal integration enhances survival and performance of *Potentilla anserina*, suffering from partial sand burial on Ordos plateau, China. In: Stuefer JF, Erschbamer B, Huber H, Suzuki JI (eds) Ecology and evolutionary biology of clonal plants: proceedings of clone-2000. An International Workshop held in Obergurgl, Austria, 20--25 August 2000. Springer Netherlands, Dordrecht, pp. 81–96. https://doi.org/10.1007/978-94-017-1345-0_5
Yu F, Dong M, Krüsi B (2004) Clonal integration helps *Psammochloa villosa* survive sand burial in an inland dune. New Phytol 162(3):697–704
Yu FH, Wang N, He WM et al (2008) Adaptation of rhizome connections in drylands: increasing tolerance of clones to wind erosion. Ann Bot 102(4):571–577
Zhang W-M, Yang T-Y, Qu J-J et al (1994) The extension and hazard of desertification disaster in China (in Chinese with English abstract). J Nat Disast 3(3):23–30
Zhao WZ, Li QY, Fang HY (2007) Effects of sand burial disturbance on seedling growth of *Nitraria sphaerocarpa*. Plant Soil 295(1–2):95–102
Zheng Y, Xie Z, Yu Y et al (2005) Effects of burial in sand and water supply regime on seedling emergence of six species. Ann Bot 95(7):1237–1245
Zheng MQ, Zheng YR, Zhou GS et al (2009) Effects of watering regime and depth of burial on seedling emergence of four dominant psammophytes in the Mu Us sandy land, Inner Mongolia, China, and relevance to revegetation of a desertified region. Ann Appl Biol 154(1):87–96
Zheng M, Lai L, Jiang L et al (2012) Moderate water supply and partial sand burial increase relative growth rate of two *Artemisia* species in an inland sandy land. J Arid Environ 85:105–113
Zhou H-S, Xiang Z, Shen J-J et al (1998) An overview of desertification disasters in China (in Chinese with English abstract). J Catastrophol 13(3):67–71
Zhu Z-D, Chen G-T (1994) Land aeolian desertification in China (in Chinese with English abstract). Science Press, Beijing

Zhu Z-D, Liu S, Di X-M et al (1989) Aeolian desertification and rehabilitation in China (in Chinese with English abstract). Science Press, Beijing

Zhu X-W, Huang Z-Y, Chu YU et al (2004) Effects of burial in sand and seed size on seed germination and seedling emergence in two leguminous shrubs in the Otindag Sandland, China. Israel J Plant Sci 52(2):133–142

Zhu Y, Dong M, Huang Z (2009) Response of seed germination and seedling growth to sand burial of two dominant perennial grasses in Mu-Us Sandy Grassland, semiarid China. Rangel Ecol Manag 62(4):337–344

Zhu Y, Yang X, Baskin CC et al (2014) Effects of amount and frequency of precipitation and sand burial on seed germination, seedling emergence and survival of the dune grass *Leymus secalinus* in semiarid China. Plant Soil 374(1–2):399–409

Part II
Measures to Combat Aeolian Desertification

Chapter 7
Goals and Principles for Combating Aeolian Desertification

Xian Xue

Abstract Land degradation neutrality (LDN) is the overall goal proposed by the United Nations Convention to Combat Desertification (UNCCD) for combating desertification. Based on the framework of LDN, this chapter proposes sustainable development, poverty eradication, and land management as subgoals for restoring desertified lands globally. These three subgoals have not yet been achieved, although decades of efforts have been implemented. The relevant principles have not been implemented, and this lack of implementation may be one of the fundamental causes of not achieving these subgoals. To better understand these principles, this chapter provides a detailed introduction to aeolian desertification and restoration in the Minqin Oasis, which is located in northwestern China. Based on the experience and lessons obtained from Minqin and other regions of China, the author recommends nine principles for achieving the three subgoals and LDN. These principles are also suitable for controlling and restoring areas experiencing aeolian desertification.

Keywords Aeolian desertification · Combating · Goals · Principles · LDN

7.1 Introduction

The history of combating desertification can be traced back to 1977 when the United Nations Conference on Desertification (UNCOD) adopted a Plan of Action to Combat Desertification (PACD). However, 15 years of efforts have not adequately

X. Xue (✉)
Northwest Institute of Eco-Environment and Resources, Chinese Academy of Sciences, Lanzhou, Gansu, China

Key Laboratory of Desert and Desertification, Chinese Academy of Sciences, Lanzhou, Gansu, China

Drylands Salinization Research Station, Northwest Institute of Eco-Environment and Resources, Chinese Academy of Sciences, Lanzhou, Gansu, China
e-mail: xianxue@lzb.ac.cn

T. Wang et al. (eds.), *Combating Aeolian Desertification in Northeast Asia*, Ecological Research Monographs, https://doi.org/10.1007/978-981-16-9028-0_7

addressed desertification due to the lack of substantive action (UNEP 1991). Therefore, in 1992, the United Nations Conference on Environment and Development (UNCED) held in Rio de Janeiro proposed the requirements for new and integrated approaches to combat desertification. Promoting sustainable development at the community level was emphasized as the essential aim of combating desertification at that time (UNCED 1992). Based on the proposal, effort, and preparation of the UNCED, the United Nations Convention to Combat Desertification (UNCCD) was adopted in Paris on 17 June 1994. In the strategic plan and framework of the UNCCD, the clear vision is to reverse and prevent desertification by forging global partnerships (UNCCD 1994). In response to the UNCCD, 196 countries and the European Union had been Parties of the Convention (COP) through August 2018. Therefore, the adoption of UNCCD is undoubtedly a milestone in the global effort to combat desertification. The UNCCD also supports the development and implementation of national and regional policies that contribute to the reduction of poverty (UNCED 2014). After world leaders concluded the 2030 Agenda for Sustainable Development at the SDG Summit in September 2015 (UNGA 2015), achieving the "land degradation neutrality (LDN)" described in target 15.3 of the 17 Sustainable Development Goals (SDGs) became an important task and vision of the UNCCD (Chasek et al. 2015; UNSD 2016; Stavi and Lal 2015). To date, over 120 countries have committed to setting LDN targets, and many countries have secured high-level government commitments to achieving LDN. The involvement and commitment of high-level governments are particularly important because land management practices supported by policymakers are critical to the implementation of LDNs (Safriel 2017).

Throughout history, the main goals of combating desertification are related to sustainable development, poverty eradication, and the pursuit of productive (UNCCD 2012). The restoration of aeolian desertification also needs to achieve the three goals (Wang et al. 2015). Determining the goals is not difficult, but determining how to achieve these goals without reverting back to initial states is difficult. This process has been difficult specifically in drylands. The author believes that the policies and measures implemented to achieve these goals must follow some principles. These principles might not be specific recommendations, but they should provide guidelines or rule so that aeolian desertification control efforts will not deviate from the expected results. Minqin Oasis, located at the end of the Shiyang River Basin in northwestern China, had experienced severe aeolian desertification before 2010, and then, the aeolian desertification has been reversed to a great extent since 2010. This chapter presents the Minqin case study to help describe the process of aeolian desertification and restoration. Based on the experience and problems that have emerged in the process of aeolian desertification control in Minqin and other regions of China, the author proposes essential principles for achieving the above goals. These principles are, to no small extent, preliminary discussions, with flexible concepts or theories.

7.2 Case Study on Aeolian Desertification and Restoration in Minqin Oasis

An oasis is a fertile area in deserts or drylands with a continuous water supply (Krevisky and Jordan 1989). The water supply of oases usually comes from inland rivers due to the sparse rainfall. Inland rivers flow into lakes or deserts instead of the ocean. The inland rivers in China are mainly distributed in the northwest where there is an arid climate. Except for the Yellow River, the rivers shown in Fig. 4.5 are inland rivers. These inland rivers originate from high mountains with glaciers and snow cover. The precipitation, snow, and glaciers in the mountains provide water to the inland rivers. From the mountains, these inland rivers flow into the premountain alluvial plains. On the plains, oases are distributed on both sides of the rivers. The inland rivers finally input into lakes such as Juyan Hai at the end of the Heihe River and Lop Nur at the end of the Tarim River (Fig. 4.5).

7.2.1 Aeolian Desertification Process in the Minqin Oasis

The Shiyang River is the easternmost inland river of China and originates in the Qilian Mountains (Fig. 4.5). Along the Shiyang River, the Gulang Oasis, Liangzhou Oasis, and Minqin Oasis are distributed from south to north (Fig. 7.1). Annual precipitation decreases from 700 mm in the mountains to 120 mm in the Minqin Oasis. Therefore, the Minqin Oasis has a typically arid continental climate. Local agriculture mainly depends on irrigation water from the Shiyang River and groundwater. The annual average runoff of the Shiyang River flowing out of the Qilian Mountains is approximately 13.7×10^8 m^3, and then, it supplies the Gulang Oasis, Liangzhou Oasis, and Minqin Oasis. In the 1950s, the water flowing into the Minqin Oasis was approximately 5×10^8 m^3. After the 1950s, with the rapid increase in irrigated croplands in the Gulang and Liangzhou Oases, the water flowing into the Minqin Oasis gradually decreased and was only 0.98×10^8 m^3 in 2003.

The increase in irrigated croplands resulted from the rising population. The total population of Gulang and Liangzhou oases increased from 59.1×10^4 people in 1950 to 140.3×10^4 people in 2011. In addition to the naturally occurring increase, policy-induced immigration also greatly contributed to the rapidly increasing population. In 1977, the Chinese Communist Party Central Committee approved the Hexi Corridor as one of the ten Commodity Grain Bases of China and encouraged local governments to rapidly develop agriculture and use abandoned lands. In response, a large number of people immigrated to the Gulang and Liangzhou Oases, which are an important part of the Hexi Corridor. In addition, the actual sown area significantly increased from 118.04×10^3 ha in 1950 to 175.77×10^3 ha in 2011 (Xue et al. 2015). Increasing cropland areas requires additional irrigation water. The Liangzhou Oasis consumed almost 80% of the total water resources in the river basin. Therefore, the water volume from the Shiyang River flowing into the Minqin Oasis decreased dramatically.

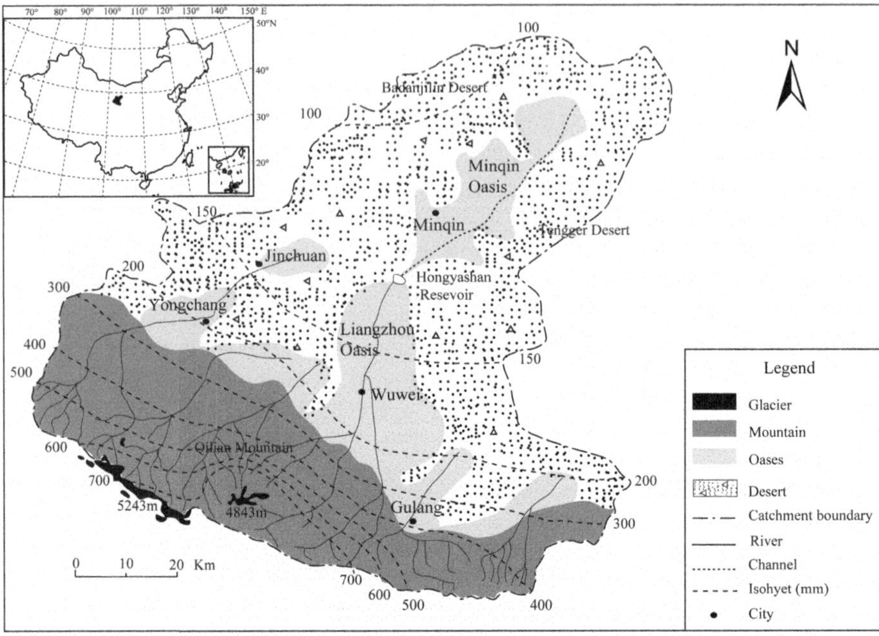

Fig. 7.1 Map of the Shiyang River Basin and Minqin Oasis (from Xue et al. 2015)

In contrast to the decrease in surface water, the population and croplands in Minqin Oasis significantly increased from 1950 to 2011. To address the shortage of surface water, the local government built the Hongyashan Reservoir in 1965 to store water in the flood season and recharge croplands during the irrigation season through artificial cement channels. Because all water was used to irrigate croplands, no water flowed into Qingtu Lake (lake at the end of the Shiyang River) to provide water to the natural vegetation that protect the oases from wind-sand disasters. Furthermore, groundwater was exploited to compensate for the shortage of surface water. Since the mid-1970s, 11,000 wells have been drilled in the Minqin Oasis, among which 250 wells have depths of 300 m. The average groundwater extraction volume is 6×10^8 m^3 per year, and the overexploited volume was 3×10^8 m^3 per year from 1998 to 2011 in the Minqin Oasis. The overexploitation of groundwater resulted in the groundwater table decreasing by an average of 6.84 m (Xue et al. 2015).

The other negative result of the overuse of groundwater is the deterioration of groundwater quality and soil salinization. The Minqin basin is a well-confined hydrologic basin in which the equilibrium of groundwater is mainly regulated by evaporation and transpiration. Due to the low circulation rate in this area, the overuse of groundwater has already induced increased levels of total dissolved solids (TDS). The average value of the TDS in the groundwater in Minqin Oasis significantly increased by 0.1–1.01 g l^{-1} from 1908 to 2011. Of the parts of the Minqin Oasis, the northern part, adjacent to the Tengger Desert, has the worst groundwater quality because of a lack of surface water recharge and overuse of groundwater. The TDS of well water along the desert margin can reach 10 g l^{-1} (Xue et al. 2015). Flood and furrow irrigation with saline or high-sodium water can change pH values, soil exchangeable sodium percentages (ESPs), and soil water-holding capacity (WHC), resulting in the formation of alkaline soil and damage to soil structure (Table 7.1). In the Minqin Oasis, the average salt content is as high as 16715.18 mg kg^{-1} and 2 or 2.5 times that in the Gulang and Liangzhou Oases. Before 2007, over 80% of the cropland experienced different intensities of salinization. In the northern part of the Minqin Oasis, minimally and moderately salt-tolerant crops such as wheat and cotton can no longer grow. The main crops have become fennel, alfalfa, and sunflower, which are highly salt-tolerant crops. Even some cropland had to be abandoned due to severe salinization (Fig. 7.2).

The decreased groundwater table, deteriorated water quality, and soil salinization directly caused a large area of natural and artificial vegetation to disappear. Research shows that natural vegetation cover in the Minqin Oasis decreased from 44.8% in the 1950s to 15% at the end of the 1990s (Xie and Chen 2002). Because the Minqin Oasis is adjacent to desert, which is covered by active dunes, shelter forests are especially important to protect the oasis from wind erosion and sand burial. Vegetation degradation directly resulted in oasis exposure to wind and sand disasters. Since the 1950s, sand dunes in the north have encroached southward by 50–70 m into the oasis and have destroyed 400 ha of farmland (Fig. 7.2). Desert areas located in the west have moved eastward by 30–60 m and have destroyed 467 ha of farmland (Zhang et al. 2004; Sun et al. 2005). Based on remote sensing imagery, aeolian desertification in the Minqin Oasis has rapidly occurred, and 70% of farmland has been affected by aeolian desertification (Xue et al. 2015). Severe aeolian desertification has resulted in the Minqin Oasis being a major source of sandstorms in China before 2010 (Wang et al. 2004; Dong et al. 2010).

7.2.2 *Restoration Process in the Minqin Oasis*

Minqin has become well-known in China and the world based on its severe aeolian desertification and corresponding refugee problems. After 2007, the Chinese national government implemented a series of environmental policies and invested over 4 billion RMB in the Shiyang River Basin and Minqin to combate aeolian desertification. These policies included water allocation inside the basin, agricultural water management, agricultural structural adjustments, and construction of shelter forest systems.

First, based on the average annual water allocation plan, the amount of water flowing into the Minqin Oasis must account for 22% of the total water volume of the river flowing out of the mountains and can change in high and low years depending on the mountain water volume. Furthermore, some water from the Yellow River is transported into the Minqin Oasis directly by channel. In 2019, the total surface water volume that flowed into the Minqin Oasis was from the Shiyang River, and the

Table 7.1 Soil salinity accumulation after farmland was abandoned in the Minqin Oasis

Sampling site	Depth (cm)	Total salt content (%)	Ion content (mg 100 g^{-1})							
			CO_3^{2-}	HCO_3^-	Cl^-	SO_4^{2-}	Ca^{2+}	Mg^{2+}	K^+	Na^+
Irrigated farmland	0–20	0.281	–	0.27	0.44	3.14	2.37	0.83	0.18	0.78
	20–48	0.971	–	0.22	0.41	13.68	10.38	2.85	0.26	1.44
	48–60	1.390	–	0.22	1.01	19.79	12.18	4.49	0.28	3.58
	60–100	1.395	–	0.20	1.52	19.39	10.41	5.29	0.23	5.04
Abandoned irrigated farmland (10 years)	0–10	10.715	–	0.24	80.63	89.72	15.62	37.3	1.77	117.17
	10–20	7.957	–	0.22	36.74	83.32	12.82	24.0	1.59	88.26
	20–40	1.791	–	0.14	3.92	22.91	11.22	6.41	0.28	10.00
	40–60	2.129	–	0.16	4.23	28.36	12.66	5.45	0.23	12.17
	60–100	2.035	–	0.13	6.48	33.71	14.42	3.85	0.36	13.48

Fig. 7.2 Sands encroaching a settlement (left) and farmland (middle) and soil salinization (right) in the northern part of the Minqin Oasis (from Xue et al. 2015)

Fig. 7.3 Qingtu Lake in the 1950s (left), 2010 (middle), and 2012 (right) (**a**) is provided by Minqin County government, (**b**) and (**c**) are taken by Xian Xue

Yellow River reached approximately 3×10^8 m^3. In addition to being used for irrigation, a certain amount of surface water flowing into the Minqin Oasis recharges Qingtu Lake at the end of the Shiyang River Basin. Qingtu Lake disappeared for over 40 years starting in the 1950s and reappeared with a maximum size of over 30 km^2 in 2019 (Fig. 7.3).

In terms of water use and management, the local government regulates the average cultivated farmland area for each person and the water consumption amount for a given unit of farmland. This policy was implemented better in the initial period after its establishment. Each person could own only 2.5 mu irrigated farmland. Therefore, the total farmland area was reduced from 900,000 mu (1 ha = 15 mu) to 600,000 mu in the early 2010s in the Minqin Oasis. However, in a recent survey, we found that many local farmers have moved to places outside the Shiyang River Basin to utilize more abandoned land to compensate for their losses in their hometowns. Furthermore, many scattered privately owned arable lands, abandoned lands, and sandy lands are leased to enterprises or owned by large landowners for intensive management. Farmers benefit by working for these farmland renters. This kind of agricultural model not only reclaimed abandoned farmland in a short time but also opened newer abandoned lands and sandy lands.

In contrast to farmland management, water use control and management has always remained relatively strict in the Minqin Oasis. Surface water is considered the priority source for irrigation water, especially in the northern part of the Minqin Oasis, where groundwater quality is poor. Under the unified management of the Water Resource Bureau, surface water is sent to farmlands in the different periods of

Fig. 7.4 Drip irrigation in the initial stage (left) and drip irrigation substituted by flood irrigation (right) (provided and taken by Xian Xue)

the growing season of crops (usually four to five times per year, including spring or winter irrigation to leach salt downward). Groundwater is mainly used to compensate for the lack of fresh water. After 3161 wells were closed in the early 2010s (Xue et al. 2015), new or closed wells seldomly were drilled or reopened. The amount of groundwater in the irrigation water volume was controlled by an electricity card. Farmers can use more groundwater by purchasing electricity, which is also the reason farmers rent their farmland to large landowners or enterprises. Enterprises have enough funds to pay for irrigation water to open more abandoned or sandy land.

To save water resources, the local government suggested and implemented tube irrigation and drip irrigation, but the effort failed. By the end of 2010, drip and tube irrigation systems had been set up in half the total irrigated farmland area of the Minqin Oasis, including 15,000 ha of drip irrigation and 3700 ha of tube irrigation in open fields and 1827 ha of drip irrigation in greenhouses. By 2020, for several reasons, almost all drip irrigation and tube irrigation systems were abandoned (Fig. 7.4). First, local farmers did not implement this method because they believed that more water would produce more yield. Second, the higher initial investment hindered the popularization of this method. Initially, the central government invested substantial funding into drip irrigation. When the investment stopped, local governments and farmers could not afford it and had to stop using this method. Finally, the local unsustainable technology and lack of technicians prevented adoption of drip irrigation. Farmers did not know the right time and amount for drip irrigation or how to cope with dripper clogs caused by poor water quality. The scattered farmland plots also made implementation of drip irrigation use and management difficult. The newly emerging large-scale agriculture operated by enterprises and large landlords has gradually made drip irrigation possible. On the one hand, these enterprises have the initial investment and the ability to cooperate with scientific research departments. On the other hand, the large farmland area makes drip irrigation easy to apply and manage.

Adjusting the agricultural structure is another important measure used to save water and improve water use efficiency in the Minqin Oasis. Fruit trees such as red dates and red wolfberries with high water use efficiency, vegetables in greenhouses,

Fig. 7.5 Google map (left) and photo (right, taken by Xue et al. 2015) of a *Haloxylon ammodendron* forest protecting Minqin Oasis

and stable animal husbandry have become popular and are replacing traditional grain crops (Xue et al. 2015). However, fruit trees have not resulted in the expected benefits. Red dates and red wolfberries are not Minqin's specialty products. The former in Xinjiang and the latter in Ningxia have been planted over a long period and account for the main portion of the market. Greenhouse vegetables are mainly tomatoes, eggplants, and peppers, which provide more benefits than traditional crops. However, their competitive ability in the market is still limited. High-value and specific crop, vegetable, and economic tree species need to be explored.

Since 2007, the Chinese central government has invested a large amount of funds in Minqin to establish the shelterbelt system to protect the oasis. The local government organizes public servants to plant trees voluntarily in the spring each year. By 2020, most of the northern and western edges of the Minqin Oasis had been protected by the shelterbelt system (Fig. 7.5). The current shelterbelt system mainly contains a checkerboard with *Haloxylon*. Although some scholars have questioned whether dense *Haloxylon* forests consume a large amount of soil water and whether a single checkerboard approach can effectively prevent dune encroachment and wind erosion, the effectiveness of the shelterbelt system is still obvious over a short time.

Based on the above description, although the measures to address aeolian desertification did not meet initial expectations and even had some negative effects, aeolian desertification reversal and restoration of aeolian desertified areas obviously occurred in the Minqin Oasis. Vegetation coverage increased. The groundwater table ascended. The frequency and intensity of sand dust storms have obviously decreased. The next step is to consider how to sustainably restore areas experiencing aeolian desertification. To achieve this aim, it is important to evaluate the current measures and find smart ways to balance livelihoods and environmental protection. In the following sections, we will discuss this aim.

7.3 Principles for Achieving Sustainable Development

Achieving and maintaining sustainable socioeconomic development and environmental protection are the main goals in combating desertification. This target of the UNCCD has received increasing attention since it was proposed in the 1970s. Compared with the increasingly severe desertification in the twentieth century, vegetation coverage in some countries and regions gradually increased after the beginning of the twenty-first century (Fig. 7.6). Among these countries and regions, China and India have been restored and have contributed significantly to global greenness (Chen et al. 2019). Desertification, therefore, is considered to have already been reversed in some countries or regions (Piao et al. 2020). However, does increasing greenness levels indicate that the SDG related to desertification has been achieved or does it promote sustainable development in these areas? Before answering the question, some principles must be identified in terms of how the restoration of aeolian desertification can be scientifically assessed.

7.3.1 Principle 1 Building a Scientific Assessment System

Determining suitable indicators and conducting scientific assessments are especially important steps to achieve the SDGs related to combating desertification. Currently, vegetation coverage is one of the most common indicators used to evaluate the restoration of degraded land because it is visible and easy to monitor. However, vegetation coverage changes seasonally and yearly in arid areas. Substantial

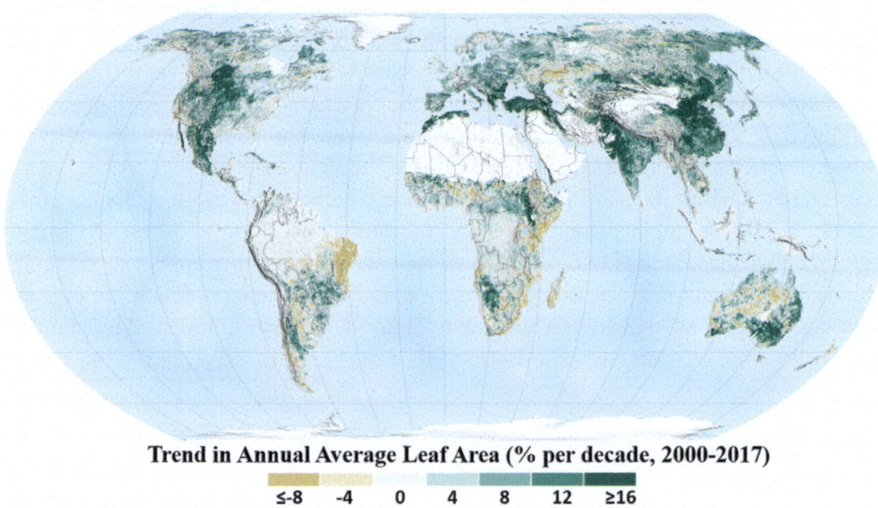

Fig. 7.6 Trend in growing season mean leaf area index (cited from https://earthobservatory.nasa.gov/images/146296/global-green-upslows-warming)

precipitation variability is the main climatic characteristic of arid climate zones. Sudden intense precipitation can rapidly initiate the growth of xerophytic plants or annual grasses, therefore increasing vegetation coverage over a short period. However, heavy rainfall is often followed by prolonged drought, which decreases vegetation cover and dries the land. Instability in vegetation coverage has recently become more common due to global warming. Rapid increases in vegetation coverage are also due to afforestation. Planting trees or shrubs in drylands with annual average precipitation less than 200 mm can increase vegetation coverage for a few years through irrigation, as occurred in the Minqin Oasis. Plants, however, will die from drought after the groundwater table descends because a large amount of transpiration consumes the locally limited water resources. Some research has shown that *Haloxylon* will die when the density of *Haloxylon* is greater than 70 trees per mu. It is worth noting that the disappearance vegetation is not limited to planted vegetation but includes native natural vegetation as well.

Therefore, considering vegetation coverage, such as the NDVI, as an important assessment indicator is extremely risky in arid regions. The temporary increase in vegetation coverage tends to result in the misjudgment or exaggeration of the degree to which aeolian desertified lands have been restored. To a certain extent, this scenario can lead to unsustainable restoration of aeolian desertified areas. In comparison to vegetation coverage, vegetation community structure and soil fertility change slowly and can better reflect the characteristics of the ecosystem. Assessment indicators should consider these more stable variables and ignore or focus less on rapidly changing variables such as coverage. Gibbs and Salmon (2015) reviewed four kinds of assessment methods for land degradation at the global scale, and these were expert opinion, satellite-derived net primary productivity, biophysical model, and abandoned cropland map methods. The results show a large uncertainty in estimating the areas and distributions of degraded land. There is still a large gap between the theoretical and actual situations in evaluating aeolian desertification and restoration.

7.3.2 Principle 2 Evaluate the Trend, Not the State

Evaluating restoration of aeolian desertified land by its direction rather than its existing state is the second principle to achieving sustainable development. If vegetation community structure, soil fertility, and the other stable variables are assessment indicators, then the restoration of aeolian desertification would need to be viewed over a longer temporal scale (Reynolds et al. 2007). Research and experience at the southeastern edge of the Tengger Desert in China have shown that significant soil fertility increases on the surface occurred at least 50 years after mobile dunes were fixed by a wheat straw checkerboard grid and vegetation (Li et al. 2003). An approach that requires half of a century to determine whether a degraded ecosystem has been restored cannot address the likelihood of achieving neutral land degradation goals by 2030. However, considering this issue from a different perspective may provide a different understanding. That is, sustainable development and restoration of aeolian desertified lands are regarded as dynamic processes rather than static states. Severely aeolian desertified land may no longer be in the degradation process and might be restored from a very severe state. In contrast, moderately or lightly aeolian desertified land might be able to be restored from a better state and thus may be in the aeolian desertification process rather than the restoration process. Considering the development direction over a limited temporal scale can address issues related to stable variables, and in comparison to other approaches, this approach is also more scientific and objective.

7.3.3 Principle 3 Balanced Trade-Off Between Livelihood Improvement and Land Restoration

A balanced trade-off between livelihood improvement and land restoration is also an important principle in achieving the sustainable restoration of aeolian desertified land. Improving livelihoods and restoring degraded land are often considered nonexclusive and can be mutually reinforcing. This perspective is the basis for understanding sustainable development in terms of combating aeolian desertification. This understanding is undoubtedly true to some extent. A high level of income is always consistent with advanced scientific, technological, and cultural levels, including understanding land restoration. Healthy and stable ecosystems have high levels of production and are beneficial to the improvement of livelihoods. However, this ideal state depends on existing livelihoods much more. There is a large gap between developed and developing countries or regions in terms of basic human needs. Making a living means survival for some people, but it likely refers to vacation times per year at tourist resorts for other people.

In attempting to narrow this large gap, it is important to note that it is impossible to prevent impoverished people from exploiting resources and destroying the environment, even if they know that the damage leads to even greater poverty. The case in Minqin Oasis has proved this point. Between the short-term and long-term benefits, as well as the trade-off between improving livelihoods and protecting the environment, people prefer improving their livelihoods because of the existing inequality. This inequality is not limited to wealth but also exists in different ecosystems. Land productivity is less, and environmental conditions are poorer in arid areas than in areas with abundant precipitation and fertile soil, but both areas have the same amount of people. This kind of inequality means that achieving the same levels of livelihood improvement results in more environmental degradation in arid regions than in humid regions. Figure 7.7 conceptually illustrates the nonlinear relationship between livelihood improvement and restoration of aeolian desertified lands. Under the threshold, the restoration of aeolian desertified lands does not mean improvement in livelihood. Under these conditions, the trade-off between improving livelihoods and restoring aeolian desertified lands is substantial. The answer to resolving this problem depends on the second goal, which is poverty eradication.

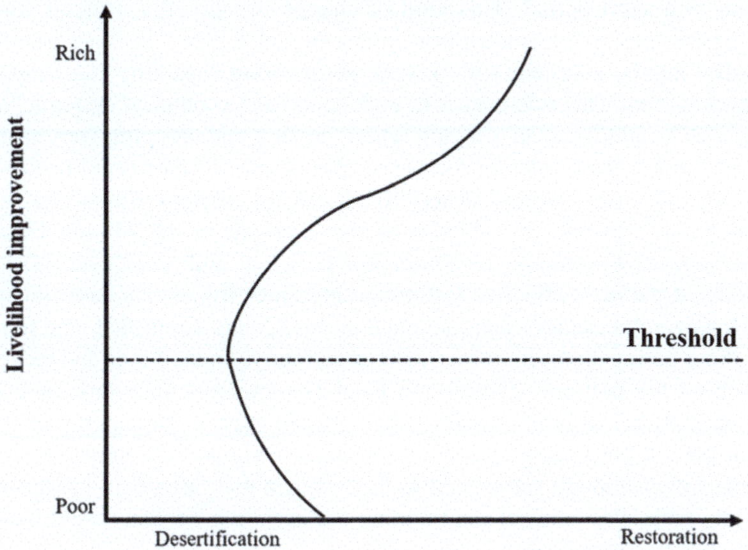

Fig. 7.7 Conceptual model of the relationship between livelihood improvement and aeolian desertified land restoration

7.4 Principles for Achieving the Poverty Eradication Target

Figure 7.7 shows that only when the state of aeolian desertification is above the threshold can restoration of aeolian desertified lands and livelihood improvement be mutually reinforcing. In most drylands, especially in the developing world that experiences substantial poverty, the current aeolian desertification state is below the threshold. This situation means that aeolian desertification control might lead to further poverty. Does this imply that the sustainable restoration of aeolian desertified lands and achieving LDN in these areas are impossible? The answer is no. When poverty is eliminated and the inequality gap becomes narrower to a certain extent, the relationship between livelihood improvement and aeolian desertified land restoration will cross the threshold and become a mutually promoting relationship. The key problem aeolian desertified regions are facing is how to reach the threshold line quickly, that is, to eradicate poverty. Therefore, the eradication of poverty has become the second main target of restoring aeolian desertified lands. Thus, how can poverty eradication be achieved, and what principles need to be practiced to eradicate poverty?

7.4.1 Principle 1 External Inputs

External inputs are the most important factor in eradicating poverty. It is difficult for a system whose society, economy, and environment are unsustainable to change its status quo in a short period. This kind of change is especially difficult when the system experiences unfair competition from external sources. Without healthy and high quality inputs, a system with a weak competition capability can change to a lower-level system. This rule occurs in both social and ecological systems. We can imagine what would occur in a region where grazing and cropping were forbidden without any inputs from external sources so that desertified lands could be restored.

As a typical case, restored desertified lands in northern China can help us understand this principle. In addition to the example in the Minqin Oasis, the *Grain for Green*, *Forbidding or Rotational Grazing*, and *Sand Control* projects are considered the most effective restoration measures for aeolian desertification in China. However, the achievement obtained in this aeolian desertified land relied not only on these policy inputs but also, more importantly, on the funds and technology inputs behind the policies. Central and local governments have provided a great number of subsidies to farmers and herders, usually with funds ranging from 50 to 200 yuan per mu per year with even more funds to support the implementation of these measures. Although some surveys have shown that farmers or herders cannot control all funds, the external contributions to restoring aeolian desertified lands cannot be underestimated.

The other external contributions were from NGOs, enterprises, communities, and private companies outside the degraded system. Alxa SEE Ecological Associates, a well-known NGO in China, is continually active in western Inner Mongolia. The SEE mainly solicits funds from entrepreneurs throughout China to contribute to the eradication of poverty by supporting the restoration of aeolian desertified lands in drylands. In the last 20 years, the main activities of SEE have included guiding local farmers and herdsmen to adjust their land use practices, introducing highly beneficial crops and animals, providing solar stoves to reduce the use of wood, financially subsiding afforestation, and training local people. To date, SEE has become the main team implementing restoration of aeolian desertified lands except for local governments, and SEE has high-level authorities in local society. Therefore, external inputs play the most important role in the achieving the restoration of aeolian desertified lands in China. The active and healthy connection between desertified regions and the outside world is an important principle and a necessary way to eradicate poverty.

7.4.2 Principle 2 Improving Internal Capacity

Although the elimination of poverty through external inputs can restore aeolian desertified lands, improving the internal capacity of the system is also crucial. There is never continuous support and help from outside entities. After the receiving system has been improved or reached certain level of stability, the improvement of a society and an economy and the continued restoration of aeolian desertified lands will mainly rely on the system. This improvement of system capabilities includes many aspects, such as the average education level of citizens, scientific and technological capabilities, women's status and rights, community management, medical conditions, and awareness of environmental protection. If these aims cannot be achieved, then the prosperity and recovery of an aeolian desertified system will be temporary. Once external inputs cease, the system will quickly return to being impoverished and degraded. However, in the current social, economic, and ecological situation of degraded dryland systems, the improvement of these internal capabilities is difficult to achieve without the support of external inputs.

The second principle, therefore, is closely related to the first principle. The principle of developing internal capabilities indicates that external inputs should not be limited to material inputs but should also focus on cultural, educational, and ideological inputs. Although globalization is becoming mainstream, the principle is not easy to implement due to many constraints, such as social systems, race, religion, and traditions. The true eradication of poverty and the restoration of aeolian desertified lands greatly depend on the construction and improvement of internal capacity and resilience of aeolian desertified systems. Correctly understanding and addressing this principle will help to objectively recognize the challenges in achieving poverty eradication and restoration of aeolian desertified lands and in eliminating blind optimism.

7.4.3 Principle 3 Periodic Evaluations

Carrying out necessary periodic evaluations during the implementation of poverty eradication is also an important principle. Based on the evaluation results, the strategic direction and specific implementation measures can be adjusted to obtain the best results. Two important factors determine the implementation this principle. First, the effectiveness of any measure cannot be objectively evaluated before its implementation as no one knows whether it is appropriate or not. In particular, a complex socioeconomic-ecological integrated system has extremely high complexity. Different traditional customs, religious understanding, and the environmental protection awareness may all have unpredictable results in terms of implementing strategic measures. It is even more likely that the consequences will be the opposite of the expected goal. It is particularly important to adjust the strategic direction and implementation plan at any time by conducting a staged evaluation of the implementation effect. Second, both the socioeconomic system and the ecological environmental system dynamically develop. Even if external inputs and intervention measures of the system are correct in terms of the strategic direction and implementation process, the internal response of the system will change with time and its development process. By scientifically evaluating the development direction and speed of the system after receiving external inputs and the level of its capacity building, continuously adjusting the method and intensity of inputs will help the system to develop steadily and rapidly and accelerate the eradication of poverty.

In summary, eradicating poverty is an extremely complicated and difficult goal. In the process of achieving this goal, the above three principles should not be ignored.

7.5 Principles for Improving the Land Productivity Target

According to the UNCED (2012), desertification occurs because of reduced land productivity. Therefore, the most intuitive result of restoring desertified land is the improvement of land productivity. However, how can the relationship between land productivity improvement and the restoration of desertified land be quantified? In other words, what level of land productivity can be considered as restoring desertified lands? Second, how can stable and sustainable improvements in land productivity occur? These two issues are closely related to the following three principles.

7.5.1 Principle 1 Evaluating Carrying Capacity

Scientifically understanding and assessing the carrying capacity of the aeolian desertified land to be restored is the primary principle. Carrying capacity includes three aspects: past, present, and future capacities.

Before aeolian desertification occurred, what kind of ecosystem existed, and at what level was the carrying capacity? The answer largely determines the restoration target for aeolian desertification. For example, if the land was previously a typical steppe or desert steppe, then it is unscientific to restore it to a forest or shrubland. Based on most of the current restoration measures in aeolian desertified lands, we have determined that this principle is being neglected. Afforestation is becoming popular in some regions where no forests existed before. Agriculture is also prosperous in the boundary regions where desert is located. The above land use activities that neglect this principle have resulted in the land being at risk of aeolian desertification again, although the measures restore aeolian desertified lands for a short time.

The second consideration is related to the carrying capacity of current aeolian desertified lands. Even though the land had been a typical grassland ecosystem, after years of degradation, the groundwater level might have fallen sharply, soil fertility may also have disappeared, and mobile dunes have been densely distributed. In this case, is it possible to restore the former grassland ecosystem? In other words, people need to find the optimal land use activity that fits the current carrying capacity of the water and land resources and not those the previously undesertified ecosystem. It is easy to understand that it is difficult to restore severely aeolian desertified regions to their previous state. Therefore, the definition of restoration must not mean going back to the same states that previously existed. Of course, past conditions and future changes to the current ecosystem need to be considered and evaluated to determine the land use method and intensity.

Finally, future changes in climate and human activities also need to be better understood and projected, which is undoubtedly a critical issue. If the local climate develops to become increasingly warm and humid, then the aeolian desertified land is expected to be restored, and it may even be restored to an ideal state. Conversely, if the local climate changes toward increased droughts, then the restoration of aeolian desertified lands will become increasingly difficult, and more efforts will be needed. Therefore, people should continue to adjust land use activities and intensities to adapt to future climate conditions. Similarly, human activity changes are the other determining factor through which land use can be adjusted with time. For example, technologies such as saving water and desalinizing saline water can change the local water supply situation, thereby increasing the land's water resource carrying capacity and land productivity and ultimately effectively enabling recovery from aeolian desertification.

A scientific understanding of the carrying capacity of land resources in different periods is important for the restoration of aeolian desertified land. When the national or local government formulates and implements land restoration strategies and measures, the unsustainability of restoring aeolian desertified lands might largely take place if the carrying capacity assessment of the land resources in different periods is ignored.

7.5.2 *Principle 2 Improving Land Management*

After a scientific assessment and determination of the land carrying capacity in aeolian desertified areas, the next principle that people must consider is how to improve land management. Land management has always been an important issue in the field of aeolian desertification research. A land carrying capacity assessment is part of land management, but because of its importance, we treat it as a separate principle and provided a detailed description in the section below. In addition to the land carrying capacity assessment, land management includes many aspects.

At the macroscale or regional level, land management includes formulating land uses and periodic monitoring and evaluating land use effects. The implementation of land use policies such as land acquisition management and environmental protection subsidies also falls into the macroscale category of land management. Land management varies in different regions. For example, water resource allocation is more important in the catchment areas of inland rivers than in pastoral regions. In areas with substantial land use conversion, people are concerned about the stability and duration of land policies and subsidies. In ecological resettlement areas, especially immigration areas, basic security measures such as health, education, entertainment, and religion of the immigrants are particularly important to macroscale land management. At the microscale, land management mainly implies the improvement of soil fertility, the maintenance of soil moisture, the prevention of erosion, and the improvement of saline-alkali soil.

The case study of the Minqin Oasis introduced in Sect. 7.2 can help us understand the importance of improving land management to sustainable restoring aeolian desertified lands. Due to the large-scale reclamation of abandoned land in the middle reaches of the Shiyang River in the last half of the twentieth century, the surface water entering the downstream area gradually decreased from approximately

500 million m^3 in the 1950s to 80 million m^3 in 2000 (Xue et al. 2015). Under a severe lack of surface water, the increase in the area of cultivated land in the lower reaches mainly depended on the exploitation of groundwater. Long-term overexploitation of groundwater caused the groundwater table to drop rapidly. The descending groundwater table caused natural vegetation to grow in the boundary zone between the irrigated oasis and the desert, where the vegetation started to wither and die because they cannot absorb enough groundwater. The ecological function of the natural vegetation zone is protecting oases from wind erosion and sand disasters. The loss of natural vegetation, therefore, led dunes to move into oases from the desert. Many villages and farmlands were destroyed and buried (Fig. 7.2). Sandstorms have also affected surrounding cities and even those farther east of China (Wang et al. 2004). On the other hand, groundwater is repeatedly used to irrigate farmlands. With strong evaporation, soil salinity continued to increase. In 20 years, soil salinization gradually degraded wheat farmlands with high productivity levels to low productivity farmlands, which only supported crops with high salt tolerances, such as fennel and alfalfa.

Since 2000, severe aeolian desertification in the Minqin Oasis has attracted global attention. Chinese national and local governments have implemented many policies and measures to improve the level of land management and reverse aeolian desertification. These policies and measures include water resource allocation in the whole catchment, land use conversion from crops to economy forest-agriculture and from high-water consumption crops to low-water consumption crops, construction of farmland protection forests, and the fixation of active dunes. These measures have had some good effects, and the environment has improved to a large extent. However, the abandonment of farmland and increased water supply to lakes have raised the groundwater table rapidly, which has resulted in more severe salinization problems. This series of environmental changes shows the importance of scientifically based land management to control and restore aeolian desertified lands. Land management includes monitoring, evaluation, policy formulation, market regulation, and many other aspects. No land management policies or measures can resolve aeolian desertification by themselves.

7.5.3 Principle 3 Dependence on Science and Technology

The improvement of land productivity depends on not only the scientific evaluation of land carrying capacity and sustainable land management but also the continuous progress of science and technology. Science and technology are double-edged swords because they can promote the restoration of aeolian desertified land, but they can also lead to aeolian desertification. The use and popularization of motor pump wells provide a typical case of technological improvements leading to aeolian desertification. The case of the Minqin Oasis shows that motor pump wells in arid areas are important because they use deepwater, greatly improve land productivity, and reduce poverty. However, the existence of motor pump wells has also promoted the overexploitation of groundwater, which is one of the causes of aeolian desertification in arid regions. However, the overuse of motor pump wells does not mean that science and technology cannot restore aeolian desertified lands. In contrast, the rational use of science and technology is an indispensable factor in aeolian desertification control. If farmers better grasp drip irrigation technology in the Minqin Oasis, then the unsustainable use of water resources can be resolved.

Implementation of the previous principles depends on science and technology. For example, the application of high temporal and spatial resolution remote sensing technology makes it possible to monitor and assess aeolian desertification control effects in a timely manner. Saline water irrigation and desalination technologies have greatly increased the available water resources in drylands with limited freshwater. Biogenetic technology has effectively improved land yield by changing the drought and salt resistance ability of crops. Additionally, the improvements in material, chemical, and biological technologies have significantly increased the efficiency of sand disaster control in harsh climate regions. Chapter 8 will specifically introduce the use of these sand control technologies.

It is worth noting that the application of biotechnology in aeolian desertified areas is gradually being promoted. As mentioned above, the improvement of soil fertility is a long process in arid areas, especially in areas where severe aeolian desertification is occurring. However, biotechnology can promote soil formation by rapidly changing the soil structure and increasing the soil organic matter content. For example, in northwestern China, researchers are now screening indigenous root microorganisms, such as bacteria and fungi, that promote plant growth to expand them to plant roots for inoculation to increase drought and salt tolerance while increasing soil fertility to quickly restore aeolian desertified land. At the same time, algae or other plant extracts can be developed into water-retaining agents and sand-fixing agents and are widely used as environmental protection materials in the field of aeolian desertification control.

7.6 Summary

Although the UNCCD set the overall goal of achieving LDN by 2030, there are too many problems or difficulties preventing the achievement of this overall goal. This chapter proposed three sublevel targets to achieve the restoration of aeolian desertified lands and the principles that need to be followed to achieve them. However, the facts show that to date, these principles have not been scientifically recognized or implemented. Therefore, the overall goal of achieving LDN has a long path forward and requires more effort and scientific exploration. Finally, we note that there are still more subtargets and principles in addition to the contents presented in this chapter that need to be considered in terms of restoring aeolian desertified lands.

References

Chasek P, Safriel U, Shikongo S, Fuhrman VF (2015) Operationalizing Zero Net Land Degradation: the next stage in international efforts to combat desertification? J Arid Environ 112:5–13
Chen C et al (2019) China and India lead in greening of the world through land-use management. Nat Sustain 2:122–129
Dong ZB, Man DQ, Luo WY et al (2010) Horizontal aeolian sediment flux in the Minqin area, a major source of Chinese dust storms. Geomorphology 116(1–2):58–66
Gibbs HK, Salmon JM (2015) Mapping the world's degraded lands. Appl Geogr 57:12–21
Krevisky J, Jordan L (1989) Webster's encyclopedic unabridged dictionary of the English language. Gramercy Books, New York
Li X, Zhou H, Wang X et al (2003) The effects of sand stabilization and revegetation on cryptogam species diversity and soil fertility in the Tengger Desert, Northern China. Plant Soil 251:237–245
Piao S, Wang X, Park T et al (2020) Characteristics, drivers and feedbacks of global greening. Nat Rev Earth Environ 1:14–27
Reynolds JF et al (2007) Global desertification: building a science for dryland development. Science 316(5826):847–851
Safriel U (2017) Land Degradation Neutrality (LDN) in drylands and beyond – where has it come from and where does it go. Silva Fennica 51:1650. 19 p. https://doi.org/10.14214/sf.1650
Stavi I, Lal R (2015) Achieving zero net land degradation: challenges and opportunities. J Arid Environ 112:44–51
Sun D, Dawson R, Li H, Li BG (2005) Modeling desertification change in Minqin County, China. Environ Monit Assess 108:169–188
United Nations Conference on Environment and Development (UNCED) (1992) Earth summit, Rio de Janeiro
United Nations Conference on Environment and Development (UNCED) (1994) United Nations convention to combat desertification in those countries experiencing serious drought and/or desertification, Particularly in Africa. Paris
United Nations Conference on Environment and Development (UNCED) (2012) zero net land degradation-a sustainable development goal to Rio+20
United Nations Conference on Environment and Development (UNCED) (2014) Rio declaration on environment and development. Habitat.igc.org. Archived from the original on 2 Apr 2003. Retrieved 4 Aug 2014
UNEP (1991) Status of desertification and the implementation of the United Nations plan of action to combat desertification. UNEP, Nairobi
United Nations General Assembly (2015) Transforming our world: the 2030 agenda for sustainable development. Retrieved from http://www.un.org/ga/search/view_doc.asp?symbol=A/RES/70/1&Lang=E
United Nations Statistics Division (2016) SDG indicators global database. Retrieved from https://unstats.un.org/sdgs/indicators/database/
Wang XM, Dong ZB, Zhang JW, Liu LC (2004) Modern dust storms in China: an overview. J Arid Environ 58:559–574
Wang T, Xue X, Zhou LH, Guo J (2015) Combating aeolian desertification in northern China. Land Degrad Dev 26:118–132
Xie YW, Chen FH (2002) Study on the change of Minqin Oasis since recent twenty years on digital Rs images. Arid Zone Res 19(1):69–74
Xue X, Liao J, Hsing Y-T et al (2015) Policies, land use, and water resource management in an arid oasis ecosystem. Environ Manag 55(5):1036–1051
Zhang KC, Qu JJ, Liu QH (2004) Environmental degradation in the Minqin oasis in Northwest China during recent 50 years. J Environ Syst 31(4):357–365

Chapter 8
Engineering Measures to Combat Aeolian Desertification

Xian Xue, Kecun Zhang, and Weimin Zhang

Abstract Engineering measures (EMs) are methods to prevent and control aeolian disasters that do not involve plants. Usually, EMs are the preferred prevention and control methods in arid areas, where planting trees is not always possible. In semiarid and dry subhumid areas, EMs are sometimes used in conjunction with other measures, especially in the early stages of biological measures (BMs), for immediate effects. EMs can be divided into four types based on function: sand blocking, sand fixing, sand transporting, and sand guiding. These types of EMs have different forms, such as upright, semiupright, flat, and buried. The materials used in EMs are diverse and include crop straw, degradable nylon mesh, gravel, clay, and synthetic materials. This chapter mainly introduces four forms of EMs and their materials and functions. Chemical measures (CMs) are usually used in regions where both EMs and BMs are not available. CMs can immediately fix the sand surface by spraying chemical materials. Due to their high price and potential pollution, CMs are not extensively used and are introduced at the end of the chapter.

X. Xue
Northwest Institute of Eco-Environment and Resources, Chinese Academy of Sciences, Lanzhou, Gansu, China

Key Laboratory of Desert and Desertification, Chinese Academy of Sciences, Lanzhou, Gansu, China

Drylands Salinization Research Station, Northwest Institute of Eco-Environment and Resources, Chinese Academy of Sciences, Lanzhou, Gansu, China

K. Zhang (✉) · W. Zhang
Dunhuang Gobi and Desert Ecological and Environmental Research Station, Northwest Institute of Eco-Environment and Resources, Chinese Academy of Sciences, Lanzhou, Gansu, China

Key Laboratory of Desert and Desertification, Chinese Academy of Sciences, Lanzhou, Gansu, China

Northwest Institute of Eco-Environment and Resources, Chinese Academy of Sciences, Lanzhou, Gansu, China
e-mail: kecunzh@lzb.ac.cn

T. Wang et al. (eds.), *Combating Aeolian Desertification in Northeast Asia*,
Ecological Research Monographs, https://doi.org/10.1007/978-981-16-9028-0_8

Keywords Engineering measures · Function · Forms · Material · Chemical measures

8.1 Introduction

The influences of aeolian desertification mainly include wind erosion, sand burial, and sand dust storms. These activities are usually called aeolian disasters because of the resulting decreased land productivity and damage to buildings, bridges, roads, and human health. In northern China, many measures have been implemented to control aeolian disasters. These measures include engineering, chemical, and biological measures. Biological measures (BMs) mainly involve reducing wind erosion and sand burial by increasing vegetation cover, such as through afforestation and grazing management. Chemical measures (CMs) prevent wind erosion by spraying chemical materials on the sand surface and are rarely used due to their high price. Engineering measures (EMs) refer to those that prevent and control windblown sand disasters using crop straw, gravel, pebble, soil, fiber, and other nonliving materials.

EMs widely appear in different regions because they are not limited by water, soil, and climatic conditions. They are most extensive in arid areas because drought prevents large areas of afforestation. On the Tibetan Plateau with its cold climate, EMs also play essential roles. In semiarid and dry subhumid areas, EMs are usually used to aid biological measures, especially when the plants do not result in beneficial effects. EMs utilize the physical characteristics of aeolian sand to set up a measure to prevent windblown sand and the forward movement of dunes. The basic principles of EMs are as follows: (1) prevent the movement of sand particles by inhibiting wind erosion on the surface; (2) reduce sand deposition by accelerating the movement of wind-sand flows; and (3) reduce the overall forward movement of dunes by changing the direction of the wind-sand flows (Wang 2011).

There are many types of EMs. EMs can be divided into four types based on their functions: sand blocking, sand fixing, sand transporting, and sand guiding. Their forms can be upright, semiupright, flat, and buried. The materials used in EMs are diverse and include crop straw, degradable nylon mesh, gravel, clay, and synthetic materials. This chapter mainly introduces the characteristics and functions of different EMs based on their forms and materials used. Finally, the use of chemical measures is briefly introduced.

8.2 Four Forms of EMs

EMs are designed to inhibit sand movement and weaken wind erosion. However, no single measure can achieve a good effect. Effective prevention and control of wind-sand disasters are usually achieved by a comprehensive prevention system. The system consists of upright, semiupright, flat, and buried measures from upwind to downwind. Upright measures are located at the forefront of the entire system. These measures minimize the amount of sand entering the downwind area. This type of measure reduces the wind speed, blocks the sand material, or leads the sand away from the protected area by changing the wind direction. Semiupright measures are usually set in the downwind direction of upright measures. Sometimes, there is a blank buffer between the two. The functions of semiupright measures include blocking sand from the upwind direction by reducing the wind speed and fixing local sand by covering the surface. The functions of flat measures are to force upwind sand to move downwind quickly, thereby reducing the local deposition of sand. The buried measures trap sand from the upwind area with a trench. After implementation of the four measures, there is very little sand material that can move downwind and cause aeolian disasters.

8.2.1 Upright EMs and Their Effects

Upright measures usually involve a standing fence or wall. The height, density, porosity, and arrangement of the fence or wall are different, depending on the actual requirement. Usually, the functions of upright EMs include decreasing wind speed, changing the wind direction, and blocking sand particle or dune movement. The section presents three examples to help describe their characteristics and functions.

8.2.1.1 Sand-Blocking Fence

Sand blocking means intercepting sand particles by setting walls and fences perpendicular to the main wind direction. These fences are usually deployed in areas upwind of the protected areas, and the fences reduce the amount of sand moving into the protected area (Dong et al. 2007). A sand-blocking fence is a high vertical barrier made of nylon mesh, wood plate, corn straw, or cement (Fig. 8.1). As a typical sand-blocking engineering measure, it prevents sands from moving downward to protected areas by intercepting sand sources. The interception effect of the

Fig. 8.1 Sand-blocking fences with different porosities and heights (Photos courtesy of Yanping Yu and Tao Wang)

fence depends on its shape parameters, such as porosity, height, and thickness (Bitog et al. 2009).

Porosity is the ratio of the open area to the total area of the fence. It is the most significant factor in controlling aerodynamic performance and affecting protection efficiency (Qu et al. 2002; Li and Sherman 2015; Wu et al. 2018). Although different studies show a variety of results, the general trend is that the protective effect of a high-porosity fence is lower than that of a low-porosity fence. For example, a fence with a porosity of 20% can reduce the wind velocity leeward of the fence (Raine and Stevenson 1977). A fence with a porosity of 30% is the most effective in controlling dust emissions (Chen et al. 2012).

The protective effects of fences with different heights are also very different. With increasing height, the starting point of wind erosion gradually moves from near the fence to its upwind area, which means that the protective scope increases. When conditions provide for a good sand source, both erosion and deposition coexist in the leeward area of the fence, and the erosion strength increases with the height of the fence (An et al. 2011). Therefore, when building a fence barrier with EMs, the sand-control effect, the protection area, and the cost of materials should be considered to obtain the maximum ecological and economic benefits (Zhang et al. 2011).

8.2.1.2 Windbreak Wall

A windbreak wall is also a kind of sand-blocking measure that uses local materials and is usually set in the upwind area of the sand fixation area (Fig. 8.2). In contrast to a sand-blocking fence, a windbreak wall has no porosity. When the airflow reaches a windbreak wall, the wind speed is quickly reduced to a minimum, creating a low-speed area on the leeward side of the windbreak wall. The sand is deposited near the windbreak wall because the wind speed in the low-speed area is lower than the threshold wind speed of the moving sand. The energy is not enough to keep the sand in the previous movement state. The windbreak wall, therefore, can trap sand

Fig. 8.2 Windbreak wall (left) and sand-prevention dike (right) (Photos courtesy of Tao Wang and Zishan An)

and reduce the sand entering the sand-control area. This aeolian disaster control system in the downwind direction can maintain long-term efficiency.

There are many types of windbreak walls, such as pull-type, "L"-shaped, soil embankment, and plate-type windbreak walls. Among them, the structure of the pull-type windbreak wall is relatively complicated. However, its stability, safety, and reliability are sufficient, its cost is relatively low, and the symmetry of the wall is not affected by the wind direction. The "L"-shaped windbreak wall has a flexible structure and is small and of high quality, but it is relatively expensive. A soil embankment windbreak wall can be built with local materials and is convenient to construct with a simple structure, good stability, and easy maintenance and inexpensive, but these walls are large. In general, plate-type and oblique plate-type windbreak walls are more widely used in areas with abundant sand in the Gobi Desert. Currently, the numerical simulation method is often used to determine parameters such as height, air permeability, and slope of the windbreaks in the process of construction (Xue et al. 2011; Zhang et al. 2011; Shi and Jiang 2014).

8.2.1.3 Sand-Prevention Dike

A sand-prevention dike is another upright measure used in the sand-blocking zone of a protection system. The shape of a sand-prevention dike is usually triangular or trapezoidal. The dike is made of soil or sands. The main function of this measure is to intercept sand by reducing the wind speed of the saturated wind-sand flow. According to the characteristics of different flow fields, the surrounding area of a sand-prevention dike can be divided into five areas: the low-speed area at the foot of the windward slope, the wind-lifting area, the wind-acceleration area, the wind-deceleration (settlement) area, and the dissipation recovery area. The low-speed zone at the foot of the windward slope and the deceleration zone at the leeward side can effectively reduce the wind speed and block sand flow. Wind tunnel simulation studies have shown that the trapezoidal area has a larger range of low-speed areas at the foot of the windward slope formed by the triangular dike. In comparison to the trapezoidal dike, the triangular dike has a more adequate protection distance. The protection range of the triangular dike decreases with increasing wind speed. The effective protection distance formed by the triangular dike increases with increasing inclination of the windward slope. The range of the low-speed zone at the foot of the windward slope decreases with increasing inclination. A model of the triangular dike with an inclination angle of 40° forms the maximum significant protection distance under each group of wind speeds. The effective protection distance on the leeward side of the dike is positively related to the height of the dike. In addition to its sand-blocking function, a dike with the appropriate direction can change the local wind direction, control sand activities, and play a sand-guiding function (Li et al. 2012).

8.2.1.4 Sand-Guide Fence

In addition to blocking sand, upright measures also guide sand. Sand-guiding measures can change the direction of sand movement by fences or walls built obliquely to the main wind direction so that the sand particles accumulate in expected areas (Wang 2011). These measures usually supplement the other measures and are suitable for areas with simple wind directions.

8.2.2 Semiupright EMs and Their Effects

Semiupright EMs usually include different kinds of checkerboard barriers with heights of approximately 20–30 cm above the surface. The checkerboard barriers are located downward of upright sand-blocking measures. They block the interaction of air-sand particles on the bed surface and inhibit sand movement by partially covering the surface of the sand.

The checkerboard barriers are the most common horizontal sand-control barriers and are usually set up in dual- or multiwind-direction areas. Barrier materials such as crop straw, gravel, clay, or nylon mesh are put on or inserted into the sands to form a checkerboard shape. The barrier has a height of approximately 20 cm above the ground, therefore increasing the surface roughness and lowering the wind speed near the surface. When the wind flow carrying sand goes through the barriers, the wind velocity decreases because the barriers and sands in the wind flow unload near the barrier. After the sand is unloaded, the wind accelerates behind the barriers, which causes wind erosion inside the checkerboard. The ideal checkerboard should ensure the deposition and erosion reach a dynamic equilibrium in a short time, which means that the deposition amount near the barrier equals the erosion amount inside the checkerboard. The dynamic equilibrium can produce a smooth and stable concave face at the bottom of the boundary, which is beneficial to maintaining long-term equilibrium (Fig. 8.3). To achieve a stable erosion-deposition equilibrium, the

Fig. 8.3 A checkerboard barrier at the preliminary stage and at the mature stage with a concave face (Photos courtesy of Yang Gao and Yanping Yu)

Fig. 8.4 Gravel and clay checkerboard sand barriers on the Qinghai-Tibetan Plateau (Photos courtesy of Yanping Yu)

barrier material should be elastic and permeable, which allows the wind to pass through the barrier quickly and avoid erosion near the barrier (Shi and Jiang 2014).

Field observations and experiments in wind tunnels have shown that checkerboard barriers with small grid sizes have a high sand-control effect. However, the cost will increase as the size becomes smaller. Considering the effects and costs comprehensively, a checkerboard barrier with a grid size of 1 m × 1 m will be ideal (Qu et al. 2007; Xue et al. 2011; Zhang et al. 2011). However, in an actual situation, grid sizes change in different regions, and the size mainly depends on the wind-field conditions and topography, such as the slope, height, and dune shape. In strong wind erosion areas with higher and steeper dune slopes, a small grid size of 1 m × 1 m is recommended. In weak wind areas with flat sands, grid sizes with areas of 1.5 m × 1.5 m or 2 m × 2 m are also efficient (Wang 2011). In addition, the direction of the checkerboard barrier must be vertical to the main wind direction to achieve the best sand-blocking effect. Although checkerboard barrier measures have a good sand-fixing effect, some research in wind tunnels and in the field has found that the effect will decrease with increasing wind velocity (Fig. 8.4).

8.2.3 Flat EMs and Their Effects

Using clay, gravel, reeds, and straw to cover the surface of dunes or sandy lands can effectively inhibit the interaction between wind and sand. Therefore, a typical flat measure is to change the underlying surface by wholly or partially covering the sand with different materials. However, different materials, coverages, sizes, arrangements, and even wind velocities can result in varied functionality. Flat mechanisms are very complicated and involve the wind erosion process, sand deposition process, and sand transportation process. In arid or cold regions, gravel is used extensively because wheat straw, reeds, and clay are not always available. Furthermore, gravel is efficient in terms of preventing and controlling wind-sand disasters, resulting in it being researched more than other materials.

Gravel coverings provide multiple functions in terms of preventing aeolian desertification and aeolian disasters. First, gravel mulch can reduce the evaporation of soil

Fig. 8.5 A shrub and watermelon crop with gravel cover in Zhongwei in Ningxia Province, China (Photos courtesy of Hanchen Duan and Liming Ma)

moisture and help the soil retain moisture. Planting crops and shrubs on lands with gravel cover has occurred for at least 300 years in the arid area of northwestern China (Fig. 8.5). Our survey in Zhongwei in Ningxia Province showed that the crop yield in a gravelly cropland was higher than that in a cropland without gravelly cover, especially in drought years.

Second, gravel cover plays a critical role in influencing sand movement (Liu et al. 1999; Zhang et al. 2004). Gravel cover mainly affects sand movement by changing the erosion and deposition process. The process of erosion and deposition is closely related to the degree of gravel coverage, gravel height, and gravel shape (Liu et al. 1999; Xue et al. 2000). Wind conditions and sand supply lead to different erosion process intensities on the surface so that gravel cover has the dual functions of "catching sand" and "transporting sand" (Bagnold 1941; Cooke et al. 1993). Thus, if the erosion process is stronger than the deposition process, then the function of gravel cover measures is transporting sand. In the opposite scenario, the function of gravel is to catch or trap sand. Understanding the principles of wind erosion and deposition conversion on a gravel-covered surface and establishing a reasonable dynamic balance of erosion and deposition through a good configuration of gravel cover with various coverage and particle sizes are crucial issues to achieving the best prevention and control of aeolian disasters.

Research shows that gravel coverage is a crucial indicator affecting the wind erosion process. The portable wind tunnel experiment conducted at the top of the Mogao Grottoes shows that the erosion rate decreased exponentially with increasing

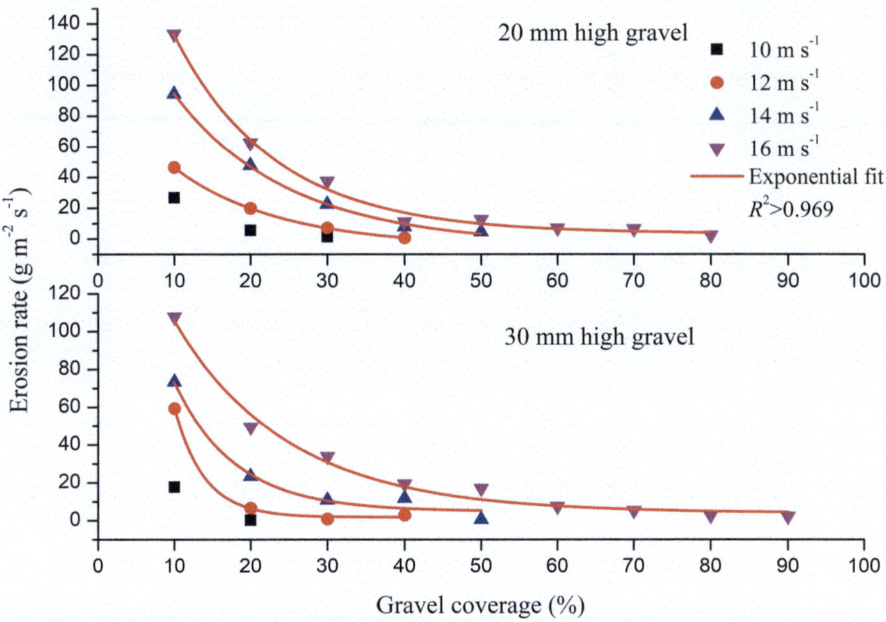

Fig. 8.6 Wind erosion rate with a gravel cover surface varies with the coverage under different experimental wind speeds (according to Zhang et al. 2014)

gravel coverage (Fig. 8.6). Under a wind velocity of 10 ms^{-1}, 2–3-cm-diameter gravel with 30% coverage resulted in no erosion. With an increase in wind velocity and gravel coverage, no erosion events occurred. However, the efficiency of preventing wind erosion is not significant when gravel coverage was greater than 50%. Furthermore, the different effects of gravel diameter became nonsignificant when the coverage was 50%. Therefore, 40% is recommended as the best gravel coverage at the top of the Mogao Grottoes (Zhang et al. 2014; Tan et al. 2016).

Similar to the erosion process, the sand deposition rate was also controlled by the experimental wind velocity and gravel coverage. Figure 8.7 presents a group of data from the gravel cover experiment at the top of Mogao Grottos. The research showed that no sand deposition occurred when gravel coverage was less than 40% under strong wind (12 m s^{-1}). With increasing gravel coverage, the sand deposition rate had a decreasing trend (Tan et al. 2012). For 3-cm-diameter gravel, when the coverage was 40–80%, the maximum sand deposition rate occurred. For 4-cm-diameter gravel, the maximum wind velocity occurred when the coverage reached 40–60%. This result is consistent with the wind tunnel experiment result conducted by McKenna Neuman (1998).

Based on the comprehensive evaluation of sand blocking and sand transportation by gravel coverage, 30–40% is recommended as the ideal gravel coverage at the top of the Mogao Grottoes (Zhang et al. 2014; Tan et al. 2016).

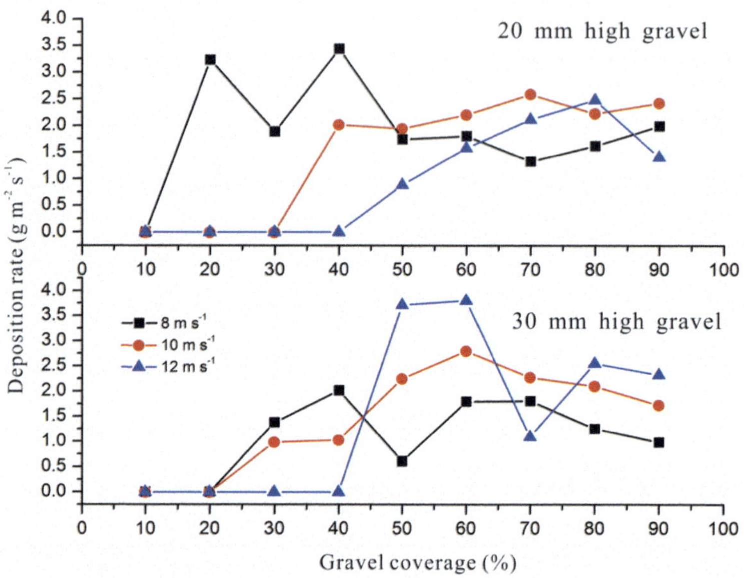

Fig. 8.7 Wind deposition rate of the gravel cover surface varies with the coverage under different experimental wind speeds (according to Zhang et al. 2014)

8.2.4 Buried EMs and Their Function

Compared with research on the previous three EMs, research on buried EMs is rare, and they are not widely useful. A sand ditch is considered the primary buried measure. When the wind carrying sand passes through the ditch, sand moving slowly along the ground surface drops into the ditch. Sand carried by the wind is thus reduced. After the sand in the sand-carrying flow is unloaded into the ditch, the wind speed will increase significantly and quickly pass through the protected area. Through this process, less sand is deposited in the area.

8.3 Main Materials Used in EMs

In combating aeolian desertification and aeolian disasters, the sand-control efficiency of the engineering measures greatly depends on the characteristics of the materials used. Ideal materials should have good elasticity, plasticity, wind erosion resistance, and environmental friendliness properties. A low cost, easy access, and a long life are also important characteristics of the materials. At present, there are diverse types of EM materials. These materials can be generally divided into traditional materials, synthetic materials, and new materials.

8.3.1 Traditional Materials

Traditional EM materials are mostly raw materials obtained locally, such as crop straw, willow branches, reeds, clay, gravel, and pebbles (Fig. 8.8). These materials have been used for a relatively long time. As the primary material of checkerboard barriers, wheat straw has been used for nearly half a century and is still one of the most common sand-barrier materials. Wheat straw is widely used because wheat is the most extensively distributed crop in northern China and the raw material market is vast. Additionally, wheat stalks have perfect elasticity, flexibility, and porosity, which can effectively reduce the incoming wind speed. In addition to wheat straw,

Fig. 8.8 Wheat straw (**a**), clay (**b**), cotton straw (**c**), corn straw (**d**), willow twig (**e**), and reed (**f**) checkerboard sand barriers (**a**, **e**, **f** from Wang 2011; **b**, **c**, **d** provided by Xian Xue)

cotton and corn straw is also common in the sand-control field of northwestern China. Cotton and corn straw is taller and stronger than wheat straw and therefore is used for not only horizontal checkerboard barriers but also vertical sand-blocking fences and semivertical checkerboard barriers. Compared with cotton straw, corn straw has higher flexibility and plasticity. It is thus more widely used. In recent years, the crop straw market has decreased gradually, and prices have also increased because more straw is consumed for papermaking and returning to fields as organic fertilizer. Compared to other materials, crop straw easily rots and has a shorter service life, so other materials gradually replace it.

Tamarix is a kind of saline-tolerant and drought-resistant native plant widely distributed in the arid regions of northwestern China. In recent years, it has become an important shelter forest tree species at the edge of the oasis. During the pruning of tamarix, a large number of tough and relatively hard branches are produced, and they provide material for setting up engineering measures to control aeolian desertification. Initially, people usually lay tamarix branches on the sand surface in a checkerboard pattern, but this method has been shown to be less effective in reducing wind and sand. In a later approach, tamarix branches were cut short and inserted vertically into the sand to form a vertical grid or a vertical strip-shaped fence. Reeds are not only the main halophyte at the end of an inland river in an arid region but also widely distributed in the low interdune areas. Therefore, it is also used as a sand-control material to address the shortage of raw materials in the market.

Climate change and human activities have resulted in the gradual shrinkage and disappearance of terminal lakes at the end of inland rivers. The large amount of lacustrine clay deposited at the bottom of lakes is transported to dune fields to cover and fix the mobile dunes. Inland rivers, which originated from the high mountains, brought a large amount of gravel from the mountain areas, deposited in the middle and lower reaches of riverbeds, and the gravel was polished over the long term by running water to form pebbles. Therefore, these materials are used as alternative materials for dune fixation when crop straw is scarce. However, clay barriers are not strong and are susceptible to erosion (Sun et al. 2012), and the actual sand-fixing effects of gravel and pebble barriers are also not effective enough, mostly due to their high transportation costs, which have limited the scope of their application.

8.3.2 Synthetic Materials

In recent years, the Chinese government has strengthened control measures for aeolian desertification and aeolian disasters, which has rapidly increased the construction of engineering measures. The supply of crop straw is not suitable to meet increasing demand. It is also difficult to expand the use of clay and gravel barriers due to transportation problems. Therefore, different synthetic materials have begun to appear and have been used in sand-control fields. The region using synthetic materials is currently smaller than that using traditional materials. However, there is a possibility that synthetic materials will replace traditional materials because a sand-control barrier with synthetic material can be manufactured and constructed inexpensively with a high level of production.

Plastic and nylon mesh are the most common sand-control materials that are constructed with high-density polyethylene (HDPE). HDPE has strong UV resistance, extreme temperature resistance, and aging resistance. Therefore, nylon mesh has the advantages of producing no pollution and being inexpensive, reusable, and easy to construct in sand. It is especially suitable for some regions with severe aeolian desertification and disaster. The HDPE material was first successfully developed and used in the prevention and control of wind-sand hazards on the southeastern coast of China in 2000 by Jianjun Qu. Currently, HDPE nylon horizontal checkerboard barriers and vertical fence have been widely applied in different areas of China, and even areas of Africa and Central Asia (the right photo in Fig. 8.3). Research shows that the HDPE mesh checkerboard sand barrier can efficiently reduce the wind speed and control the wind-sand flow. However, it is easily buried and eroded if no sand-block measures occur upwind (Liu et al. 2011; Qu et al. 2014).

Because HDPE nylon is not degradable and pollutes the environment, polylactic acid (PLA) fiber material is attracting attention in the field of engineering measures. PLA is a new biodegradable polymer material. Its raw materials are wheat, corn, sweet potato, potato, sugar beet, and other starch-containing crops. After a period of time, PLA material can convert into CO_2 and H_2O without causing secondary pollution to the environment. The construction of a PLA barrier includes three steps. PLA fiber is first made into tubular bags with a diameter of approximately 15–20 cm. Then, these bags are filled with sand after being transported to the construction area. Finally, the PLA fiber sandbags are laid on the ground surface in a grid shape to fix dunes. Compared with traditional sand barriers such as wheat straw, clay, and gravel, PLA checkerboard sand barriers have the advantages of weighing less, being easy to transport, being able to be constructed rapidly, not weathering, and not polluting the environment (Wei et al. 2017).

The other new synthetic material used in engineering measures is clay-based, sand-fixing composite material (CSCM). The main component of CSCM is red clay and loess. Accessory materials are environmentally friendly polymers such as CMC and PVA and biopolymers such as plant straw. Because of the addition of polymers, in comparison to ordinary clay, CSCM has a higher cohesive force and a stronger resistance to wind erosion. At the same time, due to its low porosity level, CSCM water resistance and aging resistance are also high. CSCM is usually laid directly on a dune or sandy surface to fix sand, and sometimes, it is made into different shapes of sand barriers, such as hexagonal grids, square grids, plum grids, and arch grids, to reduce wind speed and fixed dunes (Fig. 8.9).

Gypsum board (GB) is a kind of sand-fixing material made of gypsum composite material. The raw material in GB is gypsum, which is inexpensive, has good cohesiveness, and does not cause pollution. GB can not only effectively prevent wind erosion but also absorb moisture and reduce soil moisture evaporation. Therefore, it is used at a large scale, especially on the Tibetan Plateau, with fewer other materials.

Fig. 8.9 Clay-based sand-fixing composite material sand barriers (Photos courtesy of Ziqiang Lei)

8.4 Chemical Measures

Chemical measures involve spraying a chemical substance with a certain gelling property on the surface of loose sand. The water quickly penetrates the sand layer. The chemical gelling substance stays in the gap between sand layers of a certain thickness, gelling the single grain of sand into a layer of protection. The protection layer can improve the threshold velocity of sand particles moving and inhibit sand particles from entering the atmosphere. Thus, chemical measures can effectively reduce the concentration of sand in airflows. Additionally, the layer with chemical bonding materials also maintains the soil moisture content by reducing evaporation. Therefore, chemical measures can effectively control erosion and dust release (Han et al. 2004; Zhou et al. 2012). Due to the advantages of their simple structures, low costs, fast effects, and rapid improvements in aeolian desertified soil, chemical measures are increasingly used in engineering for aeolian desertification prevention.

The effectiveness of chemical measures is related to the properties of sand particles, such as their chemical and mechanical composition, and to the physical and chemical properties of the chemical material, such as its structure, molecular size, viscosity, and adsorption force. In the process of aeolian desertification prevention, chemical materials must be nontoxic, nonpolluting, and beneficial to plant germination and growth. In addition, the adaptability of chemical materials to the climate and their low cost, efficiency, and durability should also be considered. According to their source and function, chemical materials include inorganic, organic, and organic-inorganic composite chemical materials. The most common chemical materials for preventing aeolian desertification are cement slurry, sodium silicate, petroleum products, biomass resources, and synthetic polymers. The sand-fixing functions of chemical materials include bondage, surface cover, hydration, precipitation, and polymerization (Ju et al. 2019).

8.4.1 Inorganic Chemical Materials

Tie et al. (2013) divided inorganic chemical, sand-fixing materials mainly into cement slurry, water glass, and gypsum and determined the advantages and disadvantages of these materials.

After being sprayed on the sand surface, the cement slurry-based, sand-fixing material plays a role in condensation and solidification to form a covering layer. Due to the dry and hot climate in aeolian desertified areas, the sand surface temperature is high in summer, and the water in the cement slurry sand fixer will easily evaporate quickly. Therefore, the cement slurry, sand-fixing agent lacks sufficient moisture to be completely hydrated and can only form a thin consolidation layer with low strength on the surface of the sandy land. This consolidation layer is not flexible and is easily affected by dune migration and severe weather. It therefore loses its sand-fixing and water-holding effects in a short time. Due to the above defects, cement slurry-based, sand-fixing materials are rarely used alone (Ding et al. 2003; Tie et al. 2013).

Water glass slurry, sand-fixing materials have a long history of use in preventing aeolian desertification. In the past, this kind of material mainly consisted of water glass and acidic reactants. Under strong alkaline conditions, gelation and solidification occur, the gelation time is limited, the permeability is poor, and the curing reaction is incomplete; thus, the strength of the consolidated layer is not high. Additionally, the composite material of water glass and acidic reactants is easily damaged by external forces and affected by strong alkalinity. After the silica colloid product gradually dissolves, the water resistance becomes poor, the durability decreases, and the environment experiences secondary alkali pollution. To resolve these problems, in the current field of aeolian desertification prevention, people are committed to the study of various modified water glass slurries. The main aim is to obtain a kind of liquid composite water glass slurry, sand-fixing material suitable for spraying by adding organic substances such as glyoxal and vinyl carbonate. Then, bentonite can be mixed with the water glass to obtain porous materials by solidification.

Gypsum is a widely used industrial and building material. It is inexpensive, strong, and nontoxic; has a high water absorption and retention capacity; and does not pollute the environment. Therefore, gypsum is one of the materials used for preventing aeolian desertification. Gypsum is also beneficial to the growth of plants. It is commonly used to maintain the survival of plants by mixing it with bentonite and fertilizers in dry environments. Due to the brittleness of gypsum, pure gypsum is easily pulverized in arid climates, and its durability is also low. Thus, gypsum is mixed with other materials to improve its toughness and make it an excellent sand-fixing material (Wang 2008).

8.4.2 Organic Chemical Materials

Organic chemical sand-fixing materials mainly include petroleum products, synthetic polymer-based materials, biomass resources, and modified plastic waste.

At present, cementitious petroleum materials used for sand fixation are mainly byproducts of the petrochemical industry. These include asphalt emulsions, high-resin petroleum, rubber emulsions, and oil-rubber emulsion mixtures. Asphalt emulsion is a representative petroleum-based sand-fixing agent and is currently the most widely used material in chemical sand fixation measures around the world. Asphalt is solid or semisolid at normal temperature, has a high freezing point and melting point, and has a high viscosity. These physical characteristics make asphalt a traditional bonding, waterproof, and anticorrosive material. Asphalt emulsions made by using emulsifying equipment under the action of an emulsifier can be divided into cationic, anionic, and nonionic types. As a soil improver, asphalt emulsions can prevent soil erosion, improve soil hydrothermal conditions, increase temperature and soil moisture, reduce fertilizer and pesticide loss, and improve fertilizer efficiency. As a sand-fixing material, asphalt emulsions can be used for sand fixation alone or in combination with plants and mechanical sand barriers. When combined with plants, asphalt is nontoxic to plants, does not affect vegetation germination and growth, and can be fixed to the ground for a longer time. The former Soviet Union began to study the use of asphalt emulsions for sand fixation as early as 1935. In the early 1970s, the Tashkent Institute of Railway Transportation Engineering researched this area and achieved positive results. The Ministry of Railways and the former Lanzhou Desert Research Institute of the Chinese Academy of Sciences conducted sand fixation by spraying emulsified asphalt and planting along the Baotou to Lanzhou Railway from the 1960s to 1990s. Although asphalt emulsions are an ideal chemical material for preventing aeolian desertification, they cannot be widely used alone because of some shortcomings, such as the high demand for them, their high cost, and their resulting reduction of water seepage into soil. Therefore, in the process of aeolian desertification control, asphalt emulsions are often mixed with other inorganic substances to improve performance and reduce costs (Lai et al. 2017).

Synthetic polymer-based, sand-fixing materials are new chemical sand-fixing materials that have been developed since the 1960s. The synthetic high-viscosity polymers can seal sand particles, making them change into cemented joints and no longer unconnected or weakly connected particles. The formation of a strong and elastic cover can effectively fix the sand surface (Wang et al. 2003). Popular synthetic polymer-based, sand-fixing materials include polyvinyl alcohol (PVA), polyvinyl acetate (PVAC), polyacrylamide (PAM), and SH sand-fixing agents. Using high-molecular-weight polymers to fix active dunes or sands has some advantages, such as a simple and easy construction process and short construction period. The effectiveness of synthetic polymer-based materials is, therefore, more significant and stable than that of other chemical materials.

Using biomass resources to prevent aeolian desertification is becoming the primary prevention mechanism. Lignin is the most widely used biomass sand-fixing material and is produced by chemical modification of lignin in pulping waste liquid. There are a large number of active groups in the lignosulfonate molecule, such as the phenolic hydroxyl group, the alcoholic hydroxyl group, the carboxyl group, and the carbonyl group. After being sprayed on the sand surface, lignosulfonate binds to sand particles by creating electrostatic attraction, hydrogen bonding, and chemical complexation between sand particles. Additionally, the three-dimensional network structure of the lignosulfonate molecular chain makes the sand particles combine to form a dense, strong consolidation layer. Notably, lignosulfonate has poor water erosion resistance because of its strong solubility. This disadvantage hinders its wide application in practice. Researchers are trying to modify lignin sand-fixing materials to address the desired requirements using grafts, polycondensation, etc. For example, the new sand-fixing agent obtained by grass-lignosulfonate grafting modification with acrylic acid and acrylamide monomers in a hydrogen peroxide environment significantly improves sand-fixing efficiency and waterproof resistance (Li and Song 2002). Zhao et al. (2015) produced a water-soluble dust inhibitor with good sand fixation performance by modifying urea-formaldehyde resin with sodium lignosulfonate. Using methylcellulose fiber-pulp waste liquid as the main raw material, Jin et al. (2006) synthesized a new type of biodegradable biomass material through methylation and condensation reactions under specific conditions.

Discarded plastic products severely pollute the environment. Currently, some scholars are studying how to produce sand fixation materials by modifying plastic waste. The research is still in the preliminary phase and limited to indoor research. The sand-fixing performance of modified plastic waste in the field needs to be further investigated and studied (Lai et al. 2017).

8.4.3 Organic-Inorganic Composite Chemical Sand-Fixing Materials

Because of the poor mechanical properties and weak water-retention performance of inorganic sand-fixing materials, organic components are added to inorganic sand-fixing materials to form organic-inorganic composite chemical sand-fixing materials. Composite sand-fixing materials not only compensate for shortcomings but also integrate advantages, thereby enhancing the performance of sand-fixing materials. For instance, Xu (2003) developed high-performance polymer cement-based, sand-fixing materials by adding emulsified asphalt and sodium polyacrylate superabsorbent resin to cement mortar. The new composite sand-fixing material is stronger and tougher than traditional cement mortar. Li et al. (2010) obtained a polymer gel-fixing binder by adding an aluminum chloride curing agent to sodium silicate. The mechanical strength and bonding effect of the composite material are greatly improved compared with those before modification.

8.5 Conclusion

EMs are essential and useful methods to prevent aeolian desertification and aeolian disasters. However, how to maximize the effectiveness of these measures depends on many factors. At present, many EMs in northwestern China have not achieved the desired targets. The most important reason for this is an insufficient understanding of the mechanisms of these measures. First, any single measure cannot wholly control wind erosion or sand deposition without the assistance of the other measures. A comprehensive system including sand blocking, sand guiding, sand transporting, and sand fixing is necessary. Second, the material, size, porosity, and elasticity of the EMs affect their disaster prevention functions. The relationship between these features and functions is not static but changes with terrain, sand sources, and wind fields. Therefore, before constructing any engineering measures, it is necessary to research the local environment and wind-sand characteristics. Based on the research results, a prevention system with different EMs is designed to ensure its functions are maximized. However, many of the current measures do not address the above two points, resulting in EMs eroding or being buried over a short time. Therefore, understanding the protection mechanisms of EMs is essential for preventing aeolian desertification and aeolian disasters.

References

An ZS, Zhang KC, Tan LH, Li JG (2011) Process of aeolian sand erosion and deposition and sand controlling effect of fences with different heights. Bull Soil Water Conserv 31:37–41 (in Chinese with English abstract)

Bagnold RA (1941) The physics of blown sand and desert dunes. Morrow & Company, New York, pp 167–171

Bitog JP, Lee IB, Shin MH, Hong SW, Hwang HS (2009) Numerical simulation of an array of fences in Saemangeum reclaimed land. Atmos Environ 43:4612–4621

Chen GH, Wang WW, Sun CF, Li JL (2012) 3D numerical simulation of wind flow behind a new porous fence. Powder Technol 230:118–126

Cooke RU, Warren A, Goudie AS (1993) Desert geomorphology. UCL Press, London

Ding Q, Xu XJ, Chen YZ, Hu SG (2003) The research advances of chemical sand-fixing materials. J Wuhan Univ Technol 25:27–29 (in Chinese with English abstract)

Dong ZB, Luo WY, Qian GQ, Wang HT (2007) A wind tunnel simulation of the mean velocity fields behind upright porous fences. Agric Forest Meteorol 146:82–93

Han ZW, Wang T, Dong ZB, Zhang WM, Wang XJ (2004) Main engineering measurements and mechanism of blown sand hazard control. Prog Geogr 23:13–21 (in Chinese with English abstract)

Jin YC, Liu J, Yang YQ, Li ZZ (2006) Preparation of biomass sand stabilization material from straw pulp effluent and its application on vegetation restoration structural characteristics of biomass sand stabilization material. China Pulp Paper Indust 27:75–78 (in Chinese with English abstract)

Ju YF, Qiu XM, Zhu JK, Zhang JM, Zhou Y (2019) Advances in sand-fixing material research in China and the application prospect. J Arid Land Res Environ 33:138–144 (in Chinese with English abstract)

Lai JH, Zhang K, Wang WS, Wang YK, Xu XL (2017) Research advances and prospect in chemical sand-fixing materials. J Desert Res 37:644–658 (in Chinese with English abstract)
Li B, Sherman DJ (2015) Aerodynamics and morphodynamics of sand fences: a review. Aeolian Res 17:33–48
Li JF and Song ZQ (2002) Study on lignosulfonate and its grafted polymers as sandy soil stabilizers. Chem Indust Forest Prod 22:17–20 (in Chinese with English abstract)
Li T, Deng XY, Li JB, Zhang HT, Zhang W (2010) Effect of concentration of aluminum salts on stability of aluminosilicate sand stabilization material. Bull Chin Ceramic Soc 29:721–727 (in Chinese with English abstract)
Li JG, Qu JJ, Li F, An ZS, Han QJ (2012) Wind tunnel simulation on flow fields around different types of sand dikes. J Desert Res 32:291–299. (in Chinese with English abstract)
Liu LY, Liu YZ, Li XY, Wang JH, Peng HM (1999) Effect of gravel mulch restraining soil deflation by wind tunnel simulation, J Desert Res 19(1):60–62. (in Chinese with English abstract)
Liu HJ, Wang JH, Li Y, Ma R, Sun T (2011) Effects of plastic checkerboard sand-barrier on wind-sand flux of leeward of crescentic dune. J Soil Water Conserv 25:26–34 (in Chinese with English abstract)
McKenna Neuman C (1998) Particle transport and adjustments of the boundary layer over rough surfaces with an unrestricted, upwind supply of sediment. Geomorphology 25(1):1–17
Qu JJ, Ling YQ, Liu XW, Lei JQ, Li F (2002) The effects of an A-typed nylon fence on aeolian sand prevention. J Lanzhou Univ (Nat Sci) 38:171–176. (in Chinese with English abstract)
Qu JJ, Ling YQ, Jing ZF, Zhang KC, Zu RP (2007) Interaction between sand blown activity and protection system in Shapotou section of Baotou-Lanzhou Railway. J Desert Res 27:529–533 (in Chinese with English abstract)
Qu JJ, Yu WB, Qin XB (2014) Wind-protecting efficiency of HDPE functional sand-fixing barriers. J Desert Res 34:1185–1193 (in Chinese with English abstract)
Raine JK, Stevenson DC (1977) Wind protection by model fences in a simulated atmospheric boundary layer. J Wind Eng Indust Aerodyn 2:159–180
Shi L, Jiang FQ (2014) Numerical simulation on parameters optimization of the oblique inserting plate retaining sand wall design. J Desert Res 34:666–673 (in Chinese with English abstract)
Sun T, Liu HJ, Zhu GQ, Zhang YH, Ma R (2012) Timeliness of reducing wind and stabilizing sand functions of three mechanical sand barriers in arid region. J Soil Water Conserv 26:12–22 (in Chinese with English abstract)
Tan LH, Zhang WM, An ZS, Li JG (2012) Response of wind velocity gradient at boundary layer to gravel coverage. J Desert Res 32:1522–1527 (in Chinese with English abstract)
Tan LH, Zhang WM, Qu JJ, Yin DY, An ZS, Bian K (2016) Quantitative simulation of erosion rates of Gobi with different gravel coverages. J Desert Res 36(03):581–588 (in Chinese with English abstract)
Tie SN, Jiang X, Wang CA (2013) Advances in chemical sand-fixing materials. Materials, 27(3): 71–75 (in Chinese with English abstract)
Wang YM. (2008) Study on mechanism of chemical desert control. J Catastrophol 23:32–35 (in Chinese with English abstract)
Wang T (2011) Deserts and aeolian desertification in China. Science Press, Beijing (in Chinese with English abstract)
Wang YM, Sun GP, Zhan WW, Ding L (2003) Strength characteristics of sand fixated by SH. Chin J Rock Mechan Eng 22:2883–2887 (in Chinese with English abstract)
Wei YJ, Zuo XF, Wang J, Dang XH, Liu XJ (2017) Summarization of application of mechanical sand barrier in desertification control. J Inner Mongolia Agric Univ (Nat Sci Ed) 38:86–93 (in Chinese with English abstract)
Wu TG, Zhang P, Zhang L, Wang JY, Yu MK (2018) Relationships between shelter effects and optical porosity: a meta-analysis for tree windbreaks. Agric Forest Meteorol 259:75–81 (in Chinese with English abstract)
Xu XJ (2003). Study on the polymer modified cementitious sand-fixing material, Wuhan University of Technology (in Chinese with English abstract)

Xue X, Zhang WM, Wang T (2000) Wind tunnel experiments on the effects of gravel protection and problems of field surveys. Acta Geograph Sin 55:375–383 (in Chinese with English abstract)

Xue CX, Jiang FQ, Cheng JJ, Pang QD (2011) Research on sand preventing benefit of sand retaining wall in a strong wind sector along Lanzhou-Xinjiang Railway. J Glaciol Geocryol 33:859–862 (in Chinese with English abstract)

Zhang WM, Wang T, Wang WF et al (2004) The gobi sand stream and its control over the top surface of the Mogao Grottoes. China Bull Eng Geol Env 63(3):261–269

Zhang JP, Wang YS and Jiang FQ (2011) Numerical analysis on the features of sand flow movement around the embankment of Lan-Xin Railway in gobi region. China Railway Sci 32:14–18. (in Chinese with English abstract)

Zhang W, Tan LH, Zhang GB et al (2014) Aeolian processes over gravel beds: field wind tunnel simulation and its application atop the Mogao Grottoes, China. Aeolian Res 15:335–344

Zhao H, Zhang LD, Han CY, Kong WQ, Zhang N (2015) A new sand fixing agent: lignin sulfonate grafted urea-formaldehyde resin. J Beijing Univ Chem Technol 42:60–64 (in Chinese with English abstract)

Zhou MJ, Zhou YS, Song JL, Liu S, Liu JH (2012) Research and application status of sand-fixing materials in China. Mater Rev 26:332–334 (in Chinese with English abstract)

Chapter 9
Biological Measures to Combat Aeolian Desertification

Xueyong Zhao, Xiaoan Zuo, Yuqiang Li, Xinping Liu, Shaokun Wang, Juanli Chen, and Rui Zhang

Abstract Aeolian desertification is a process where wind and solid (mainly sand) material movement lead to vegetation degradation, and soil erosion due to climate change and overland uses in arid, semiarid, and subhumid regions. Biological measures have been tested as long-term effective measures to moderate or even control aeolian desertification, and using or reutilizing restored land resources, including soil, plants, and water, has also been considered. This chapter focuses on biological measures, such as planting trees, shrubs, and grasses; constructing cropland protection systems around oases; afforesting; and enclosing areas around houses, villages, or targeted areas to restore aeolian desertified land and degraded habitats. In addition, some successful case studies in northeastern Asia, particularly in China and Mongolia, are introduced.

Keywords Aeolian desertification · Biological measures · Combating · Measures · Restoration

---

X. Zhao (✉)
Urat Desert Grassland Research Station, Northwest Institute of Eco-Environment and Resources, Chinese Academy of Sciences, Lanzhou, Gansu, China

Key Laboratory of Desert and Desertification, Chinese Academy of Sciences, Lanzhou, Gansu, China
e-mail: Zhaoxy@lzb.ac.cn

X. Zuo · Y. Li · X. Liu · S. Wang · J. Chen · R. Zhang
Key Laboratory of Desert and Desertification, Chinese Academy of Sciences, Lanzhou, Gansu, China

Naiman Desertification Research Station, Northwest Institute of Eco-Environment and Resources, Chinese Academy of Sciences, Lanzhou, Gansu, China

T. Wang et al. (eds.), *Combating Aeolian Desertification in Northeast Asia*, Ecological Research Monographs, https://doi.org/10.1007/978-981-16-9028-0_9

9.1 Introduction

In most cases, aeolian desertification is a process that leads to vegetation degradation and soil erosion due to climate change and overland uses in arid, semiarid, and subhumid regions (Zhu et al. 1981; Zhu and Chen 1994). However, aeolian desertification is not only an aeolian process but also more often a mixed process of soil, vegetation, and near-ground air trait changes due to water and aeolian erosion when land cover changes (Wang et al. 2016). It is therefore necessary to note that combating aeolian desertification is a compound process to restore degraded vegetation and soil systems (Zhao et al. 2009, 2015). That is, biological measures to combat aeolian desertification refer to various types of strategies adopted for revegetation or establishment of land cover in shifting and semishifting sand areas or shelterbelts to specifically protect against sand and dust risks in arid, semiarid, and subhumid regions.

The principal biological measures for combating aeolian desertification include the conservation of relic plant species patches and the adoption of revegetation measures, including planting trees, bushes, and grasses or even establishing soil biological crusts (www.forestry.gov.cn 2017; Li et al. 2016). Biological measures for combating aeolian desertification were first developed in the former Soviet Union for vegetation restoration and cropland protection around the Aral Sea and along the Amu River and then first introduced and developed in China in combination with straw-squared checkerboards for the protection of the Zhongwei section of the Baotou-Lanzhou railway in the Ningxia Autonomous Region in western China (Zhu et al. 1981). Then, biological measures were expanded to northern China and recently to northeastern Asia to restore degraded land.

Protecting remaining vegetation, also called relic vegetation, is of crucial biological importance in combating aeolian desertification, particularly in the large area where annual precipitation is in the range of 200–500 mm in northern China. The practical doctrine for combating aeolian desertification emphasizes protection, restoration, and utilization by moderating the relation between ecological benefit and development. Several successful ecological projects are also introduced in this chapter to describe the achievements and lessons.

9.2 Protection of Relic Vegetation

Relic vegetation refers to the different scales of vegetation remaining after habitat changes that occur due to climate and/or land use change. In most cases, relic vegetation is composed of native plant species with relatively strong adaptability to high temperature, salinity stress, drought, or cold stress.

In deserts in arid and semiarid regions or sandy lands in the transitional areas between semiarid and subhumid areas, relic plant species are often used in biological measures. In practice, materials for straw-squared or rectangular checkerboard or checkerboard-like structures for aeolian desertified land restoration are mainly from native bushes and/or grasses. One of the important means of biological measures is to protect relic vegetation, either as a living source of biological materials or as plants for revegetation in aeolian desertified lands. According to a report, there are 24 first-rated National Nature Reserves and 7 second-rated National Nature Reserves for valuable species and habitat protection in Inner Mongolia, China, for example, the Daqinggou National Natural Forest Reserve for the protection of trees in the Horqin Sandy Land (Liu et al. 1996; Liu and Zhao 2003) and the Urat *Equus hemionus* and *Haloxylon ammodendron* Nature Reserve for *E. hemionus* and *Haloxylon ammodendron* protection and for the protection of *Gymnocarpos przewalskii*, *Potaninia mongolica*, and the dryland corridor habitat along the border between Mongolia and Inner Mongolia in China too (Sun et al. 2000).

9.2.1 National Reserves for Relic Species Protection

Natural reserves or national parks are highly effective means to conserve valuable plant species. In the northwestern part of China, these national reserves and parks, in combination with natural reserves at the province, prefecture, county, and even township or individual levels, have protected many relic or economic, religious, ecological, and medical plant species. Most of these species could be adopted as biological materials for revegetation or use in squared checkerboards to fix dunes or moderate aeolian process impacts. Here, we take the Daqinggou National Natural Forest Reserve in eastern Inner Mongolia and Urat *E. hemionus* and *H. ammodendron* Nature Reserve in western Inner Mongolia as examples.

9.2.1.1 Daqinggou National Natural Forest Reserve

The Daqinggou National Natural Forest Reserve is located in the southern part of the Horqin Sandy Land (122°07′25″–122°15′42″ E, 42°43′40″–42°49′18 N″, 300–350 m.a.s.l.) in eastern China. The annual mean precipitation is in the range of 300–550 mm, and the reserve covers approximately 8183 ha, with a central area of 1322.4 ha, which is approximately 16.16% of the total area of the nature reserve. The buffer area is 2082.2 ha, which is approximately 25.45% of the total area. The revegetation area is 4778.4 ha, which is approximately 58.39% of the total area.

This reserve has 529 plant species that are well conserved, and they belong to 359 and 106 families. These species include some rare species in the semiarid regions, such as *Fraxinus mandshurica*, *Phellodendron amurense* Rupr., *Ulmus macrocarpa* Hance, *U. pumila* L., *Quercus mongolica* Fisch. ex Ledeb, *Manchurian walnut*, and *Ailanthus altissima*. Some of these plants, such as *F. mandshurica* and *P. amurense* Rupr. and *Tilia amurensis*, are on the national red list. Some plant species, such as *Artemisia halodendron*, *Caragana microphylla*, and *Salix gordjevii*,

Fig. 9.1 Natural vegetation in Daqinggou National Natural Forest Reserve: vista of the vegetation in central Daqinggou (right); vegetation in the buffer zone (left); photos taken by Xueyong Zhao

in the buffer area of this reserve have been well protected and widely used as living materials for biological fixation of sand dunes and movable or semimovable dunes.

Approximately ten mammals and ten species of birds, such as *Vulpes corsac*, *Eutamias sibiricus*, *Lutra lutra* Lin., *Canis lupus* Linnaeus, *Lepus europaeus*, *cinereous vulture*, and *Milvus Korschun*, live inside the reserve.

F. mandshurica, *P. amurense* Rupr., and *U. macrocarpa* Hance are the dominant forest vegetation in the valley, and *U. pumila L.* and *Q. mongolica* Fisch. ex Ledeb are dominant in the scattered grassland outside of the valley, as typical and native sandy land vegetation. The buffer area is dominated by *U. pumila L.* and *P. sevistris var. mongolica* and associated with *A. halodendron*, *C. microphylla*, and *S. gordjevii* (Fig. 9.1).

The habitats in the Daqinggou National Natural Forest Reserve are also valuable. As indicated by the name of the reserve, it is located in a valley, approximately 16 km long from the south to the north and 23 km wide from the east to the west and 50 m deep in the Horqin Sandy Land. The formation of the valley is unclear. One perspective is that it was formed by an earthquake, and the other consideration is it was formed by water erosion. Undoubtedly, the annual averaged air temperature is 5.5–7.0 °C, and the minimum air temperature is −29 to −30 °C in this region; however, the spring water runs along the valley banks through the valley all the year and has sustained the animals and plants for thousands and millions of years (Cao et al. 1980).

9.2.1.2 Urat *Equus hemionus* and *Haloxylon ammodendron* Nature Reserve

More often than not, rare species are distributed in specific habitats and play valuable roles in ecosystem succession and related processes. The Urat *E. hemionus* and *H. ammodendron* Nature Reserve (106°15′–108°E; 41°50′–42°27′N, 840–960 m.a.s.l., 140 km from east to the west and 22 km from north to south) was officially

Fig. 9.2 Small-scale *U. pumila* shelterbelts for natural grassland protection (upper right), *Caragana microphylla* shelterbelts for desertified land restoration (upper left), *H. ammodendron* community in the protection area (lower right), and *Cistanche deserticola* Ma community (lower left). Photos taken by Xueyong Zhao

established in 1985 and promoted as one of the national nature reserves in 2001. It is located administratively in Urat Hou County along the border between Inner Mongolia, China, and the Republic of Mongolia, with an area of 131,800 hm^2.

The reserve is in an arid region, the annual mean precipitation is 50–150 mm/a, and the potential evapotranspiration is in the range of 2300–2800 mm/a. The annual average air temperature is 6.5 °C, and the average air temperature is −14 °C in January and 34.3 °C in July. The annual mean precipitation is 90 mm and ranges from 50 to 200 mm, of which, 70–80% falls in July and August. The number of gale days is 60–70, and the maximum is 120 days. The windy and dusty days occur on approximately 25 days/a, with an average velocity of 5.4 m/s and a maximum speed of 24 m/s. The habitat is extreme for rare plants and animals.

The zonal soil is gray sand and pebble desert soil, and the azonal soil is desert soil and salinized wetland soil. Plants are mainly distributed along rivers (mostly seasonal) or even in seasonal gullies (Sun et al. 2000; Jia et al. 2017).

In addition to protecting animals of biological importance, such as *E. hemionus*, *Capra sibirica*, *Gazella subgutturosa*, and *Aquila chrysaetos*, approximately ten plants on the red list of China, such as *H. ammodendron*, *G. przewalskii*, *P. mongolica Maxim*, *Tugarinovia mongolica* Iljin, *Prunus mongolica* Maxim, *Ammopiptanthus mongolicus* Maxim, and *Cistanche deserticola* Ma, are well preserved (Fig. 9.2 lower left).

Haloxylon., *P. mongolica* Maxim, and *A. mongolicus* Maxim have been used to establish biological checkerboards to fix shifting sand in severely or relatively severely desertified lands and to revegetate gobi or sand-covered lands in arid regions in northwestern China and southern Mongolia. *C. deserticola* Ma is increasing in economic value due to its high value as a medicine for health and dietary needs, and the cropping area of this species is increasing.

G. przewalskii, *T. mongolica* Iljin, and *P. mongolica* Maxim have been used as core accompanying species for the revegetation of desertified land and for the restoration of degraded vegetation in arid regions due to their tolerance to drought, salinity, and high temperatures as native species in this region (Fig. 9.2 upper panels).

Relic vegetation protection also protects habitats and related resources (Table 9.1). These preserved resources include soils, rivers, lakes, mosaic grasslands, and bush lands as well as small wetlands in the relatively lower parts of dunes and various dune landforms, and these resources provide special or sometimes unique habitats for organisms, such as *A. chrysaetos* and *Tetraena mongolica* Maxim and *E. hemionus*, which are all on the top of the national red list of rare plants and animals (Sun et al. 2000).

Rivers and lakes are valuable to human activities, providing life-supporting services for animal husbandry and cropping industries and wetlands for moderating extreme climate change. Small wetlands play a core role in animal, bird, and plant protection, and most wetlands are severely desertified due to intensive grazing by livestock and frequent vehicle traffic through these areas. Traffic is not as heavy at this kind of site as at other sites, but given the vulnerability of this area, traffic is already too extensive.

Geomorphologically, this reserve covers gobi, mountains, and deserts, dotted with lakes and rivers running through, and it provides diversified habitats and sustains rare species due to its relatively extreme soil, water, and landform diversification.

Ecologically, the grassland, bushland, and even small patches of wetlands systematically provide preventive land cover and reduce further land degradation from aeolian desertification as well as sand and dust processes. This area is part of the ecological screen in the northern part of China.

Recent investigations have found that protected rivers, lakes, and even lower lands are areas of local people and wildlife, and overuse of resources is becoming increasingly popular. This is a practical problem for local governments and communities because many people recognize the problem but do not have a good solution.

9.3 Construction of the Oasis Forest Shelter Belt

An oasis is a fertile or green area in an arid region (such as a desert) or an area that provides refuge, relief, or pleasant contrast. Ecologically, an oasis refers to heterogeneous small patches of landscape with microclimatic traits and relatively stable

Table 9.1 Comparison of the main characteristics of the two nature reserves (Cao et al. 1980)

Daqinggou	*Precipitation (mm)*	*Temp. (°C)*	*Soils*	*Rivers (number)*	*Lakes*	*Community*	*Land*
	300–500 (min. 200 to max. 550)	5–7 (min. 5 to max. 40)	Chestnut, sand	2 annual rivers	3 perennial and 7 seasonal	*Ulmus* scattered forest and grassland	Sand land or degraded grassland
Urat	150–220 (min. 50 to max. 300)	5–7 (min. 5 to max. 40)	Gray desert soil, salinized soil, sand	45 flood gullies or seasonal rivers	5 to 9 seasonal lakes	*H. ammodendron* + *A. mongolicus*	Desert and gobi or desertified grassland
Comments	Annual	Growing season	Recently deposited soil	One flooding in a year	<1 * 1 km	Most drought-tolerant bushes and grasses	Seasonal accumulation of water in lower lands

Fig. 9.3 Bird's-eye view of cropland shelterbelts (left) and shelterbelts in intensively managed cropland (right) in the Horqin Sandy Land (photos by Xueyong Zhao)

biocommunities in a large desert matrix in arid and semiarid lands. Per its definition, an oasis is identified as a refuge for life and life-sustaining services. Oases in China are mainly distributed in the deserts in Xinjiang, Ningxia, and western Inner Mongolia and provide habitats, food, water, and related resources for approximately 70 to 90 million people and billions of livestock and wildlife species. Oases in China are characterized by extreme temperatures, climate, salinity, and drought and near-to limit land use pressure. Oasis protection is economically and ecologically imperative.

Oasis shelterbelts refer to the barrier systems of trees, shrubs, and grasses that provide protection for croplands, woodlands, river systems, transportation systems, and resident areas and project sites from wind and storm impacts in the oasis.

Oasis shelter forest belt systems are often established in and/or around oases in desert regions, especially in the desert regions with seasonal rivers running into them from neighboring or surrounding mountains. In the semiarid region, the oasis shelter forest belt is mainly composed of *Ulmus pumila*, *Salix* spp., and *Populus* spp. (Zhang et al. 2018), and in the arid region, the oasis shelter forest belt is composed of more diverse plant species, such as *Salix* spp. and *Populus* spp.; *H. ammodendron*, *P. euphorbia*, and *Tamarix* spp. occur on the salinized soils of oases (Jia et al. 2017).

The arrangement of an oasis shelter forest belt is very important, and in most cases, shelterbelts are arranged in between an oasis and surrounding desert; then, in the oasis, shelterbelts are arranged in different squares or oblong shapes that are 200 m wide and 500 m long. The long edges are laid out against the prevailing wind in the winter and in line or in a checkerboard pattern in the spring. In the drier region, shelterbelts are in patterns of 200 × 200 m, 300 × 300 m, and 300 × 500 m. Alternatively, shelterbelts are arranged along irrigation canals, roads of different slopes (Fig. 9.3 right and Fig. 9.4), or surrounding croplands (Fig. 9.3 left). In some areas, the size of the checkerboard patterns may be larger than 500 × 500 m due to irrigation canals cutting across them and well-vegetated canal banks providing protection for the oases (Fig. 9.3).

The establishment of oasis shelterbelts is related not only to protected targets but also to the abilities of different species to adapt to different soils. Thus, proper plant

Fig. 9.4 Compound shelterbelts along roads, around a village, and cropland in combination with naturally restored bush + grassland (left) in the Horqin Sandy Land and village garden in western Inner Mongolia supported by the project "Beautiful China." Photos by Xueyong Zhao

combinations and patterns are important to implement in the target area for restoration. In the northeastern part of China, shelterbelts are mainly composed of *P. simonii*, *Salix matsudana*, and *Pinus sylvestris* L. *var. mongolica* Litv. However, in the western part of China, bushes, such as *H. amondendron*, *Populus alba* var. *pyramidalis.* Bge., and *P. euphratica*, play important roles in these protective systems.

Twenty years ago, shelterbelts used 2- to 3-year-old seedlings of *Salix* spp., *Populus* spp., and *Tamarix* spp. It took approximately 3 to 5 years for shelterbelts to effectively protect croplands or fruit yards. Until approximately 2000, the seedlings used for shelterbelts were 3 to 5 years old, and the basal diameter of the main stems grew to 5 to 10 cm. This technique is called high-stem tree planting (HSTP). When starting HSTP, seedlings with root systems that are 20 to 50 cm deep are dug up, branches are cut off, and stems with root systems of approximately 150 to 200 cm are planted in scale-lined holes of 50 × 50 × 60 cm or 60× 60 × 80 cm, covered up to 5 to 10 cm from the edge with soil from the holes, and then watered (20 to 30 kg/hole) (Fig. 9.4). This planting effort can be implemented by machine or manual labor.

Water is a limiting factor for revegetation efforts in arid, semiarid, and subhumid regions. Examples of efforts are those for larger biological system construction for revegetation, such as the Three-North Shelterbelt (TNS) projects, which have been carried out in approximately 623 counties in the eastern, northern, and western parts of China since 1987, and those for smaller tree plantings for special protection purposes. In recent years, based on research on the water consumption of 13 plant species in northern China, revegetation has been optimized in terms of plant density, pattern, and species composition in the drylands in China.

9.4 Sandy Area Enclosure for Afforestation and Vegetative Sand Fixation

Sandy lands are mainly distributed in Inner Mongolia, including the Mu Us Sandy Land, Otindag Sandy Land, Horqin Sandy Land, and Hulun Buir Sandy Land. Sandy lands also include various patches of sandy land covered with sands in semiarid and subhumid areas and even in arid lands due to aeolian processes, such as deposited patches of sand along rivers, around lakes, and near villages or pastoral households due to long-term trampling by livestock.

The four sandy land areas in eastern Inner Mongolia are distributed in semiarid and subhumid regions in eastern China and were once the most aeolian desertified lands in China from the 1950s to 1980s. These four sandy land areas have been well restored since the middle of the 1980s; for example, the Horqin Sandy Land was once a desertified land area covering 42.3 thousand km^2 in 1959, and this amount increased to 51 thousand km^2 in 1987 and then decreased to 49 thousand km^2 in 2015 (Zhao et al. 2015).

Overgrazing, cropping, and fuelwood harvesting are driving forces of aeolian desertification in these four sandy land areas. However, benefiting from the semiarid climate and natural restoration processes, it is quite possible for local populations and governments to implement natural restoration by mitigating land use pressure. Enclosures (fenced areas) are a direct and effective means to promote the restoration of various aeolian desertification-impacted grasslands. The Horqin Sandy Land is the first place where enclosures were tested, and they were successful.

Enclosures are effective for all kinds or degrees of aeolian desertification and restoration and are often implemented for relic tree and/or bush vegetation restoration and bush-scattered grass vegetation restoration. Enclosures are also often used for sand fixation along transportation routes.

The Naiman Desertification Research Station, Chinese Academy of Sciences, started an experiment on the effects of grazing and no grazing on *U. pumila* scattered grasslands in the middle part of the Horqin Sandy Land in 1992 and supported a Sino-Japan joint research project. At the beginning of the project, a severely desertified *U. pumila* grassland that was 100 Chinese mu (1 mu = 1/15 hm^2) disappeared due to overgrazing and was divided into four parts, one as a control, one as a slightly grazed treatment, one as a moderately grazed treatment, and one as a severely grazed treatment. Until 1996, after 5 years of treatments, vegetation in the control, slight grazed treatment, and moderately grazed treatment was gradually restored, the plant number increased from three to five species to seven to nine species, and the *U. pumila* returned. The severely grazed treatment plot changed into shifting dunes, with eroded soil patches, dunes, and small patches of overgrazed grassland, and was defined as the most severely desertified land due to overgrazing and aeolian erosion. Since 1993, the whole experiment has been enclosed without grazing and restored. The experimental site became a *U. pumila* scattered grassland again, near native grassland status (Fig. 9.5).

Fig. 9.5 Revegetation for shifting sand dune fixation and cropland protection (left) and natural restoration of sandy land by an enclosure (right) at the Naiman Desertification Research Station. Photo taken by Xueyong Zhao

In practice, enclosures have been widely adopted as an effective means to promote the restoration of desertified lands with different aeolian impacts and land changes. The results could be totally different without proper management. Successes and/or failures can be found in China, Mongolia, and other countries.

9.4.1 Enclosure for Afforestation

In combination with enclosures, afforestation ensures high restoration success levels in semiarid regions in northwestern China and southeastern Mongolia. Afforestation refers to tree and bush plantings to restore grasslands.

Large enclosures and afforestation have been carried out since 1987, when the "National Ecological Three-North Forest Protection Project" was implemented. Before 1996, most of the sites for afforestation were along rivers (irrigation systems) and roads and around croplands, villages, and houses and yards as well as between the transitional areas between oasis and surrounding desert or desertified land that mainly occurred due to aeolian processes. The planted tree density was approximately 2$\times$ 3 m, and the bush density was approximately 2 $\times$ 2 m. After 1996, the planted tree density decreased to 3 $\times$ 3 m or 3 $\times$ 4 m, and the bush density decreased to 2 $\times$ 3 m or 3 $\times$ 3 m. The drought from 1999 to 2009 pushed governments, researchers, and farmers to reduce the density continuously to 3$\times$ 5 m to 4 $\times$ 5 m, or even 5 $\times$ 5 m, according to the water availability and water consumption rates of individual trees or the population of trees. Then, the planted areas were protected with enclosures or marked by GPS sites for natural restoration with bushes and grasses.

Plant species for afforestation included *Populus* spp., *Salix* spp., and *Larix* spp. in the beginning and then shifted to *Populus* spp. and *Salix* spp. in the sandy lands with different degrees of desertification (Fig. 9.5), and *Larix* spp. were in the hilly or mountainous region. After 2000, the drought from 2001 to 2010 indicated the importance of changing the plants and density of afforestation (Zhao et al. 2009).

Due to some existing patches of relic vegetation, fences and GPS-identified markers were used as protection systems to encourage near-natural vegetation restoration. With this mixture of protection systems, vegetation can be restored sustainably. Enclosures are a direct and simple protection means and have played an important role in the past in enhancing natural or near-natural restoration of plant communities. The disadvantages of enclosures, such as blocking animal movements and transportation routes, have also been discussed.

9.4.2 Restoration

Restoration is (1) the act of restoring or the state of being restored to a former or original condition, place, etc.; (2) the replacement or giving back of something lost, stolen, etc.; (3) something restored, replaced, or reconstructed; or (4) a model or representation of an extinct animal, a landscape of a former geological age, etc.

However, aeolian desertified land or degraded ecosystems can never be restored to their status before aeolian desertification or degradation occurred because the climate, soil, water, and plant community have changed. Here, the definition of restoration consists of the acts or processes based on scientific research that rehabilitate the composition or structure of targeted damaged ecosystems to a status near to that before the impacts occurred in the short term or to a status adapted to the current synergistic condition of the climate, soil, and water. Then, natural or near-natural restoration can occur to restore desertified lands, including aeolian desertified woodland, grassland, and cropland.

The Horqin Sandy Land is as an example of restoring an aeolian desertified grassland to *U. pumila* scattered grassland vegetation. Before planting the trees, a shelterbelt-like facility was established with branches 50 and 60 cm long from *A. halondendron*, *C. microphylla*, and *S. gordgevii*, planted on shifting or semishifting dunes or on fluctuating gentle sandy lands in a pattern of 1 × 1 m or 2 × 2 m to form a checkerboard system in the rainy season from July to August. The branches were buried into the sandy soil for approximately 20 to 30 cm deep, and the remaining parts of the branches were maintained at 30 to 20 cm above the land surface.

The rainfall was generally enough for the branches to grow as long as the rain was not less than 50% of the normal rainfall amount in the growing season, and then, the grasses and semibushes, bushes, and even *U. pumila* naturally invaded and grew (Fig. 9.5). A mixed protection system was established, and the desertified grasslands were restored. This kind of tree planting, in combination with grass and bush revegetation, is becoming more accepted in ecological projects because it is cost-saving, effective, and sustainable.

In summary, near-natural restoration can be divided into two key steps. First, the shelterbelt-like barriers with the local plants are established, and second, the regrowth of local grasses, semibushes, bushes, or even trees plays a role in near-natural vegetation restoration.

9.4.3 Sand Fixation with Biological Measures

Sand-fixation measures can be broadly divided into two categories: biological measures and nonbiological measures. The nonbiological measure includes chemical measures and physical measures (mechanical measures). The nonbiological measures can be combined with biological measures at an earlier stage of sand fixation to prepare for later plant revegetation.

Either biological or nonbiological measures can be implemented in severely desertified regions or for larger shifting dune fixation along the leeward sides of rivers and lakes. Moderately and slightly desertified land could be naturally restored, particularly in the semiarid sandy lands in the eastern part of China.

Initially, nonbiological measures, such as straw checkerboards in the shifting sand area, were introduced from the former Soviet Union in the 1950s–1960s to reduce aeolian sand movement and provide a protection system for plant colonization on the shifting sandy lands or dunes for later revegetation. Then, based on the protection provided by the nonbiological measures, selected plants were planted inside checkerboards of different sizes. The planted plants could be used as auxiliary measures to enhance the protection effects provided by the nonbiological measures and later became the predominant protection system (Fig. 9.6).

With the development of science and technology, additional new materials and measures for aeolian desertified land restoration and revegetation have been produced and applied. Since 2005, a mixture of decomposed organic wastes with clay, sand, and chemical fertilizers has been developed by researchers from the Naiman Desertification Research Station, Chinese Academy of Sciences. The wastes from leftover feed grasses and crop residues and human activities can be collected, and

Fig. 9.6 Dune fixation systems of *P. simonii*, *C. microphylla*, and *P. sylvestris* var. *mongolica* in the Horqin Sandy Land (left) and multifunction shelterbelts for oasis cropland protection from shifting sands in western China (photos from Tao Wang)

Fig. 9.7 Decomposition of residual materials from agricultural activities (left) and application of the decomposed materials on shifting dunes in checkerboards for dune fixation (Wang et al. 2016). Photos by Xueyong Zhao

these materials are then decomposed with the addition of fiber decomposers infiltrated from sandy land soil or even dune soil (Fig. 9.7).

When the material decomposes or semidecomposes, it is mixed with topsoil and covered with nylon nets. If there is rainfall, then this material will be thoroughly mixed with the soil and result in improved water holding capacity, and biological soil crust will gradually cover the shifting sand dunes. This cover provides the opportunity for plants to colonize the dune area quickly and produce households and villages with less sand.

To evaluate the quantitative nutrient return in the cropland soil, soil organic carbon (SOC) and total nitrogen (TN) were collected and analyzed at the end of the field experiment. The results showed that maize straw turnover significantly increased SOC. SOC was significantly higher at 13.92% and 34.08% in the soil under net bags without (ND) and with (WD) selected decomposers than in the control soil (CK), respectively. TN did not show a statistically significant difference between ND and CK. However, TN was significantly higher in WD than in ND and the CK. The soil organic carbon and nitrogen ratio (C:N) increased significantly by 13.06% and 16.80% in ND and WD, respectively, compared to that in the CK. The C:N ratio did not show a significant difference between ND and WD (Fig. 9.8). The results indicated that our selected decomposing fungi could not only accelerate the straw turnover rate but also increase the soil organic carbon and nitrogen content and potentially promote soil fertility and soil health on the sandy cropland.

9.5 Conclusion

Establishing protection systems is an effective way to enhance degraded vegetation restoration, but practically optimized land management could produce long-term restoration results. In the Horqin Sandy Land, the local governments engaged local

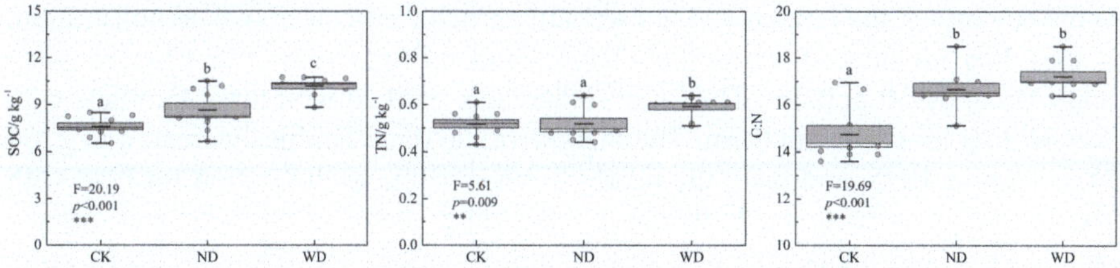

Fig. 9.8 Influence of straw turnover (with and without efficient cellulose decomposers) on soil organic carbon (SOC), total nitrogen (TN), and the carbon and nitrogen ratio (C:N) in the field decomposition experiment. *CK* bare soil without straw, *ND* soil under straw net bags without selected decomposing fungi, *WD* soil under straw net bags with selected decomposing fungi (Wang et al. 2016)

Fig. 9.9 Two Mongolian households living in shifting sand areas and protecting their house and yards with trees, bushes, and restored grasslands (left) and a local intercropped area with a water-saving irrigation system for a cropland in Inner Mongolia, China (photos by Xueyong Zhao)

interest in combating Aeolian desertification and adopted the "Law of Combating aeolian Desertification of the People's Republic of China" and "Law of Forestry of the People's Republic of China," with increasing transparency and in-depth dissemination of research findings, to ensure the location population understood the importance of improving the environment and growing economy as beneficial, providing the win-win result of environmental improvement and economic development for the locals (Fig. 9.9).

Since 1985, researchers have suggested that local governments and village committees reduce rain-fed cropland and proportionally increase irrigated cropland to sustain cropland productivity and adopt livestock yard feed practices to increase income levels. These suggestions were gradually accepted and resulted in good economic outputs. By 2000, aeolian desertification was reversed in 20% of the total Aeolian desertified land. The annual income increased to 6000 CY/person in 2019, which was 1000 CY/person less than that in the demonstration village (Zhao et al. 2009, 2015).

The Naiman Desertification Research Station, Chinese Academy of Sciences, has established a model village called Yaoledianzi village in 1985 to demonstrate research findings for combating Aeolian desertification and cropland management.

At the beginning, the farmers did not want to participate in the above demonstration, as they had no interest when their annual income was only 178 Chinese yuan (approximately 30 dollars). Then, the researchers worked with three families, who were willing to work with them. The income of the three families increased to 2100 CY/person in 1995, and then, the nearby farmers came to the station for help with crop species introductions, fertilization (including chemical fertilizer and organic manure use), soil management, water-saving irrigation, and livestock health care. Improved cropping management (Fig. 9.7 right) and animal care greatly increased income and enthusiasm for cooperating with the researchers. Income levels increased to 10,000 CY/person in 2019 (Zhao et al. 2009, 2015).

An investigation found that small scale of eco-economic sites have been established as models for the national "Green Great Wall" program to prevent aeolian desertification and dust storms, and these sites run through the middle part of Mongolia from the eastern grassland to the western gobi and desert (Nyamtselen 2014). In addition, the investigation also identified some failures in this program due to misuse of plants for revegetation.

China provides many successful ecological stories for inside and outside this area, but the techniques supporting the stories should be improved in China to support other countries, and the countries supported must know how to use these techniques, their advantages in China, and their challenges when transferred to targeted countries.

When combating aeolian desertification or any kind of land degradation, it is imperative for researchers, decision-makers, or any stakeholders related to this issue to adopt the attitude and understanding that it is very difficult to win the battle against land degradation caused by climate change without extra materials or energy inputs, and we can combat aeolian desertification mainly caused by human activity with properly implemented measures, particularly with the active participation of local people and local governments (Zhao 2012).

China has made great progress in combating aeolian desertification and at the same time faced the great challenge of water availability reductions. To some extent, it is predictable and even likely that this challenge will become more threatening with climate change and growing land use pressures. China, including Mongolia, must develop more efficient measures, either biological or nonbiological, to win the war against aeolian desertification and lift local people out of poverty, and then, we will have a win.

References

Cao XS et al (1980) Vegetation in daqinggou nature reserve of Inner-Mongolia. Inner-Mongolia Peoples' Press, Hohhot

Jia KF, Cui ZR et al (2017) Atlas of Urat Houqi county in Bayannaoer prefecture. Inner-Mongolia People's Press, Hohhot

Li XR, Zhang ZS, Liu YB et al (2016) Fundamental ecohydrology of ecological restoration in Sandy Desert regions. Science Press, Beijing, China

Liu XM, Zhao HL (2003) Research in comprehensive prevention of the eco-environment in Horqin Sandy Land. Gansu Science and Technology Press, Lanzhou
Liu XM, Zhao HL, Zhao AF (1996) Wind-sandy environment and vegetation in the Horqin Sandy Land. Science Press, Beijing
Nyamtselen M (2014) Land degradation and desertification process in Mongolia
Sun CH, Jia KF et al (2000) Investigation of the Urat wild donkey and H. ammodendron nature reserve. Chinese Forestry Press, Beijing
The fourth Bulletin of desertification and aeolian erosion in China (2017) http://www.forestry.gov.cn
Wang SK, Zuo XA, Zhao XY et al (2016) Responses of soil fungal community to the sandy grassland restoration in Horqin Sandy Land, northern China. Environ Monit Assess 188(1):1–13
Zhang CM, Zhun HM, Shen ZX et al (2018) Atlas of Naiman county in Horqin Sandy Land. Inner-Mongolia People's Press, Hohhot
Zhao HL (2012) Desert ecology. Science Press, Beijing
Zhao HL, Zhao XY, Zhang TH et al (2009) General restoration ecology. Scientific Press, Beijing
Zhao XY, Wang SK, Luo YY et al (2015) Toward sustainable desertification reversion: a case study in Horqin Sandy Land of northern China. Sci Cold Arid Region 7(1):23–28
Zhu ZD, Chen GT (1994) Sandy desertification in China. Science Press, Beijing
Zhu ZD, Liu S, Di XM (1981) Regionalization of desertification process and prevention in the northern part of China. China Forestry Press, Beijing

Chapter 10
Integrated System to Combat Aeolian Desertification and Disasters

Weimin Zhang, Lihai Tan, Kecun Zhang, and Xian Xue

Abstract An increasing number of projects have been established in northern China to prevent aeolian desertification and related disasters. Although many funds, laborers, and techniques have been dedicated to these projects, the effects of these projects in terms of preventing and controlling aeolian disasters have not been satisfactory. The reason for this unsatisfactory effect is that these projects ignore the fundamental principles of aeolian desertification and disaster prevention. This chapter introduces two successful cases in detail to understand the above problem. One case occurred in Shapotou, southeast of the Tengger Desert. An integrated system controlling aeolian disasters was established in the 1950s to protect railways across high-dune fields. The system primarily consisted of a sand-blocking measure combined with sand-fixing measures. Another case occurred on the top of Mogao Grottoes in Dunhuang. An integrated system controlling aeolian disasters containing sand-blocking measures combined with sand transport measures was established at the end of the last century to prevent wind erosion and sand burial in the Mogao Grottoes. This chapter clarifies the importance of comprehensive and nature-based

W. Zhang · L. Tan · K. Zhang
Dunhuang Gobi and Desert Ecological and Environmental Research Station, Northwest Institute of Eco-Environment and Resources, Chinese Academy of Sciences, Lanzhou, Gansu, China

Key Laboratory of Desert and Desertification, Chinese Academy of Sciences, Lanzhou, Gansu, China

Northwest Institute of Eco-Environment and Resources, Chinese Academy of Sciences, Lanzhou, Gansu, China

X. Xue (✉)
Northwest Institute of Eco-Environment and Resources, Chinese Academy of Sciences, Lanzhou, Gansu, China

Key Laboratory of Desert and Desertification, Chinese Academy of Sciences, Lanzhou, Gansu, China

Drylands Salinization Research Station, Northwest Institute of Eco-Environment and Resources, Chinese Academy of Sciences, Lanzhou, Gansu, China
e-mail: xianxue@lzb.ac.cn

T. Wang et al. (eds.), *Combating Aeolian Desertification in Northeast Asia*,
Ecological Research Monographs, https://doi.org/10.1007/978-981-16-9028-0_10

measures through these two successful cases and provides some instructions for preventing and controlling aeolian disasters.

Keywords Integrated system · Mogao Grottoes in Dunhuang · Nature-based measure · Shapotou · Aeolian disaster control

10.1 Introduction

Sustainability is the fundamental prerequisite for aeolian disaster control and aeolian desertification prevention. Based on information from northwestern China, two factors are vital to ensuring the sustainability of prevention projects. First, an integrated system, including different measures with a variety of functions, is necessary. A single sand-control measure has minimal economic costs and can achieve individual results in the short term. However, from a long-term development point of view, a single measure's inherent shortcomings will gradually become prominent. The service life and prevention effectiveness will be significantly reduced. Second, an integrated system's scientific design, including the measurement arrangement, materials, size, porosity, and other indicators, should be consistent with the local conditions. Two successful integrated systems in northwestern China demonstrated these factors well. One system is in Shapotou, and it protects the railway from aeolian disasters. The other is in Mogao Grottoes, and it protects cultural heritage from aeolian disasters.

The Shapotou case presents a principle of self-recovery based on the integrated system. From the view of ecological environment construction, a single mechanical protective measure is similar to "a temporary solution but not a permanent cure." The most fundamental measure for the control of aeolian disasters is sand fixation by plants, which has a self-repairing function and improves the ecological environment. In Shapotou, the primary function of an Integrated System that Combats Aeolian Disasters (briefly as ISCAD in the chapter) has evolved from that of a purely artificial system to that of a natural fusion of artificial and natural systems. This evolution not only enhances the system functions but also improves its stability. Plants in sand-fixing areas were initially planted vegetation. With the development of plant succession, the pioneer species gradually withdrew. Through self-renewal, the vegetation community formed a relatively stable *Caragana korshinskii-Artemisia ordosica* vegetation community. Atmospheric dust deposition and the formation of soil crusts could progressively strengthen system stability and durability. The ISCAD of the Shapotou Railway has been tested for more than 50 years. The combination of sand blocking and transportation, together with the plant protection system without irrigation in semiarid areas, can ensure safe railway transportation. Even during a strong sandstorm, the railway can avoid being buried by sand.

The case of Mogao Grottoes presents the importance of adopting measures that are based on local conditions. After long-term evolution, the surface at the top of the Mogao Grottoes was in a relatively stable gravel-cover state. However, some inappropriate measures were implemented. The original erosion-deposition balance on the gravel-cover surface was interrupted, resulting in the overdeposition of sand. These unsustainable measures finally led to the failure of aeolian disaster control and even resulted in additional negative results. The failure occurred because the local wind field and gravel-cover function were ignored. The current ISCAD at the top of the Mogao Grottoes mainly seeks to recover the balance of erosion and sedimentation on the surface. Based on the local wind regime, the ISCAD is dominated by sand stabilization, combined with sand blocking and sand transport.

Based on the above facts, ISCAD with sustainable design can maximize its function. This chapter introduces two cases in the Shapotou and Mogao Grottoes to understand how to set up an integrated system that can effectively combat aeolian desertification and its related disasters.

10.2 Integrated System Combating Aeolian Disaster Along a Railway in Shapotou

The railway from Baotou to Lanzhou (BLR) has a total length of 990 km. The Zhongwei-Gantang Section (ZGS) of the BLR traverses the southeastern edge of the Tengger Desert, with a length of more than 40 km. The Shapotou section is the central part of the ZGS, with high crescent and net dunes (Liu 1987). Therefore, during the construction and early stage of BLR operation, the most crucial issue was addressing wind erosion and sand burial of the railway. In the early 1950s, the Chinese Academy of Sciences, the Chinese Ministry of Railways, and the Chinese Ministry of Forestry began to systematically research aeolian disasters along railways. The researchers designed an integrated railway protection system characterized as "focusing on dune fixation, combined with sand blocking," and "biological measures combined with engineering measures." Sixty years after the system was completed, the dunes near the railway have been completely fixed. A green belt with a width of 10 km has formed along both sides of the railway. Trains are protected from wind and sand disasters.

10.2.1 Aeolian and Environmental Characteristics in Shapotou

Shapotou is located on the southeast edge of the Tengger Desert (Fig. 10.1). Dune fields are the dominant landscape feature. The area is characterized by droughts and wind. The average annual air temperature is approximately 10.0 °C, and the annual average precipitation is 186.2 mm (1956~2012) (Li et al. 2014). The annual average potential evaporation is approximately 3000 mm. The groundwater table is below 10 m and thus cannot be used by plants. Under the dry sandy layer with a thickness of 3–20 cm, the moist layer exists with a volumetric water content of approximately

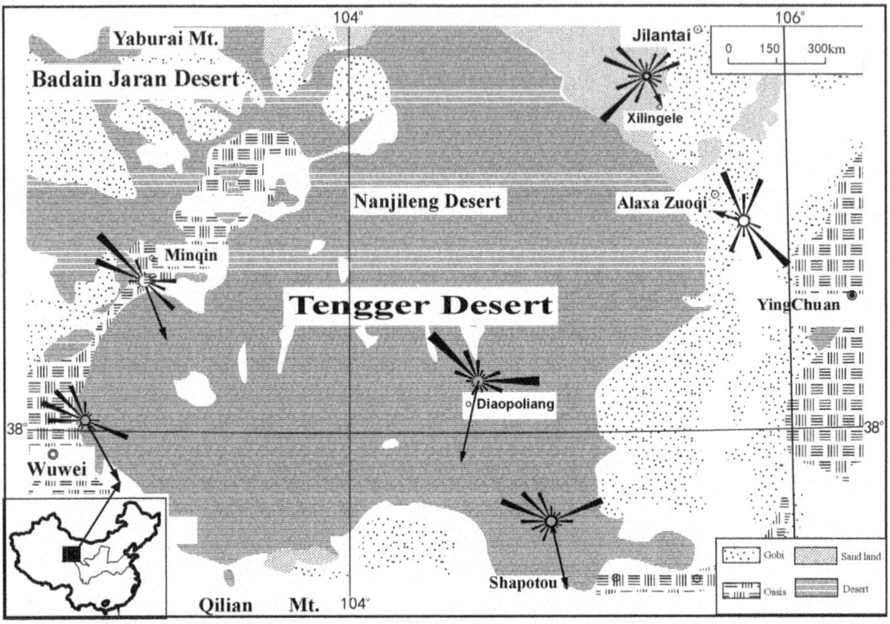

Fig. 10.1 Location of Shapotou in the Tengger Desert and China and the local wind regime (from KC Zhang et al. 2012)

2–3%. Depending on soil water and infrequent precipitation, native species such as *Agriophyllum squarrosum*, *Pugionium cornutum*, *Stilpnolepi scentiflora*, *Artemisia sphaerocephala*, and *Hedysarum scoparium* grow on dunes with an average coverage of 1–2%.

The velocity of sand-driving wind at a height of 2 m is approximately 4.5 m s^{-1}. The annual average frequency of sand-driving wind ranges from 7.2% to 12.3%. There are three dominant sand-driving wind directions. They are W-WNW, E-ENE, and S-SSW (Fig. 10.1). The westerly wind is strongest followed by the easterly wind. The two groups of wind form dense networks of dunes and dune chains. In winter and spring, the westerly wind is stronger than the easterly wind, causing the dunes to move toward the southeast by 2–4 m per year. In summer, the dune ridge swings to the northwest under the combined influence of the northeast, southeast, and southwest winds. The direction of the dune ridge swing has a 30–60° angle with the railway, which produces windblown sand. The windblown sand destroyed the railway by eroding the shoulder, abrasing the rails, and covering the railbed.

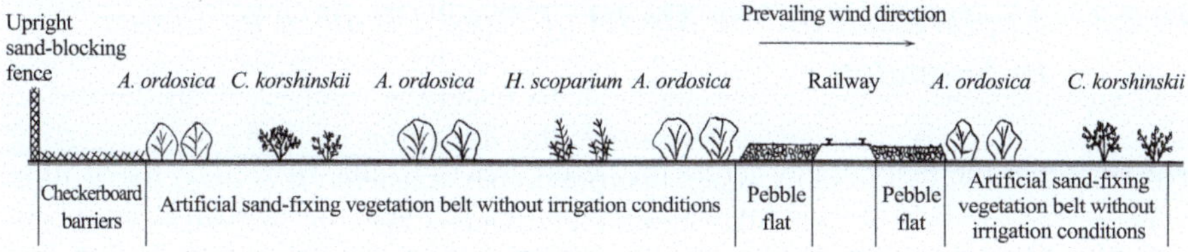

Fig. 10.2 Cross-section of the integrated system controlling aeolian disasters along the railway in Shapotou (provided by Wang 2011)

10.2.2 Integrated System Combating Aeolian Disaster Along the Railway in Shapotou

The ISCAD in Shapotou consists of three main measures (Fig. 10.2). There are sand-blocking zones, sand-fixing zones, and sand-transporting zones in sequence along the windward direction. The sand-blocking zone contains fences with a height of approximately 1 m. The fence materials include chaste twigs, poplar branches, corn stalks, and bamboo. The sand-fixing zone is the main body of the ISPAD and contains wheat-straw checkerboard barriers and xerophytic shrubs. The wheat-straw checkerboard barriers are 1 m × 1 m and 1 m × 2 m. The shrubs include *Caragana korshinskii* Kom, *Artemisia ordosica* Krasch., *Hedysarum scoparium*, *Caragana microphylla*, *Calligonum arborescens* Litw., and *Atraphaxis bracteata* A. Los. Two-year-old seedlings of these shrubs were mixed and planted inside the checkerboard at a distance of 1 m × 2 m and 2 m × 3 m in 1964, 1973, 1982, and 1992. The total length of the sand-fixing zone along the railway is approximately 16 km. The sand-transporting zone is a gravel-cover platform and is located between the sand-fixing zone and the railway.

The structure of the ISCAD reflects its comprehensive functions, including sand blocking, sand fixation, and sand transportation. The width of the ISCAD was designed according to the direction and strength of the sand-driving wind. The prevailing wind is the northwesterly wind, followed by the northeasterly wind and a south-south westerly wind. The northwesterly and northeasterly winds threaten the northern area of the railway, accounting for 55% of the annual sand-driving winds. The southern area of the railway is threatened by southerly winds, which account for 26% of the annual sand-driving winds. The theoretical widths of the ISCAD are approximately 700 m and 300 m in the northern and southern areas of the railway, respectively. The actual width changes with the sand sources and dune movement speed. The ratio is consistent with the frequency and intensity of different winds. In the early period, it was vital to maintain the function of the sand-blocking zone. After 10 years, the shrubs grew, and the importance of the sand-blocking zone gradually decreased. Throughout the ISCAD, the three measures were arranged continuously without any space to prevent the sand inside the system from eroding and moving.

10.2.3 Exploration on the Model of Sand Fixation by Plants in Shapotou

The BLR was the first railway across the desert area of China. How to fix the drifting sand on both sides of the railway was an important problem that needed to be solved urgently at that time. The Shapotou area on the southeastern margin of the Tengger Desert is a semidesert area characterized by high temperatures, drought, and a windy climate. All dunes are tall, mobile trellis dunes (Zhao 1988a). The groundwater is tens of meters deep and cannot be used by plants. The only main ecological characteristic is that the sand layer contains 1.3–2.3% effective water. However, plant sand fixation is a permanent sand fixation measure, which is the fundamental measure to prevent and control sand hazards. To address this problem, the Shapotou Desert Scientific Research Station was established in 1955 to perform positioning research on plant sand fixation under unirrigated conditions, with the mechanical sand fixation measure of a straw checkerboard barrier combination.

The experiments with sand fixation by plants in the first few years and the large-scale plantings the following years show that the effect of sand fixation by plants was far less than that of semiarid areas and desert oasis areas in China. The coverage of planted vegetation in 4–5 years was generally less than 10–15%, and wind erosion on the sand surface could not be controlled; thus, it needed to be protected by an alternating straw checkerboard barrier. In the early 1960s, it seemed impossible to fix trellis dunes with plants along the railway lines in the Shapotou area and to address sand damage in this section of the railway. The consideration at that time was to either rely on straw checkerboard barriers for a long time or build forest belts with diversion irrigation from the Yellow River. However, some of the older generations of scientists, such as Kezhen Zhu and Shene Liu, had come to SDRES many times to check the research and had great hopes for sand-fixing plants, and they thought that an experiment with sand-fixing plants with no irrigation was entirely possible; thus, it was important to persist (Zhao 1988b).

In the 30 years since the establishment of the station, dozens of trees, shrubs, and half-shrubs, including *Pinus Tabulaeformis*, *Pinus sylvestris* var. *mongolica* L., *Ulmus Pumila*, and *Robinia pseudoacacia*, have been planted, and all have died due to the unsuitable conditions of drifting sand. The local *Populus simonii* and *Populus nigra* L. var. *italica* also experienced poor survival rates and growth due to insufficient water and nutrients in the sand dunes. Even the few plants preserved at the sandy foot site with better ecological conditions did not grow well. The local adaptable native *Elaeagnus angustifolia* also grew into shrubby shapes. Other shrubs, such as *Lespedeza bicolor* Turcz. and *Amorpha fruticosa*, and semishrubs, such as *Artemisia halodendron*, also were not successful. After years of planting tests, the most suitable species of sand-fixing plants on the trellis mobile dunes were *Hedysarum scoparium*, *Caragana Korshinskii*, *Caragana microphylla*, *Calligonum mongolicum*, *Calligonum caput-medusae*, *Hedysarum laeve*, *Atraphaxis bracteata* and *Salix gordejevii*, *Artemisia ordosica*, and *Artemisia sphaerocephala* (Wang 1988).

After the determination of sand-fixing plants, a sustainable configuration and density were needed to achieve the expected benefits of sand fixing. At the beginning of the experiment, considering that minimal plantings could not control drifting sand, the chosen density of planting was too high, leading to insufficient water supply from the sand dunes and then poor plant growth (Wang 1988). Thus, during a dry year, many plants died. However, through many experiments, the semishrub and shrub strip-planting method was finally adopted, and a certain area was left in the space between the belts; in addition, the sustainable density was artificially adjusted according to the plant years. Finally, good results were obtained. Research and field practice have shown that it is feasible and effective to establish an integrated sand-control system of plant sand fixation in trellis dunes on both sides of railways without irrigation, and these systems mainly contain sand-fixing measures combined with sand-resistance measures.

10.2.4 Effect of an Integrated System Combating Aeolian Disaster Along the Railway in Shapotou

The ISCAD in Shapotou has had several important influences. First, the "fixing is primary, fixation combined with blocking" integrated system preventing aeolian disasters was proposed and applied in China. The concept of comprehensive protection has become the basic principle for combating aeolian desertification and aeolian disasters in China. Second, the checkerboard barrier size of 1 m $\times$ 1 m used in the ISCAD presents protection and economic benefits. Thus, it has become a widely promoted model in China. Third, the coverage of rain-fed shrubs used to fix dunes is suggested to be no higher than 15%, which is the best in models for arid areas. Experience in Shapotou proves the feasibility of achieving ecological restoration by planting vegetation in sandy areas (Li et al. 2016, 2017). The ISCAD model for Shapotou provides crucial guidance for the prevention and control of aeolian desertification and disasters in arid regions around the world.

Under the protection of the ISCAD, aeolian disasters along the railway in Shapotou have been effectively controlled over the past 60 years. The most significant change has been the reduction of aeolian action (Qu et al. 2007). In the sand-blocking zone, the porosity of the fence is 30%–40%, the height is 0.8–1.0 m, and the angle of the intersection with the primary wind direction is 70–90°. This kind of fence can block 5.5 m^3 of sand from the primary wind direction per year, which is 78% of the maximum possible sand transport of 6.5 $m^3 . m^{-1} . a^{-1}$ (Ling et al. 1984). As the forefront barrier, the fence blocked a large part of the sand, which is deposited into the sand-blocking zone. The significantly decreased sand transportation effectively weakened the movement of dunes and protected the sand-blocking zone. The experiments in the other regions showed that dunes moved 20–50 m per year and buried the sand fixation zone without the protection of the sand-blocking zone. The stability of the sand-blocking zone guarantees the safety of the vegetation zone and is beneficial to the growth of plants.

Soil moisture is one of the main factors that control plant succession. Changes in soil hydrological processes drive the succession of planted vegetation. Changes in vegetation composition, structure, and spatial distribution patterns are also the result of long-term adaptation to water processes (Li et al. 2014). Plants in the sand-fixing area were initially not naturally occurring vegetation. With the development of the plant succession process, some pioneer species with high water consumption abilities, such as *Calligonum mongolicunl*, *Caragana microphylla*, and *Polygonum sibiricum*, gradually withdrew. The drought-tolerant plants *Caragana korshinskii*, *Hedysarum scoparium*, and *Artemisia ordosica* increased. These drought-tolerant plants have low coverage and self-renewal ability, so they become a relatively stable *Caragana korshinskii-Artemisia ordosica* sand-fixing vegetation community that is dominated by grass under water stress. The coverage of planted shrubs increased from 3% in the second year to 35% in 15 years and then gradually decreased to 9% after 45 years (Fig. 10.3). Finally, the shrub coverage remained stable at 8~10%. The coverage of grasses decreased from 35% at the beginning to 12% after 15 years and to 30–45% after 50 years (Li et al. 2014).

In the sand-fixing zone, the combined semiupright barriers and vegetation effectively inhibited wind erosion and stabilized the surface. Blocked by barriers and vegetation, dust and nutrients deposit on the sand surface and gradually formed a stable physical crust. The appearance of physical crusts creates a suitable habitat for herbaceous plants to settle. Annual naturally occurring plants such as *Eragrostis cilianensis* Link ex Vign. Lut., *Setaria viridis* Beauv., *Bassia dasyphylla* O. Kuntze, and *Corispermum patelliforme* Iljin invaded and increased. Subsequently, cryptographic plants such as cyanobacteria, green algae, lichens, mosses, and microorganisms gradually appeared and increased. These plants produce mycelium, pseudoroots, and exudates, which can fix the soil particles and form biological crusts on the surface. The biological crust is compact and can resist strong wind erosion. In Shapotou, after 10 years of dunes being fixed, a crust dominated by cyanobacteria formed on the sand surface with a coverage of 10% to 30%. After 40 years, a mixed crust of lichen and moss appeared, with a coverage of 30% to 40%. After 50 years, the crusts were dominated by moss and lichens, with a coverage of 50% to 60% (Fig. 10.4) (Li et al. 2014).

The reduction in planted vegetation coverage does not affect the fixation of mobile dunes because the dune surface is consolidated by the crust layer, and the function of the ISCAD is not weakened. The original planted vegetation with a single community structure evolved into a complex multilayered structure and multifunctional vegetation. Restoring the biodiversity of the fixed dunes caused the original planted sand-fixing vegetation system to evolve into a stable desert ecosystem similar to those in the same climatic zone (Li et al. 2014). Therefore, the sand-control system adjusted to the natural environment, and the system functioned stably to ensure the safe operation of the railway.

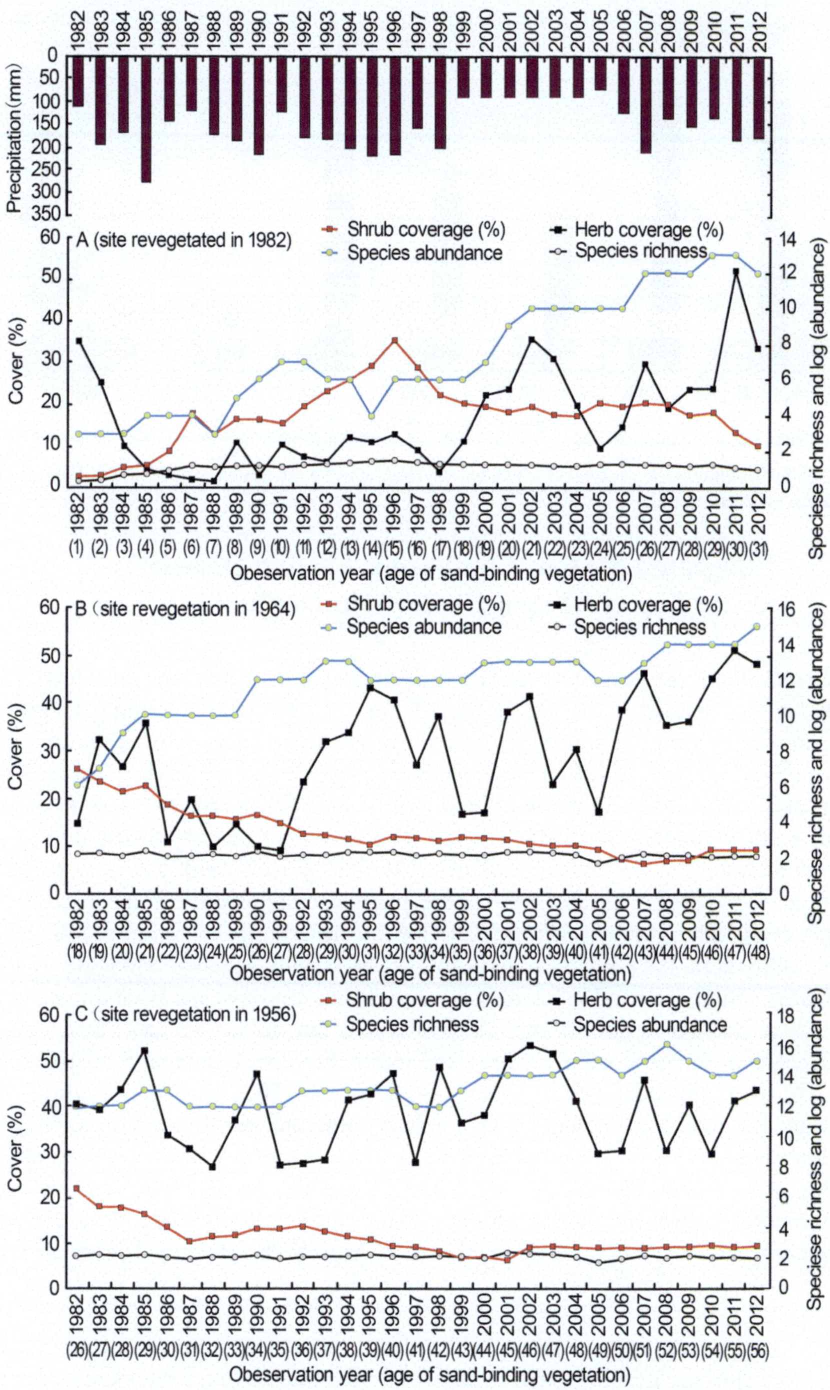

Fig. 10.3 Coverage of shrubs and grasses that are sand-fixing plants in different years, the changes in the sequence of natural settlement species, and their relationship with annual precipitation (according to Li et al. 2014)

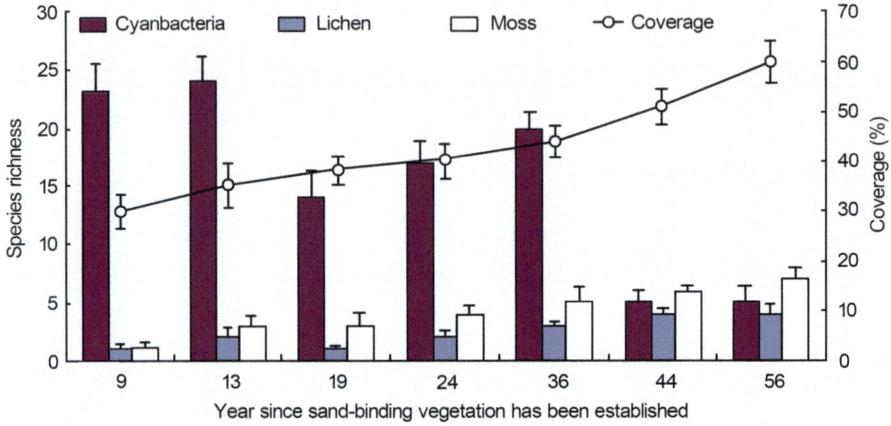

Fig. 10.4 Change in the biological crust and cryptographic plant abundance with the fixation time of dunes (according to Li et al. 2014)

10.3 Integrated System Combating Aeolian Disaster at the Top of Mogao Grottoes in Dunhuang

Mogao Grottoes are known as the "treasure of oriental art" and "the museum on the wall." It was named a "World Heritage" site by UNESCO according to the six selection criteria of the world heritage program. The Mogao Grottoes were excavated on the vertical cliff of an alluvial terrace on the west bank of the Daquan River (Fig. 10.5). The cliff is in the north-south direction. The entire cliff is 1600 m long from north to south and 10–45 m wide. The caves are densely distributed in the upper, middle, and lower layers of the cliff. The gravel, sand, and clay particles account for 70%, 25%, and 5% of the particle composition of the rock layer, respectively. The ground surface at the top of the grottoes is composed of gravel gobi, sandy gobi, flat sand, and sand Mingsha Mountain. Mingsha Mountain is a tall and large compound sand hill, which is a significant source of sand that threatens the cave area, and gravel gobi is another sand source that affects the cave area.

The top of the grottoes is a windy and multiwind-direction area. Statistics show that the average annual wind velocity is 4.3 $m.s^{-1}$. The local sand-driving wind velocity is 5 $m.s^{-1}$ at a height of 2 m. Winds with velocities lower and higher than the threshold sand-driving wind velocity account for 63.7% and 36.3% of all the winds, respectively. The primary winds in this area are northeasterly, northwesterly, and southerly winds (Fig. 10.5). Of these winds, the southerly winds occur the most frequently and account for 49.5% of all winds. Among all of the winds with different directions, northwesterly winds account for 25.3%, and northeasterly winds account for 24.5%.

Westerly winds have a strong ability to transport sand from sand mountains. Easterly winds are generally free of sand but are limited by rare sand sources in the gobi area, thus presenting a vital erosion function. In spring, the frequency of

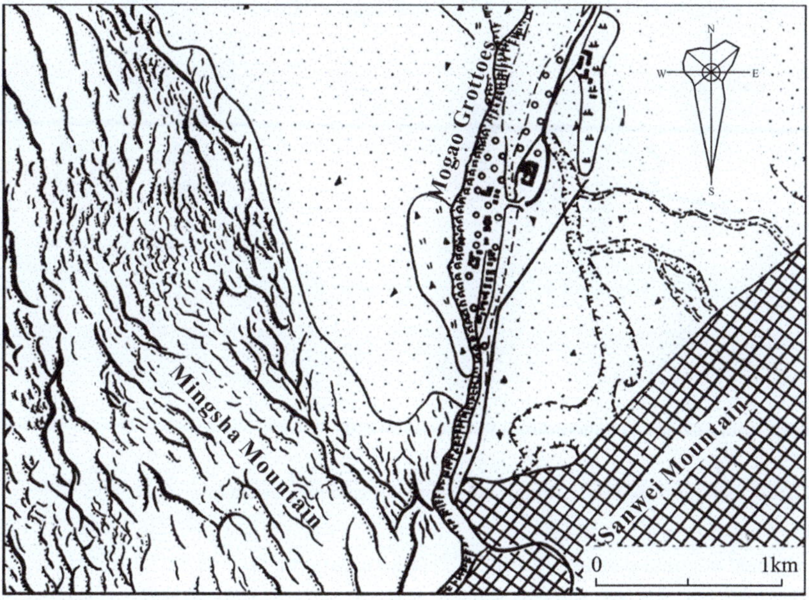

Fig. 10.5 Landscape and dynamic wind conditions on the top of Mogao Grottoes in Dunhuang (provided by WM Zhang)

northwesterly and northeasterly winds increases, and the wind is strong. Although the southerly winds flow through areas where the sand sources are sufficient and the wind frequency is high, the wind speed is relatively low, which results in the limited capacity of the southerly wind to transport sand.

10.3.1 Theoretical Calculation of Sand Transportation in the Gobi

The sand transport potential at the top of caves was calculated using the Fryberger sand transport potential eq. (Fryberger 1979): $Q \propto V^2(V - V_t)t$, where Q is the sand transport potential; V is the wind velocity at the height of 10 m; V_t is the threshold sand-driving wind velocity at the height of 10 m; and t is the wind blowing time, expressed by frequency (%). The Fryberger sand transport potential is usually the process of sand movement caused by surface wind. According to the local wind conditions, the total sand transport potential during the monitoring period (May 2008 to April 2009) was 129 VU. The combined sand transport potential was 90 VU, and the combined direction was 258°. Among the winds, the sand transport potential in the westerly wind (SW, WSW, W, WNW, and NW) was 36 VU; the sand transport potential in the easterly wind (NE, ENE, E, ESE, and SE) was 74 VU; and these two

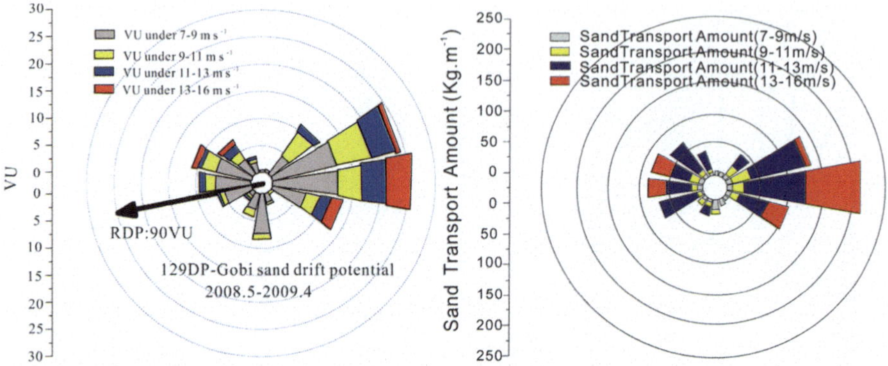

Fig. 10.6 Sand transport potential (left) and sand transport amount (right) at the top of the Mogao Grottoes (2008.5–2009.4) (According to Zhang et al. 2014)

Table 10.1 Sand transport in different wind directions/velocities (kg/m)

Wind velocities (m s^{-1})	7–9	9–11	11–13	13–16
Easterly sand transport potential (kg $m^{-1}y^{-1}$) (sand transport potential VU)	16 (7.1)	73 (31.8)	270 (23.1)	140 (12)
Westerly sand transport potential (kg m^{-1} y^{-1}) (sand transport potential VU)	15 (6.5)	39 (16.9)	203 (9.8)	62 (3)

values accounted for 28% and 57% of the annual sediment transport potential, respectively (Fig. 10.6). According to the calculations, the easterly wind transported 500 kg m^{-1} per year. The westerly wind transported 320 kg m^{-1} per year, and the total sediment transport was approximately 905 kg m^{-1} per year (Fig. 10.6).

As the westerly wind has an abundant sand supply from Mingsha Mountain, a large amount of sand is added in passing, which caused the sand transport of the westerly wind unit vector (VU) to be 1.8 times that of the easterly wind. However, from the perspective of sediment transport at different wind speeds, the easterly wind-sand transport potential was 2.1 times that of the westerly wind-sand transport. Under high wind conditions, easterly wind-sand transport dominated, resulting in easterly wind-sand transporting 1.5 times that of sand transported by westerly winds. According to the quantitative relationship between different wind directions (velocities) and sediment transport potential, the sediment transport at different wind speeds during the monitoring period of the grotto roof is shown in Table 10.1. Westerly winds transported 265 kg m^{-1} of sand to the grotto area, while easterly winds blew 410 kg m^{-1} of sand back to Mingsha Mountain (Fig. 10.6).

10.3.2 Basic Principles of the ISCAD at the Top of Mogao Grottoes

There have been two different views on how to design aeolian disaster prevention and control measures for protecting the Mogao Grottoes. The first design uses sand-blocking measures on the top of the grottoes to clear the sands deposited in the grotto area. The second design that other scholars have proposed focuses on sand transport at the top of grottoes as a more appropriate method. The easterly wind is stronger than the westerly wind. If the ground surface is an ideal sand transportation corridor, then the easterly wind can bring the sands westward and cause sand to be deposited on the sand mountain, which will effectively reduce the sand sediment in the grotto area (Zhao 1990). In addition, Zhu (1999) noted that any single measure cannot achieve the target of controlling aeolian disasters at the top of the Mogao Grottoes. Therefore, it is urgent to set up an ISCAD composed of engineering, biological, and chemical measures to fundamentally eliminate the aeolian disasters destroying the Mogao Grottoes (Zhang et al. 2000, 2004; Wang et al. 2004).

Mingsha Mountain in the west and the sandy gobi in the east are two sand sources on the top of the Mogao Grottoes. The main aeolian disasters influencing the Mogao Grottoes are wind erosion and sand deposition during sand movement. Therefore, sand-fixing measures should be set up first on Mingsha Mountain to reduce sand movement eastward. Second, maintaining the stable wind regime in the gobi (east part on the top of Grottoes) is crucial for reaching the balance between erosion and sediment (Zou et al. 1995). To maintain this balance, any upright or semiupright measures should not be placed on the gobi surface because they will result in more sand depositing near the caves. Gravel has been suggested to cover the gobi surface to reduce wind erosion and improve sand transportation ability. Under this scenario, most sands coming from Mingsha Mountain can be returned to the mountain through the gravel surface by the easterly wind.

The mean annual precipitation in the Mogao Grottoes is 39 mm. The harsh natural conditions limit the large-scale implementation of biological sand fixation measures. These conditions also indicate that the protection system must be based on engineering measures. Therefore, the ISCAD on the top of the Mogao Grottoes includes upright sand-blocking fences, semiupright sand-fixing wheat-straw checkerboard barriers on the eastern margin of Mingsha Mountain (Fig. 10.7), and a gravel-cover sand-transporting platform in the gobi area. Two simple biological measures, two shrub belts, were set up between Mingsha Mountain and gobi area (Fig. 10.7) to combine the blocking-sand and fixing-sand engineering measures. The gravel-cover measure was introduced in Chap. 8. The gravel coverage is approximately 30%, and the diameters of the gravel range from 3 to 4 cm. The gravel gobi on the top of the Mogao Grottoes formed a new dynamic balance between sand deposition and erosion and effectively reduced the damage to the Mogao Grottoes caused by wind-sand (Xue et al. 2002; Wang et al. 2004, Zhang et al. 2014; Wang 2011).

Fig. 10.7 Upright sand-blocking fences and semiupright sand-fixing wheat-straw checkerboard barriers on the eastern margin of Mingsha Mountain (provided by WM Zhang)

Fig. 10.8 Integrated system combating aeolian disaster map (left, Zhang et al. 2014) and the landscape (right, Wang 2011) on the top of Mogao Grottoes in Dunhuang

10.3.3 Overall Layout and Configuration of the ISCAD on the Top of the Mogao Grottoes

The ISCAD is generally an irregular trapezoid with the long side to the north and the short side in the south (Fig. 10.8). The length of the ISCAD from north to south is 2686 m, which is 1.7 times the grotto length. The width of the ISCAD from east to the west changes with the abundance of sand sources, wind conditions, and distance to the grottoes. The width is 2035 m at the north end and 770 at the south end. The total area of the ISCAD is 2,928,881 m^2. As mentioned in the previous section, the ISCAD is composed of an upright sand-blocking fence zone, a semiupright sand-fixing checkerboard barrier zone, a vegetation belt, a gravel-cover belt, and a blank buffer zone.

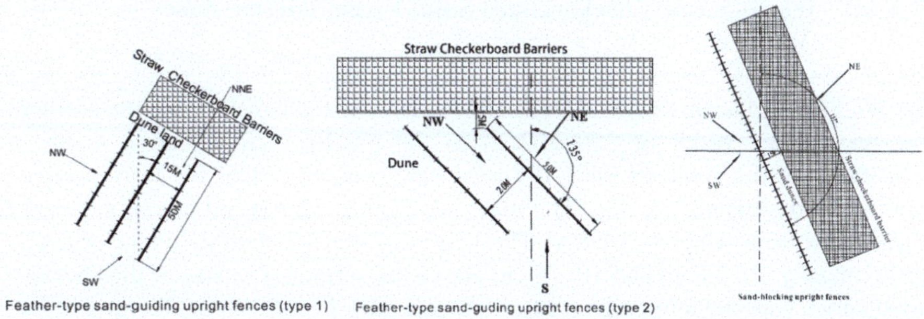

Fig. 10.9 First (left) and second (middle) kinds of feather-type sand-guiding upright fences and the sand-blocking upright fence (according to Wang 2011)

Table 10.2 Technical indicators of the three kinds of upright fences on Mingsha Mountain

Indicators	Feather-type fence I	Feather-type fence II	Sand-blocking fence
FAD	SSW-NNE, 30°	NW-SE, 135°	NNW-SSE, 157°
D (m)	15	20	–
L (m)	50–200	50–213	876
P (%)	20	20	40
DTC (m)	5	5	15
Function	Block the sands from the NW, W, and SW and guide them back to Mingsha Mountain using the strong NNE wind	Block the sands from the S and NE and guide them back to Mingsha Mountain using the W and NW winds.	Unsaturated wind-sand flow can form and transport the sand back to Mingsha Mountain

FAD represents fence arrangement direction. D represents the distance between fences. L represents the length of the fences. P is the porosity of the fences. DTC represents the distance to the checkerboard barrier zone

10.3.3.1 Upright Sand-Blocking Fence Zone (USBFZ)

The total length of the fences is 6395 m in the USBFZ, and the overall direction of the fences changes with the topography of Mingsha Mountain. There are three kinds of upright fences in the zone. Two kinds are feather-type fences. The length of the feather-type fence is 5725 m, 89% of the fence's total length. Among the two kinds of feather-type fences, one kind is arranged along SSW-NNE, and the other is arranged along NW-SE. The two kinds of feather-type fences can guide sand from the southwest and west back to Mingsha Mountain (Fig. 10.9). The third kind of fence is the blocking fence with a length of 670 m, which accounts for 11% of the total fence (Table 10.2).

10.3.3.2 Wheat-Straw Checkerboard Sand-Fixing Barrier Zone (WCSFBZ)

The WCSFBZ adjacent to the USBFZ in the east has a width of 170 m at the northern end, an average width of 390–502 m in the middle section, and a width of 352–866 m at the southern end. The total length of the WCSFBZ from north to south is approximately 2638 m. The material of the checkerboard barrier is pressed wheat straw with better flexibility. The height of the wheat-straw undersurface is approximately 10–13 cm and 10–15 cm above the surface. The size of the checkerboard varies with the topography. Most are generally 1 m × 1 m but 0.5 m × 0.5 m on the steep slopes of the dunes in the northern part. The amounts of wheat straw used in the different parts of the WCSFBZ vary because of the different sizes. The average amount is 0.65–0.7 kg/m^2, with approximately 0.98 kg m^{-2} on the top of the dune, 0.55 kg m^{-2} on the dune slope, and 0.43 kg m^{-2} in the interdunes.

10.3.3.3 Sand-Blocking Plant Zone (SBPZ)

As mentioned above, engineering measures dominate the ISCAD on the top of the Mogao Grottoes because of drought. However, the engineering measure materials can decay and degrade over time. Biological measures are thus used to assist the engineering measures. With plant growth, the effect of biological measures becomes increasingly significant, which has been shown by examples in Shapotou. The SBPZ included three shrub belts from east to west. The eastern and middle shrub belts were set up in the boundary between Mingsha Mountain and the gobi area. The western shrub belt was set up in the second easternmost interdunes of the Mingsha Mountain and mixed with the WCSFBZ (Fig. 10.10). The shrubs in the eastern, middle, and western belts were *Hedysarum scoparium* Fisch. et Mey, *Calligonum mongolicum* Turcz., and *Haloxylon ammodendron*. The length of the shrub belts from north to south was 2140 m. Each shrub belt's width was 8 m, and the space between the shrub belts was 12 m. The total width of the SBPZ was 48 m. The shrub row spacing in

Fig. 10.10 Wheat-straw checkerboard sand-fixing barrier zone mixed with the western shrub belt inside the eastern edge of Mingsha Mountain (left) and the gobi surface covered by annual natural vegetation (right) (provided by WM Zhang)

Fig. 10.11 Surface of sandy gobi (left) and the fine particles underlying the surface (right) on the top of Mogao Grottoes (provided by WM Zhang)

each belt was 1.5 × 2 m. Drip irrigation was used to support the growth of the shrubs.

10.3.3.4 Gravel-Cover Zone (GCZ)

There are three kinds of ground surfaces in the gobi region on the top of the Mogao Grottoes (Xue et al. 2002). They are gravel gobi, sandy gravel gobi, and sandy gobi (Fig. 10.11). The dust (<63 μm) content of sandy gobi is 27%, the sand content is 41%, and the gravel content is 32%. When the gravel on the surface of the gobi is destroyed, the sand and fine particles under the gravel layer will become a new sand and dust source influencing the safety of the Mogao Grottoes. Gravel cover can increase the content of gravel at the surface and inhibit the release of sand and dust due to erosion (Liu et al. 1999; Xue et al. 2000; Tan et al. 2016). Additionally, gravel cover is beneficial to sand transportation and reduces sand sediment.

Based on field experiments, the construction techniques for compacting the ground and paving the gravel were determined. Different gravel paving methods were adopted for the different surface characteristics of sandy gobi. The first step was to spray water on the surface and then let it naturally dry to a soil moisture content of 5%. Second, a layer of 0.5-cm-thick mixed soil was spread on the surface. The mixed soil was made of lime (10%), sand (50%), and gravel (40%) with a diameter of 1–2 cm. The layer was rolled twice by a roller to form a hard protective shell. Above the processed surface, gravel with a diameter of 3–4 cm was put down to a coverage of 30% (Fig. 10.12).

10.3.3.5 Blank Zone

The blank zone refers to the area where the natural surface has protective functions and does not require protective measures. The gravel gobi on the top of the grottoes

Fig. 10.12 Process of paving gravel on the sandy gobi, including watering (**a**), paving mixed soil (**b**), rolling (**c**), and paving gravel (**d**) (provided by WM Zhang)

is distributed near the cliff in the eastmost region. The gravel coverage is more than 40–60%. Gravel surface does not easily generate sand, and the protective function is more obvious under strong wind conditions. Thus, no protective measures were required. After passing through the USBFZ, WCSFBZ, SBPZ, and GCZ, most of the sand is blocked. The gravel surface can be regarded as the bed surface that has essentially reached a stable state. Moreover, the easterly wind can transport sand toward the west through the gravel surface. Therefore, to maintain its natural sand transport function, no measure was required. The blank zone not only transports sand but also reduces engineering costs.

10.3.4 Effect of the ISCAD on the Top of the Mogao Grottoes

10.3.4.1 Change in Sand Deposition Amount in the Grotto Area

Sand collectors that were 20 × 20 cm were installed in front of Grotto No. 208 on the third floor and Grotto No. 256 on the second floor to continuously monitor the changes in the amount of sand deposition. The results showed that sand deposition during March to May was 30 and 36 kg m^{-2} in 1990 when there was no protection system (Fig. 10.13). In 1991, an "A"-shaped sand-blocking nylon fence was installed on the gravel gobi near the cliff. The amounts of sand deposition were

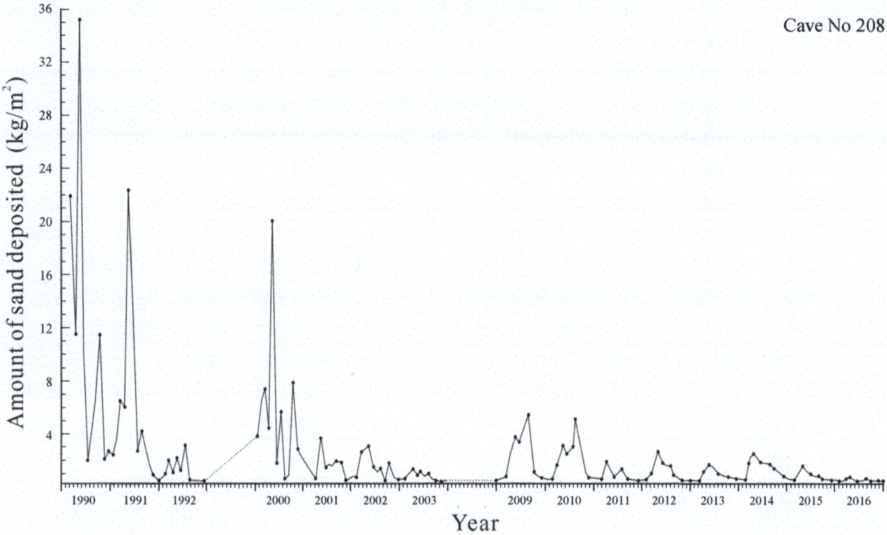

Fig. 10.13 Interannual changes in the amount of sand deposited in front of the Mogao Grottoes (according to WF Wang 2019)

10 and 8.5 kg m^{-2} in front of Grotto Nos. 208 and 256, respectively, which were 66.6% and 76.4% less than those in 1990. During 1992–1999, the Dunhuang Research Institute established eastern and middle shrub belts on the edge of Mingsha Mountain. The two shrub belts combined with the "A"-shaped nylon fence blocked a large amount of sand and reduced sand deposition in front of the grottoes by 95.1% and 88.8% in 2001.

After completing the ISCAD on the top of the grottoes by 2010, the amount of sand deposited in the front of the grottoes was significantly reduced. The sand deposition in front of the Mogao Grottoes was reduced by 97.2% in 2016 compared with that in 1990. By 2020, the ISCAD had withstood several extreme sandstorms. On April 24, 2014, when an extreme sandstorm with a westerly wind and a velocity of 27 m s^{-1} occurred, sand deposition in the grotto area was 8–10 m^3. On May 26, 2019, with a strong easterly wind with a velocity of 35 m s^{-1}, sand deposition in the grotto area was 7–8 m^3. During strong winds, the sand deposition amounts were much lower than the annual sand deposition amounts in the area during the 1960s and 1970s (3000 m^3 per year). The change in sand deposition amount fully proves the effectiveness of the ISCAD on the top of the grottoes.

10.3.4.2 Change in Sand Deposition Amount near the "A"-shaped Nylon Fence

In 1990, the Dunhuang Research Institute and the Lanzhou Institute of Desert Research of the Chinese Academy of Sciences set up an "A"-shaped nylon sand-blocking fence on the gravel gobi near the grottoes (Fig. 10.8). The fence was designed to block the sand from Mingsha Mountain and the sandy gobi in the west. The fence arrangement direction was also supposed to guide the sand away from the grotto area by moving it toward the north and south. The fence was effective in its initial stage (Ling et al. 1996). However, after a while, a large amount of sand accumulated around the northwest and southwest of the fence. The situation shows that the single fence measure moved the sand source of Mingsha Mountain to near the grottoes. The sand had to be cleared every 3–4 years, or the sand would turn over the fences and directly destroy the grottoes. The amount of sand needed to be cleared reached 30,000 m^3 each time, requiring a considerable amount of labor and funds. Additionally, the clearing efforts destroyed the gravel surface, which caused additional fine particles underlying the gravel cover to be exposed, eroded, and deposited in the grotto area. The suddenly increased sand deposition in front of the grottoes (Fig. 10.13) in 2009 and 2010 was attributed to the sand-clearing work. Since the ISCAD was set up in 2009, there has been no or minimal sand deposition around the "A"-shaped nylon fence. This situation indicates that the current integrated system has achieved an erosion-deposition balance on the top of the Mogao Grottoes. It also proves that the effect of the ISCAD is better than that of any single measure.

10.3.4.3 Improved Ecological Environment

With the establishment of the ISCAD, the environment on the top of the Mogao Grottoes has also been effectively improved. Gravel cover prevents wind erosion, enhances sand transportation, and reduces soil moisture evaporation, especially during heavy rainfall. The local rainfall was 100 mm in 2019. The annual natural vegetation coverage in the gobi area has reached 15%. These plant species include *Halogeton glomeratus* and *Salsola ruthenica* var. *ruthenica* Iljin. (Fig. 10.10). In 2020, many *Pterocles orientalis* appeared in the region. These results prove that the ecological environment on the top of the Mogao Grottoes is developing in a positive direction.

10.4 Conclusions

To achieve the expected protection effect, any sand-control project functions and objectives should be defined first, and the dialectical relationship between structure and function should be deeply understood. Given that the sand-control grades of the major national projects, such as railway traffic and nonrenewable cultural heritage sites, tend to be higher than the standard of other general facilities, their control target is much higher than that of other general projects. Thus, a comprehensive protection project should be adopted, and these control measures must not cause negative effects on the safe operation of the railway and the protection of cultural relics or pose a potential threat, which are the basic premises of sand control. Protective measures should not have any negative impact on the environment in the future (hundreds of years or more). Based on the scale of the project and the determination of a reasonable width of the sand-control system, the selection of relevant protection parameters should be based on safe operation of the railway and the safety of the cultural relics in the caves to improve the insurance parameters of the design of windblown sand-control engineering. Scientific sand-control technology is guaranteed to enable successful sand control. The key technologies in the system play a vital role in the windblown sand-control system and determine the function and effect of the sand-control engineering system. For example, the Shapotou Railway sand-control system has been tested for more than 50 years. The principle of sand fixation combined with sand resistance and the model of plant sand fixation without irrigation in semiarid areas have ensured the safety of railway transportation, even if the railway is not buried by drifting sand in the process of strong dust storms.

The dominant sand-fixing measures combined with sand-resistance measures and the successful application of the new technology of paving grave at the top of Mogao Grottoes prove that only when various single sand-control measures are organically combined into an engineering system can the overall maximum functional benefit be gained. Although the economic cost of a single sand-control protection measure is low and some results are achieved in the short term, from the perspective of long-term development, the inherent drawbacks and shortcomings of single sand-control measures will gradually become prominent, and the service life and sand-control efficiency will be greatly reduced.

This chapter presents two cases occurring in Shapotou and Dunhuang located in the eastern and western parts of China's arid areas, respectively. In the two regions, the integrated system for preventing and controlling aeolian disasters has been set up for over 50 and 20 years, respectively. These facts show the importance of and need for an integrated system, including multiple measures and functions. However, many current projects designed to protect against aeolian desertification and its related disasters in China still involve a single measure. This situation has resulted in unsustainable restoration effects and extreme waste in terms of labor and funds. Therefore, the promotion and demonstration of scientific results are even more urgent in combating desertification and disasters.

References

Fryberger SG (1979) Dune forms and wind regime. In: McKee ED (ed) A study of global sand seas, USGS professional paper, US geological survey and United States National Aeronautics and Space Administration, vol 1052, Washington, DC, pp 137–169

Li XR, Zhang ZS, Tan HJ, Gao YH, Liu LC, Wang XP (2014) Ecological restoration and recovery in the wind-blown sand hazard areas of northern China: relationship between soil water and carrying capacity for vegetation in the Tengger Desert[J]. Sci China Life Sci 57(05):539–548

Li XR, Zhou HY, Wang XP et al (2016) Ecological restoration and recovery in arid desert regions of Chin: a review for 60-year research progresses of Shapotou desert research and experiment station, Chinese Academy of Sciences. J Desert Res 26(02):247–264 (in Chinese with English abstract)

Li XR, Zhang ZS, Liu YB et al, (2017) Long-term ecological research guides ecological restoration and recovery in sandy areas of northern China. Bull Chin Acad Sci, 32(07):790–797(in Chinese with English abstract)

Ling YQ Jin J, Zou BG et al (1984) Effect of fence techniques in levelling sand accumulation around sandbreaks -case study in Shapotou District, J Desert Res (03):20–25 (in Chinese with English abstract)

Ling YQ, Qu JJ, Fan JS et al (1996) The effect analysis of sand control engineering on the cliff top of Mogao Grottoes. J Desert Res 16(1):13~18 (in Chinese with English abstract)

Liu YQ (1987) The establishment and effect of protecting system along the Baotou-Lanzhou railway in Shapotou Sandy area. J Desert Res 7(04):4–14 (in Chinese with English abstract)

Liu LY, Liu YZ, Li XY, Wang JH, Pen HM (1999) Effect of gravel mulch restraining soil deflation by wind tunnel simulation. J Desert Res 19 (1): 60–62 (in Chinese with English abstract)

Qu JJ, Ling YQ, Jing ZF, Zhang KC, Zu RP (2007) Interaction between sand blown activity and protection systemin Shapotou section of Baotou-Lanzhou railway, J Desert Res 27(4):529–533 (in Chinese with English abstract)

Tan LH, Zhang WM, Qu JJ, Yin DY, An ZS (2016) Quantitative simulation of erosion rates of Gobi with different gravel coverage, J Desert Res 36(03):581–588 (in Chinese with English abstract)

Wang KF (1988). Studies on sand dunes stabilization in Shapotou area. In: Shapotou Desert Research Station, Institute of Desert Research, Chinese Academy of Sciences. Drifting sand control in Shapotou area of Tengger Desert (2). Ningxia People's Press, Yinchuan, pp 13–26 (in Chinese)

Wang T (2011) Wind-sand disaster prevention projects in China. Science Press, Beijing

Wang WF (2019) Wind-sand disaster and prevention in the Mogao grottoes of Dunhuang. Science Press, Beijing

Wang T, Zhang WM, Wang WF, Qu JJ (2004) Study on functions of arresting and Transporting Sands of Gobi preventing system over top of Mogao grottoes. J Desert Res 24(2):187–190 (in Chinese with English abstract)

Xue X, Zhang WM, Wang T (2000) Wind tunnel experiments on the effects of gravel protection and problems of field surveys, Acta Geograph Sin, 55(3):375–383 (in Chinese with English abstract)

Xue X, Wang T, Sun QW, Zhang WM (2002) Field and wind tunnel studies of aerodynamic roughness. Bound-Layer Meteorol 104(1):151–163

Zhang WM, Wang T, Xue X, Wang WF, Guo YS, Liu JX (2000) The discuss of comprehensively preventing blown-sand system in Mogao Grottoes, Dunhuang J Desert Res 20 (4):409–414 (in Chinese with English abstract)

Zhang WM, Wang T, Wang WF (2004) The Gobi sand stream and its control over the top surface of the Mogao Grottoes,China. Bull Eng Geol Env 63(3):261–269

Zhang KC, Qu JJ, An ZS (2012) Characteristics of wind-blown sand and near-surface wind regime in the Tengger Desert, China. Aeolian Res 6:83–88

Zhang WM, Tan LH, Zhang GB, Qiu F, Zhan HT (2014) Aeolian processes over gravel beds: field wind tunnel simulation and its application atop the Mogao Grottoes, China. Aeolian Res 15: 335–344

Zhao XC (1988a) Summary of construction of railway tracks in dune area-case study at Zhongwei-Gantang section. In: Shapotou Desert Research Station, Institute of Desert Research, Chinese Academy of Sciences. Drifting sand control in Shapotou area of Tengger Desert (2). Ningxia People's Press, Yinchuan, pp 1–7 (in Chinese)

Zhao XL (1988b) Problems of revegetation on sand dunes in Shapotou area. In: Shapotou Desert Research Station, Institute of Desert Research, Chinese Academy of Sciences. Drifting sand control in Shapotou area of Tengger Desert (2). Ningxia People's Press, Yinchuan, pp 27–57 (in Chinese)

Zhao SQ (1990) Spirit of seeking truth from fact-cherishing the memory of instructions from my teacher Prof. Zhu Kezhen Scientia Geographica 10(1):6–8

Zhu ZD (1999) The desertification control must be set up a whole preventing system - a case study of blown sand preventing in Mogao grottoes, Dunhuang. In: Desert, Sandy desertification, desertification and its control and strategies. Chinese Environmental and Science Press, Beijing, pp 365–366

Zou XY, Dong GR, Wang ZL (1995) A study on some characteristics of drifting sand flux over Gobi. J Desert Res 15(4):365–373 (in Chinese with English abstract)

Chapter 11
Land Management Policies for Sand-Dust Storm and Aeolian Desertification in Source Region Countries

Lihua Zhou and Cuizhen Xia

Abstract China and Mongolia, two major sources of dust in Northeast Asia, are combating aeolian desertification. Considering the important role of the Chinese government in addressing aeolian desertification, this chapter focuses on the policies and actions taken by the Chinese government to combat aeolian desertification. First, the historical policies and activities related to aeolian desertification control at the central government level are introduced. Then, several major ecological restoration projects are described in detail. Finally, China's unique land reform system and ecological compensation mechanisms are also introduced.

Keywords Aeolian desertification · Ecological restoration project · Northern China · Sand control · Policy

11.1 Introduction

Aeolian desertification, the main type of desertification, is a serious problem in Northeast Asia, with several sand-dust source regions, mainly in China and Mongolia (Dooley 2004; Wang 2004; Wang and Xue 2010; Zhang et al. 2003). Eighteen of the 32 dust storms occurring in China in 2001 originated in southern Mongolia, accounting for 56% of the dust storms (Qin and Zhou 2002). China and Mongolia have different attitudes toward aeolian desertification control. In Mongolia, although all pastures are state-owned property, they are jointly managed by local herdsmen, with little supervision or involvement by local government (Bruegger et al. 2014; Undargaa and McCarthy 2016; Upton 2010). In China, due to the political system, the primary responsibility for combating aeolian desertification lies with the government, and the government has played an important role in the prevention and control of aeolian desertification (Wang 2009). This chapter mainly introduces the efforts of the Chinese government to overcome aeolian desertification and provides

L. Zhou (✉) · C. Xia
Institutes of Science and Development, Chinese Academy of Sciences, Beijing, China
e-mail: lhzhou@casisd.cn

T. Wang et al. (eds.), *Combating Aeolian Desertification in Northeast Asia*,
Ecological Research Monographs, https://doi.org/10.1007/978-981-16-9028-0_11

some information for readers to understand the role of the government in aeolian desertification control. This chapter first reviews the government's efforts to control aeolian desertification throughout China's history, then introduces several major ecological restoration projects organized and implemented by the central government since the founding of the People's Republic of China, and finally introduces China's unique land reform and ecological compensation system.

11.2 History of Aeolian Desertification Control in China

Historically, rotational grazing was the main land-use type in the arid or semiarid regions of northern China. Human interference on grasslands was negligible because of the small population and extremely low productivity (Wang 2014). Inappropriate farming or grazing activities have accelerated aeolian desertification since modern times and are often a result of increasing population, poverty, and state policies (Fan and Gao 2000; Fan and Zhang 2009).

Since 1949, the Chinese central government has paid attention to the problem of aeolian desertification. In 1950, the State Council established the leading group on sand control and set up the first forest protection field in Yulin, Shaanxi. In the late 1950s, with high enthusiasm for preventing and combating aeolian desertification, the first aerial seeding experiment was completed in some sandy areas, such as Yulin County and Minqin County, and the technology on aeolian desertification control continued to improve. In 1959, many scientific and technological researchers in various fields were organized by the Chinese Academy of Sciences to carry out comprehensive surveys on most of China's deserts, sandy lands, and Gobi, and they set up six comprehensive test stations and dozens of central stations, initially forming a platform for observation, scientific research, and experimental networks in the northern deserts of China. From the mid-1960s to the mid-1970s, due to the influence of far-left beliefs, the control of aeolian desertification was seriously hindered. In addition, large-scale reclamation of land led to the rapid deterioration of the ecological environment in northern China.

In 1978, the State Council officially approved a 73-year ecological construction project, the Three-North Shelterbelt Project, involving 358.3 thousand km^2 of land, which meant the beginning of China's major ecological conservation projects and increased the national awareness of ecological protection. In the 1980s, China implemented a sustainable development strategy with the goal of coordinated development of economy, society, and environmental protection. The government promulgated a series of laws and regulations, such as the Forest Law, Environmental Protection Law, Water and Soil Conservation Law, and Grassland Law, which provided legal guarantees for the protection and management of natural resources in aeolian desertified areas and consolidated the achievements of previous ecological restoration projects.

In the 1990s, efforts to prevent and combat aeolian desertification developed rapidly and constantly improved. In 1991, the State Council held the first national preventing and combating desertification conference and then issued the "Nationally Preventing and Combating Desertification Plan in 1991–2000," "Suggestions on Several Policies and Measures for Desertification Control," and other related policies, all of which provided explicit directions for subsequent work on combating desertification. The United Nations Convention to Combat Desertification, signed in October 1994, marking China's desertification control work, was officially in line with international standards. The National Desertification Combating Coordinating Group was headed by the forestry sector and composed of 19 ministries (commissions, offices, and bureaus). From the central to the local level, a multilevel, cross-field, and collective management system was gradually formed.

At the beginning of 2000, major national ecological restoration projects, such as the project to return farmland to forests and grasslands and the pilot project of the Beijing-Tianjin wind and sand source control project, started, opening a new era of aeolian desertification control driven by major national ecological restoration projects. On August 31, 2001, at the ninth National People's Congress, the law on desert prevention and transformation, the first special law on aeolian desertification control in China, was adopted, and it defined legal boundaries and laid the foundation for the rule of law. This law was also a great effort in combating desertification worldwide and won the Silver Award of Future Policy Award in 2017.

In August 2005, the State Council's decision to further strengthen the work of prevention and control of aeolian desertification was officially promulgated, and in the same year, the State Council approved the national desert prevention and transformation plan (2005–2010), which clarified the long-term goals and development direction of national efforts on prevention and control of aeolian desertification. In March 2007, the national conference on aeolian desertification control was held to clarify the three steps to combat aeolian desertification at the national level. In 2013, the National Plan for Preventing and Combating Desertification (2011–2020) was approved, and it clarified the structure, objectives, and tasks of national aeolian desertification prevention and control. Since the 18th National Congress, construction of ecological civilizations has been included in the overall structure of the five-in-one strategic desertification prevention and control effort, which is an indispensable part of the construction of ecological civilizations, resulting in unprecedented challenges and opportunities.

Aeolian desertification is a common challenge for humans, and global efforts should be implemented to combat aeolian desertification. China has continuously strengthened bilateral, multilateral, and regional international cooperation and actively led and participated in the process of developing international mechanisms, such as the Forum on China-Africa Cooperation, the Forum on Cooperation between China and the Arab States, the Forum on Cooperation between China and the Gulf Organization, the China-Japan-ROK Cooperation Mechanism, and the Northeast Asia Environmental Mechanism, which have designated desertification prevention and control a priority. In 2011, China, South Korea, and Mongolia established a subregional, network cooperative mechanism for desertification and land degradation in Northeast Asia, which played a positive role in promoting regional cooperation in combating desertification and mitigating sandstorms. China and many countries along the Belt and Road launched the joint action initiative on June 17, 2016, which promotes the joint efforts of countries along the Belt and Road to combat desertification through a variety of mechanisms, including conferences, information sharing, and project demonstration by those that signed the initiative.

11.3 Major Ecological Restoration Projects in China

At present, the Chinese government has led the implementation of major ecological construction projects, including the Three-North Shelter Forest, Natural Forest Protection, Grain for Green, and Beijing-Tianjin Sandstorm Source Control. All these projects form crucial parts of national environmental protection and construction. The main sources of investment for these projects were central financial funds and national debt funds, which supported the unprecedented coverage area, scale of investment, and construction period. The progress of these projects and relevant information are briefly introduced in Table 11.1, and the provinces covered by these ecological restoration projects are shown in Fig. 11.1.

11.3.1 Three-North Shelter Forest Project (See Appendix, Point 2)

Before 1978, there were eight deserts, four sandy lands, and the vast Gobi Desert in the Three-North (northwest, northeast, and north) area of China, with a total area of 1.48 million km^2, accounting for 85% of the total land area of aeolian desertification in China. Windblown sand damage and soil erosion were very serious; wood, fuel, fertilizer, and feed were all in short supply; and agricultural production was low and unstable. It was necessary to change the local agricultural and animal husbandry production conditions by building a protective forest system with a combination of belts, corridors, and buffers in a planned way. On November 25, 1978, the State Council approved the plan of the State Forestry Administration for the construction of large-scale shelterbelts in the key areas of sandy hazards and soil erosion in the Three-North area and officially launched the Three-North Shelter Forest Project.

The construction scope of the Three-North Shelter Forest Project involves the main ecologically vulnerable areas in northeastern, northwestern, and northern China. The construction area spans nearly half of China, with a length of 4480 km from east to the west and a width of 560–1460 km from north to south, including 551 counties (banners, cities, and districts) in 13 provinces (autonomous regions or municipalities) of Shaanxi, Gansu, Ningxia, Qinghai, Xinjiang, Shanxi, Hebei, Beijing, Tianjin, Inner Mongolia, Liaoning, Jilin, and Heilongjiang. The total construction area of the project is 4.069 million km^2, accounting for 42.4% of the total land area of the country.

Table 11.1 Brief introduction to the five ecological restoration projects (see Appendix, point 1)

Name	Three-North Shelter Forest Project	Natural Forest Protection Project	Grain for Green Project	Beijing-Tianjin Sandstorm Source Control Project	Returning Grazing Land to Grassland Project
Period	1979–2050	1998–2020	1999–2020	2002–2022	2003–2020
Objective	To protect farmland and reconstruct ecological barriers	To maintain water and soil and maintain ecological balance	To maintain soil and water, combat desertification, and restore vegetation	To resist sandstorms, stabilize soil, and recover vegetation	To protect grasslands and prevent the desertification of grasslands
Coverage	In China's "Three Norths" (northwest, north, and northeast), including 559 counties (banners, cities, districts) in 13 provinces (autonomous regions or municipalities), including Shaanxi, Gansu, Ningxia, Qinghai, Xinjiang, Shanxi, Hebei, Beijing, Tianjin, Inner Mongolia, Liaoning, Jilin, and Heilongjiang	Mainly distributed in the upper reaches of the Yangtze River and the upper and middle reaches of the Yellow River and northeastern state-owned forest enterprises (farm) including 17 provinces (districts and cities), such as Yunnan, Sichuan, Guizhou, Chongqing, Hubei, Xizang, Shaanxi, and Gansu	Covering more than 1800 counties in 25 provinces (autonomous regions or municipalities) in China, where arable land is prone to soil erosion and desertification	The first phase of the project covers 75 counties (banners) in 5 provinces (regions or cities) including Beijing, Tianjin, Hebei, Shanxi and Inner Mongolia. The second phase of the project will extend to 138 counties (banners or districts) in Beijing, Tianjin, Hebei, Shanxi, Inner Mongolia, and Shaanxi	Occurred in Inner Mongolia, Sichuan, Guizhou, Yunnan, Tibet, Gansu, Ningxia, Xinjiang, and Qinghai Provinces (autonomous regions)
Content	Construct forests for farmland protection, wind prevention, sand fixation, and soil and water conservation	Protect and reasonably operate state-owned forest farm and gradually decrease the quantity of logging and stop logging conditionally	Step by step, systematically stop farming and recover vegetation according to local conditions	Regard vegetation (forest and grass) reconstruction as a major task, including forest planting, desert mountain recovery, afforestation, and sand fixation	Construct fences and forbid grazing, implement seasonal grazing, zoned rotation, and livestock feeding in barns

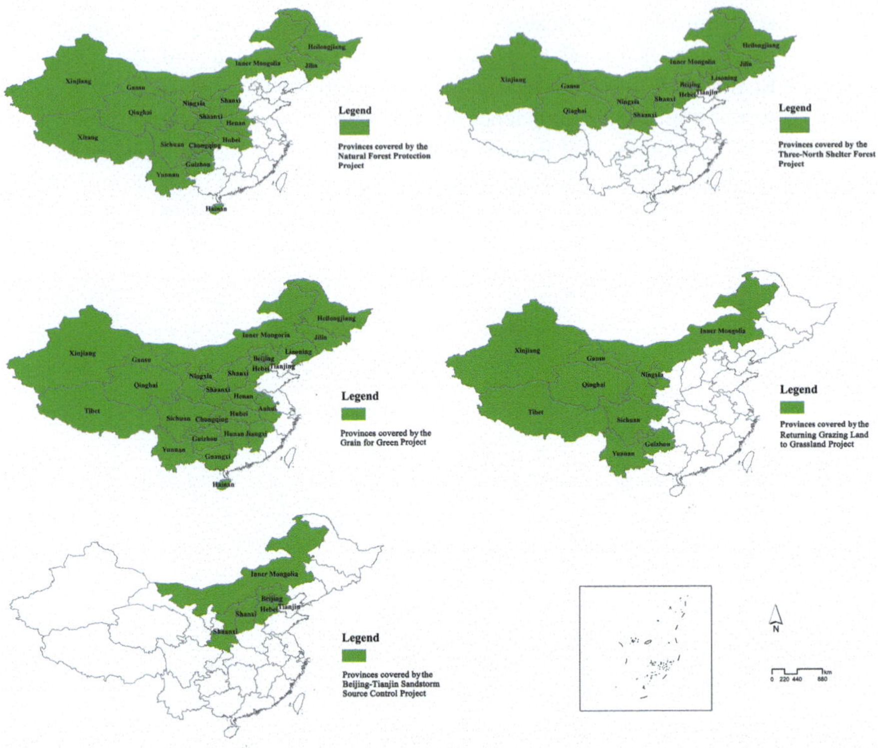

Fig. 11.1 The provinces covered by the ecological restoration projects

The planning of the Three-North Shelter Forest Project started in 1978 and will end in 2050, thus lasting for 73 years, and the planning can be divided into eight phases. The planned afforestation area of the Three-North Project is 0.35 million km^2 (including the converted area of the forest belt and forest network), of which artificial afforestation occurs on 0.26 million km^2, accounting for 75.1% of the total area; aerial afforestation occurs on 11.14 thousand km^2, accounting for 3.2% of the area; and mountain sealing and sand sealing occur on 75.98 thousand km^2, accounting for 21.7% of the area. In all, 5.24 billion trees will have been planted in the area of the project. By 2050, the forestland area should have expanded from 0.23 million km^2 in 1977 to 0.61 million km^2, and the forest coverage rate should have increased from 5.05% in 1977 to 14.95%. In the 40 years since the construction of the project, by 2018, the accumulated afforestation and conservation area of the Three-North Shelter Forest Project covered 0.30 million km^2, the forest coverage rate of the project area increased from 5.05% in 1979 to 13.59%, and the stock of standing trees increased from 740 million m^3 to 3.33 billion m^3.

11.3.2 Natural Forest Protection Project (See Appendix, Point 3)

Natural forests consist of two types of forests: primitive forests and secondary forests. In China, by 2019, there were 19.8 million km^2 of existing natural forest, accounting for 64% of the country's forest area; the stock of wood was as high as 12.96 billion m^3, accounting for 83% of the total national stock. The distribution area of natural forest in China can be summarized in four main categories: the first category is natural forest in the basic state of protection, including various types of nature reserves, scenic spots, forest parks, and undeveloped forest areas; the second category is scattered natural forest throughout the country; the third category is the natural forest distributed in the upper reaches of the Yangtze River and the upper and middle reaches of the Yellow River; and the fourth category is the natural forest distributed in northeastern Inner Mongolia and other key state-owned forest areas.

The core of this project was the classification management of forest farms; that is, the main forest area was divided into key ecological public welfare forests, general ecological public welfare forests, and commodity forest bases. With the background of strict protection of the ecological public welfare forests, the operation of commodity forests could address the gap in natural forest logging. According to the national forest protection plan, the upstream Yangtze River and the Yellow River were the key areas for the implementation of natural forest protection projects, with 0.70 million hm^2 of natural forests, accounting for 69% of the country's 0.11 billion km^2 of natural forests. Given the suspension of mining, the government increased the rate of forest construction in 36.7 thousand km^2 of the wasteland located upstream of the Yangtze River and the Yellow River and increased the forest cover ratio from 17.5% to 21.24%. The total investment in the whole project was 106.4 billion CNY, of which the financial special fund investment covered 81.2% of the total, and the State Council also stipulated that the financial debt of the financial enterprises might be reduced due to mining restrictions and suspension.

11.3.3 Grain for Green Project (See Appendix, Point 4)

China formulated a 10-year plan to return farmland to forest (grass), covering 1580 counties in 25 provinces (municipalities or autonomous regions) in China. The project planned to return 53 thousand km^2 of farmland to forest, implement 80 thousand km^2 of afforestation in suitable desert mountains, and control soil erosion on 0.36 million km^2, and control wind and sand on 0.70 million km^2.

The core content of this project was that in areas suitable for returning farmland to forest, farmers could voluntarily convert suitable sloping farmland into woodlands or grasslands, and the government provided free grain, seedlings, and cash subsidies for afforestation to farmers in accordance with uniform standards. The categories of this project also included the return of farmland to forest, grassland, and lakes and afforestation in conditional desert mountains.

In 2001, the State Forestry Administration set up a project management center for the Return of Farmland to Forest (Grass) Project, which was responsible for the entire work; relevant provinces, autonomous regions, and municipalities set up provincial-level forestry administrative departments; the forestry bureaus of each county of the project and forestry workstations in each town were responsible for implementation within the administrative areas. In 2002, the State Council formulated regulations for returning farmland to forest, which clearly defined the contents, scopes, measures, and responsibilities of various departments, turning the management of this project into the law. The compensation due to the additional costs caused by this project included compensation for farmers and local governments. To the farmers, the state provided free subsidies for food, seedlings, etc. and provided subsidies for local finance to compensate for the losses due to the reduction in farmland. In 2014, this project entered a new phase and continued to increase investments. From the start of the project to 2019, the cumulative investment has been more than 500 billion CNY, the implementation area has covered 0.34 million km^2, and the average forest coverage of the project area has increased by more than 4%.

11.3.4 Returning Grazing Land to Grassland Project (See Appendix, Point 5)

On November 16, 2002, the State Council formally approved the implementation of the Returning Grazing Land to Grassland Project in 11 provinces and cities in the western region. Starting in 2003, after 5 years, in the degraded grasslands in eastern Inner Mongolia and the eastern Qinghai-Tibet Plateau, the initial centralized management area was 6.7 million km^2, approximately 40% of the severely degraded grasslands in the western region. The Return of Grazing Land to Grassland Project has been carried out in three ways, including prohibiting grazing, delaying grazing, and rotational grazing in different regions. During the return of grazing lands, the state will provide food and feed subsidies to herders.

In the "12th Five-Year Plan" period, 0.33 million km^2 pastures were prepared for fence construction, and 0.10 million km^2 of degraded grasslands were replanted and improved. On August 22, 2011, the National Development and Reform Commission, the Ministry of Finance, and the Ministry of Agriculture issued a notice on improving the policy of returning pasture to grass, announcing that from 2011 onward, no more subsidies for grain would be provided, and the incentive mechanism of the grassland ecological protection subsidy would be fully implemented in the project area. For grasslands that have been subject to a ban on grazing, the central government provided the herders subsidies in accordance with the measurement standard of 6 CNY per mu (1 mu is equal to 666.67 m^2) per year and regarded

5 years as a subsidy period; for grasslands where grazing was suspended or rotated, the central government provided balanced rewards to herdsmen who were overloaded, at 1.5 yuan per mu per year. At the same time, the central government supplied the subsidy for supporting the construction of fences, artificial grassland construction, and shelter shed construction.

11.3.5 Beijing-Tianjin Sandstorm Source Control Program (See Appendix, Point 6)

In 1999, the Forestry Department formulated the "Beijing-Tianjin Sandstorm Sources Management Plan," which included the execution mechanism of returning desert mountain to the forest, created a desertification prevention and control system, and included ecological migration and ten other measures. The first phase of the project was launched in 2002 to maintain ecological security around Beijing and Tianjin by reducing soil erosion. The project area involved 19.58 million people, 458,000 km^2 of land, and 10.12 million km^2 of sandy lands, which were divided into four governance areas, namely, the northern arid grassland desertification management area, the Hunshandak sand land management area, sandy land management of the agropastoral transition area, and water source reserve of the Yanshan hilly mountain. The total area of the governance task was 1.5 million km^2, and the initial budget investment was 55.8 billion CNY. The project adopted comprehensive management measures based on the construction of forest and grass vegetation. Forestry measures included the return of 26.3 thousand km^2 of farmland to forest and the planting of 49.4 thousand km^2 of forest. Agricultural measures included 14.8 thousand km^2 of artificial grass, 2.9 thousand km^2 of aerial seeding grass, 27.9 thousand km^2 of enclosure, 3.4 thousand km^2 of basic grassland construction, 0.39 thousand km^2 of grass seed base, and 56.8 thousand km^2 of no grazing. Water conservation measures included 66,059 water source projects, 47,830 water-saving irrigation projects, 23.4 thousand km^2 of integrated management of small watersheds, and 180,000 ecological migrants.

The second phase of this project occurred from 2013 to 2022, with a total investment of 87.79 billion CNY. Through the construction of the second phase of the project, the outcomes in the first phase were effectively consolidated, and the sandy lands in the project area were managed, generally curbing the expansion trend. In addition to the improvement in the ecological environment and the further enhancement of ecosystem stability, sand and dust events and sand hazards were further reduced in the Beijing-Tianjin area; for example, sand and dust events in the Beijing-Tianjin area decreased from an average of 13 per year in 2000 to 2–3 in 2016. A green ecological barrier was built in Beijing and northern China.

11.3.6 Ecological Construction Planning After 2020 (See Appendix, Point 7)

On June 11, 2020, the National Development and Reform Commission and the Ministry of Natural Resources jointly issued the "National Master Plan for the Protection and Rehabilitation of Major Ecosystems (2021–2035)," focusing on the ecological barrier area of the Qinghai-Tibet Plateau, the key ecological areas of the Yellow River (including the ecological barrier area of the Loess Plateau), the key ecological area of the Yangtze River (including the ecological barrier area of Sichuan and Yunnan), the northeastern forest belt, the sand control belt in the north, the hilly and mountainous areas in the south, and the coastal belt. Nine major ecological protection and restoration projects have been deployed in these areas, and compared with the original ecological restoration projects, they focus more attention on integrity and systems.

11.4 Reform of China's Land Property Rights System

After 1949, to effectively safeguard the rights and interests of farmers, the central government actively explored the property rights of rural land (mainly including farmland, grassland, and woodland). From 1978 to 1988 was the main period of carrying out the household contract responsibility system, which started based on the attempt to implement land contracts in Xiaogang village, Fengyang County, Anhui Province. After the Third Plenary Session of the Eleventh Central Committee of the Communist Party of China, the implementation of the household contract responsibility system throughout the country was approved and allowed farmers to have the autonomous right to use land and gain the interest from production under the premise of collective ownership of land. This action greatly mobilized the enthusiasm of farmers to implement agricultural production and eliminated the constraints of the "common big rice pot" system.

After 2003, based on the excessive tax burden of farmers, the agricultural tax was completely abolished, and the reform of the transfer system of land contract and management rights started to promote the integrated development of urban and rural areas and to change the long-standing dual economic structure. In October 2008, the communiqué of the Third Plenary Session of the 17th Central Committee of the Communist Party of China stated that the next tasks were to improve the right to contract land management rights; to protect farmers' rights to the possess, use, and operate and other rights to contracted land in accordance with the law; to strengthen the management and service of the transfer of land contracts and management rights; to establish and improve the transfer market of land contract and management rights; to allow farmers to transfer land contracts and management rights in the form of subcontracting, leasing, swapping, transferring, joint stock cooperation, etc. in accordance with the principle of voluntary compensation; and to develop various forms of moderate-scale operations.

In 2014, the Central No. 1 document, "Several Opinions on Comprehensively Deepening Rural Reform and Accelerating Agricultural Modernization," clearly stipulated that farmers (herdsmen) are given collateral and security rights to the possession, use, proceeds, and circulation of contracted land under the premise of adhering to the most stringent system of farmland protection. In 2017, the Central Document emphasized that the weak link of agriculture and rural areas should be complemented, and the impetus for the endogenous development of rural areas should be activated. In 2017, the focus of the work was the reform of the rural property rights system. In fact, it was also the focus and main point of rural reform in China's "13th Five-Year Plan period." In 2018 and 2019, the Central Committee's Document No. 1 continued to include clear directions and requirements for the reform of the rural property rights system. The 19th National Congress of the Communist Party of China (CPC) proposed to keep the land contract relationship stable and unchanged for the long term and extend the contract period for another 30 years after the expiration of the second land contract period.

Considering Inner Mongolia grassland property practices as an example, grassland property rights are composed of a series of rights, such as use, transfer, operation, and regulation. The rights are related not only to the attribution of grasslands but also to use and effective use of grasslands. Without a reasonable grassland property rights system, the sustainable development of rural pastoral areas cannot be achieved. The relevant legal provisions of the current grassland property rights system can be found in the Rural Land Contract Law, the Property Law, the Grassland Law, etc. Since the reform and opening-up, the government of the Inner Mongolia Autonomous Region has carried out active exploration and practices in the grassland management area. In 1984, the government began to implement the "common grassland, contract management, livestock pricing, family feeding" responsibility system and dual-power system, that is, the implementation of grassland ownership, use rights, and contract-to-house responsibility system. Since then, for many years, this area has gradually entered the grassland "two power, one system" policy.

In 2017, the government of the Inner Mongolia Autonomous Region promulgated implementation options for improving measures to separate the rights of land and grassland contracts in rural pastoral areas, and these options were consistent with the deployment requirements of the State Council for strengthening the reform of the rural land system and showing respect for the willingness of farmers and herdsmen to retain the right to contract and transfer land. The right of grassland contracts and management was divided into contract rights, management rights, ownership, and efforts to promote the modernization of agriculture and animal husbandry. The separation of powers improved the basic management system of rural pastoral areas, conforming to the objective law of a productive relationship and adapting to the development of productive forces.

11.5 Ecological Compensation Mechanism

Financial transfer payments are the most important ecological compensation methods in China; at present, the main sources of investment for ecological restoration projects are central financial funds and national debt funds. While implementing these projects, supporting funds, grain, seedlings, and other subsidies are paid together. For example, in the Returning Grazing Land to Grassland Project, starting in 2003, with a 5-year cycle, the state has granted the necessary grassland fence construction fund subsidies and feed-grain subsidies in various regions every year. The central government bore 70% of the subsidy for fence construction, and local governments and individuals bore 30%. Since 2011, the state has raised the standard of cash subsidies, and the proportion of central subsidies for fence construction has increased from 70% to 80%. The grain subsidy was cancelled, and the grassland ecological protection subsidy incentive mechanism was fully implemented in the project areas, including three types of subsidies: grazing prohibition subsidies (7.5 CNY per mu after 2016), grass and animal balance subsidies, and performance evaluation rewards. In the Grain for Green Project, the state adopted a subsidy policy to provide grain, cash, and seedlings to farmers who have returned farmland to forests, and the state also established a special fund to consolidate the project's results by basic grain field construction and rural energy construction, ecological migration, and replanting.

Various government departments, such as the departments of land, forestry, water conservancy, agriculture and environmental protection, and local governments, were also trying to adopt flexible financial transfer payment policies to stimulate the protection and construction of the ecological environment. They implemented a series of plans to establish special funds to provide financial subsidies and technical assistance for ecological protection and construction, such as new energy construction in rural areas, ecological public welfare forest compensation, soil and water conservation subsidies, and farmland protection.

11.6 Conclusions

After decades of concerted efforts by the government and its people to fight aeolian desertification, China has performed remarkably. The Hobqi Desert almost disappeared, the aeolian desertification area of Mu Us sand land decreased sharply, and the aeolian desertification in the Hunshandak sand land and Horqin sand land was effectively controlled. However, the achievements of ecological construction still need to be further consolidated. At present, China is carrying out a new round of research on ecological policies, and on June 11, 2020, the Chinese government proposed a more comprehensive and systematic plan, namely, the "National Master Plan for the Protection and Rehabilitation of Major Ecosystems (2021–2035)." This plan deploys nine major ecological protection and restoration projects, which form a basic framework for the protection and restoration of major ecological systems and provide a guarantee for maintaining China's overall ecological security. According to the plan, by 2035, the national forest coverage rate will reach 26%, the natural forest area will remain stable at approximately two million km^2, the vegetation coverage rate of grasslands will reach 60%, and more than 75% of the land that can be controlled for desertification will be managed. Through the coordinated advancement of ecological construction projects and other societal projects such as poverty alleviation, the Chinese government is striving to achieve a balance between ecological protection, economic development, and people's livelihoods so that the achievements of ecological protection will truly benefit the people.

Appendix

1. The information in Table 11.1 is collated from the following sources:
   (a) Implementation plan of Natural Forest Protection Project in upper reaches of Yangtze River and upper and middle reaches of Yellow River. http://www.docin.com/p-20452549.html
   (b) State Forestry Administration. The implementation plan of the second phase of the Natural Forest Resource Protection Project in northeast China, Inner Mongolia and other key state-owned forests. http://www.docin.com/p-428419856.html
   (c) 360 Baike. Grain for Green Project. https://baike.so.com/doc/2870457-3029080.html
   (d) Ministry of Agriculture. The country's 13th Five-Year Plan for grassland protection, construction and utilization. (2017-01-20). http://www.moa.gov.cn/nybgb/2017/dyiq/201712/t20171227_6129885.htm
   (e) Ministry of Agriculture. Opinions on further strengthening the Implementation management of the Project of Returning Grazing land to Grassland. (2005-04-11). http://www.law-lib.com/law/law_view.asp?id=91539
   (f) The State Council. Some opinions on further perfecting the policy and measures of Returning Farmland to Forests. (2002-04-11). http://www.gov.cn/gongbao/content/2002/content_61463.htm
   (g) Baidu Baike. Three-North Shelter Forest Project. https://baike.baidu.com/item/%E2%80%9C%E4%B8%89%E5%8C%97%E2%80%9D%E9%98%B2%E6%8A%A4%E6%9E%97%E5%B7%A5%E7%A8%8B/11046362Shelter Forest Project/110463
2. Three-North Shelterbelt Construction Bureau. The general plan of the Three-North Shelterbelt Construction Project. https://www.forestry.gov.cn/sbj/5030/index.html
3. Office of Natural Forest Protection. The natural forest protection is of great merit—a review of the 20-year construction of China's Natural Forest Protection

Project. (2018-05-18). http://www.forestry.gov.cn/main/425/20180518/1103006.html
4. The History of the People' s Republic of China (HPRC). Returning farmland to forests. (2009-12-30). http://www.hprc.org.cn/gsgl/dsnb/zdsj/200912/t20091230_39594.html
5. National Development and Reform Commission, Ministry of Agriculture, Ministry of Finance. Suggestions on improving the policy of returning grazing land to grassland. (2011-08-22). https://www.gaodun.com/guoshui/649684.html
6. (a) National Development and Reform Commission. Planning of Beijing and Tianjin Sandstorm Source Control Project (2001–2010). [2007-09-28]. https://www.ndrc.gov.cn/fggz/fzzlgh/gjjzxgh/200709/t20070928_1196575_ext.html
(b) Baidu Baike. The second phase of Beijing and Tianjin Sandstorm Source Control Project. https://baike.baidu.com/item/%E4%BA%AC%E6%B4%A5%E9%A3%8E%E6%B2%99%E6%BA%90%E6%B2%BB%E7%90%86%E4%BA%8C%E6%9C%9F%E5%B7%A5%E7%A8%8B/4482355
7. National Development and Reform Commission and the Ministry of Natural Resources. National Master Plan for the Protection and Rehabilitation of Major Ecosystems (2021–2035). (2020-06-03). https://www.forestry.gov.cn/main/72/20200612/093234638407152.html

References

Bruegger RA, Jigjsuren O, Fernández-giménez ME (2014) Herder observations of rangeland change in Mongolia: indicators, causes, and application to community-based management. Rangel Ecol Manag 67:119–131

Dooley EE (2004) Asian dust storm surge. Environ Health Perspect 7:547–547

Fan SY, Gao X (2000) Innovations in the Model for Bringing Desertification under Control in China and Institutional Creation. Soc Sci China:37–44. (In Chinese)

Fan SY, Zhang H (2009) Influence of the transformation of the China's rural land system on sandy desertification changes since 1949. Arid Land Geograp 32:268–273. (In Chinese)

Qin J, Zhou S (2002) The source area and transmission path of sandstorm in China was confirmed by State Environmental Protection Administration. http://www.people.com.cn/GB/huanbao/56/20020125/655995.html

Undargaa S, Mccarthy JF (2016) Beyond property: co-management and pastoral resource access in Mongolia. World Dev 77:367–379

Upton C (2010) Living off the land: nature and nomadism in Mongolia. Geoforum 41:865–874

Wang T (2004) Study on Sandy Desertification in China −3 Key Regions for Studying and Combating Sandy Desertification. J Desert Res 24:1–9. (In Chinese)

Wang XY (2009) From "rangeland leasing" to "recentralization in rangeland conservation"-policies of rangeland conservation in North China. China Rural Surv 87:36–46. (In Chinese)

Wang T (2014) Aeolian desertification and its control in Northern China. Int Soil Water Conservat Res 2:34–41

Wang T, Xue X (2010) Aims, Principles and Measures for Combating Desertification in Northeast Asia. Glob Environ Res 14:11–16

Zhang XY, Gong SL, Zhao TL et al (2003) Sources of Asian dust and role of climate change versus desertification in Asian dust emission. Geophys Res Lett 1:5–8

Part III
Successful Stories for Combating Aeolian Desertification from Countries in Northeast Asia

Chapter 12
Successful Implementation of Measures to Combat Aeolian Desertification in China

Tao Wang

Abstract China has achieved remarkable advances in controlling aeolian desertification. Many proven practical techniques, successful demonstrations, and valuable experiences as well as extension service models to combat aeolian desertification have been developed at local and national levels, and some have been successfully applied elsewhere in the world. This chapter details examples of successful projects that have addressed aeolian desertification in China. In China, major sand control and ecological construction projects designed for regional and national development, either through predominantly mechanical or biological methods and in combination with other supplemental measures, have been comprehensively and successfully applied to control aeolian desertification.

Keywords Aeolian desertification · Comprehensive approach · Project · Implementation · Demonstration · China

12.1 Introduction

China has led a long struggle against land desertification. As early as the 1950s, the government of China organized scientific surveys and studies on affected lands and has prioritized combating desertification in seriously impacted regions. Since the 1970s, China has successively initiated and implemented aeolian desertification control and sand industry development projects in typical aeolian desertification regions and urgent national projects for eco-environmental improvement and

T. Wang (✉)
Northwest Institute of Eco-Environment and Resources, Chinese Academy of Sciences, Lanzhou, Gansu, China

Key Laboratory of Desert and Desertification, Chinese Academy of Sciences, Lanzhou, Gansu, China

Drylands Salinization Research Station, Northwest Institute of Eco-Environment and Resources, Chinese Academy of Sciences, Lanzhou, Gansu, China
e-mail: wangtao@lzb.ac.cn

T. Wang et al. (eds.), *Combating Aeolian Desertification in Northeast Asia*, Ecological Research Monographs, https://doi.org/10.1007/978-981-16-9028-0_12

important construction protection. The land use categories that are suffering most from aeolian desertification in China mainly include agricultural areas, highways, railways, roads, cities, industrial sites, and mining areas. In the following sections, the projects exemplified are as follows: the most comprehensive and far-reaching ecological construction projects in China that aim to improve the overall ecological environment in the country through a clean urban and rural environment, naturally functioning ecosystems, and healthy mountains and rivers in most regions of China; successful desertification control and sand industry development models under several typical natural conditions in China; and the moving sand-prevention project along the traffic corridors that are related to the national economy.

12.2 Successful Desertification Control and Ecological Construction Projects

The China National Plan for Eco-environmental Improvement, proposed in 1997 and adopted by the State Council in 1998, is the most comprehensive and longest-ranging plan for eco-environmental improvement. The plan is divided into three stages: the near term (1997–2010), the medium term (2011–2030), and the long term (2031–2050). By 2050, the overall ecological environment in the country will be improved, a clean urban and rural environment and a natural cycle of ecosystems will be achieved, and healthy mountains and rivers will exist in most regions.

Under the National Plan, some key desertification control and ecological construction projects were formulated and implemented, i.e., the "Three North" Shelterbelt System Engineering project; the Beijing-Tianjin Sand Source Control Project; the project of returning farmland to forests and grazing land to grassland; and the grassland desertification control project. The activities carried out during these projects mainly include (a) construction of desertified land enclosure protection zones for protecting and promoting the natural restoration of forest and grass vegetation and curbing the expansion of aeolian desertification; (b) afforestation according to natural climate characteristics of different desertification areas; (c) desertified grassland restoration by changing the mode of animal husbandry production, balancing grass and livestock, and implementing desertified grassland enclosures and house feeding; (d) catchment management and water source and water-saving irrigation engineering to create favorable conditions for returning farmland to forest and grass and enclosure protection; (e) eco-migration by constructing small towns and rural energy sources to reduce ecological pressure and to consolidate sand control efforts; and (f) sand industry development by making desertification control consistent with income increases for farmers and herdsmen and increasing the efficiency of enterprises and tax increase by the government.

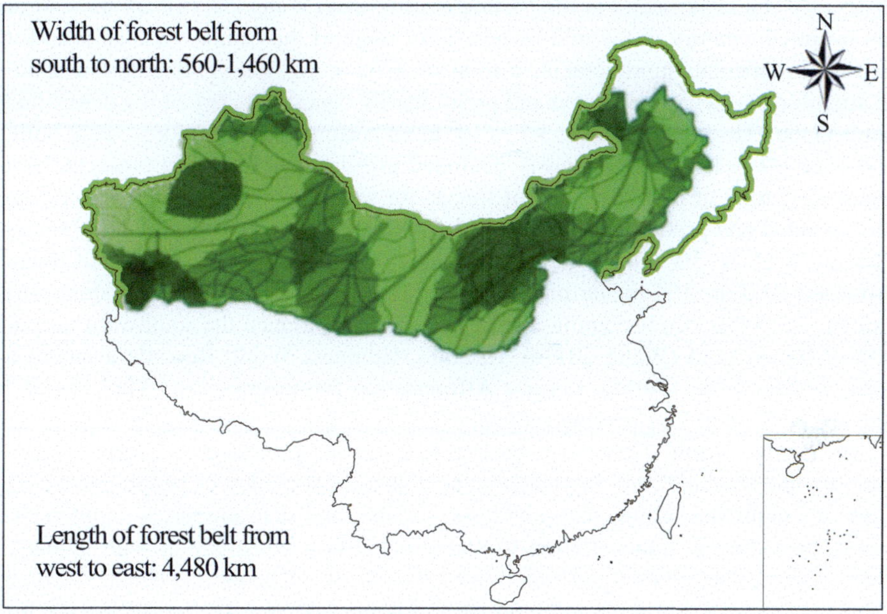

Fig. 12.1 Plan for the construction of the protection forest system in northwestern, northern, and northeastern China (Three North Office and NW Institute 1990)

12.2.1 "Three North" Shelterbelt System Engineering

The "Three North" Shelterbelt System Engineering project (running from 1978 to 2050) refers to the large-scale artificial ecological forestry system constructed in the "Three North" areas of northwestern, northern, and northeastern China that involves a total area of 4.069 million km^2, accounting for 42.4% of China's land area (Three North Bureau 1993). A well-performing shelterbelt system to coordinate the development of agriculture, forestry, and animal husbandry, which began by establishing a combination of trees, shrubs, and grasses and then included combining forest belts, forest networks, and forest patches and finally focused on the rational allocation of multiforest and multitree species, was required in the overall project plan. The plan noted that this system should be established by different afforestation methods to protect the existing forest and grassland vegetation. The forest coverage in the "Three North" area will have increased from 5.05% in 1977 to 14.95% by 2050 (Fig. 12.1).

Before the China National Plan for Eco-environmental Improvement was developed in 1997, the first phase (1978–1985) and the second phase (1986–1995) of the "Three North" Shelterbelt System Engineering project were successfully completed. During the period, the forest coverage in the "Three North" area increased from 5.05% to 8.2%, accounting for 12% of the desertified lands and protecting more than

16.0×10^6 ha of farmland in the "Three North" area (Wang 2011a). By the end of the twentieth century in the middle near-term stage of the China National Plan for Eco-environmental Improvement, a total of 4.76×10^6 ha windbreak and sand-fixation forests were constructed and preliminarily controlled 20% of the desertified lands in the "Three North" area; 3.7×10^5 ha of pasture protection forest were built, and over 1.0×10^7 ha of seriously degraded grazing land had been protected and restored. A total of 6.63×10^6 ha of forests for soil and water supply conservation were planted and preliminarily controlled 40% of the soil erosion area on the Loess Plateau; 2.13×10^6 ha of farmland shelter forests were built and effectively protected 21.3×10^6 ha farmlands. The "Three North" shelterbelt system was extended to 4480 km, becoming a "Great Wall" that stopped the southward migration of blown sand (Wang 2011a). In 2003, Guinness World Records awarded a certificate for "The World's Largest Afforestation Engineering" to "Three North" Shelterbelt System Engineering project.

The "Three North" Shelterbelt System Engineering project uses a form of "private run and state aid" and implements the construction policy of people providing labor and multichannel financing, with self-reliance and state support as supplementation, providing a demonstration of shelterbelt forest construction with Chinese characteristics that gives equal attention to ecological and economic benefits. China is leading in the world in research on the following five aspects: "aerial sowing for afforestation on sand lands," "high-yielding technology for dryland forestry," "narrow-belt and small-grid farmland shelterbelt network," "wide-belt and large-grid pasture shelterbelt network," and "mountain enclosures for forest conservation" in arid zones. As the main project of ecological environment construction in China, the state will attach importance to supporting the "Three North" Shelterbelt System Engineering project and continuously increase its capital investment.

12.2.2 The Beijing-Tianjin Sand Source Control Project

The Beijing-Tianjin Sand Source Control Project started in 2001, mainly to address problems related to wind and sand hazards in Beijing, Tianjin, and surrounding areas. The total area involved is 4.58×10^5 km^2, with an aeolian desertification area of 1.01×10^5 km^2 and a total population of 19.58×10^6 persons (Wang 2011a). The project focused on the comprehensive control of grassland desertification, the Hunshandake Sandy Land, desertified lands in agricultural and pastoral interlaced areas, and sandy lands in the Yanshan mountainous water source protection area through the implementation of afforestation, conversion of farmland to grassland, grassland protection, catchment comprehensive management, water source and water-saving facilities, etc.

The practical approaches of the project involve (a) forestry measures, including afforestation of 1.34×10^6 ha that were converted from farmlands, afforestation of 1.29×10^6 ha in badlands, artificial afforestation of 1.31×10^6 ha, aerial afforestation of 1.80×10^6 ha, and a mountain enclosure of 1.78×10^6 ha for forests;

(b) agricultural measures, including artificially planting grass on 1.48×10^6 ha, aerial sowing of plants on 0.29×10^6 ha, fencing on 2.79×10^6 ha, constructing a basic pasture on 0.34×10^6 ha, establishing a pasture seed base on 0.04×10^6 ha, prohibiting grazing on 5.68×10^6 ha, building greenhouse on 2.86×10^6 km^2, and supporting 23,100 feed machinery sets; (c) water conservancy measures, including 66,059 water source projects, 47,830 water-saving irrigation projects, and 23,445 km^2 of small watershed comprehensive management; and (d) ecological migration, including that of 1.8×10^5 persons (Wang 2011a, b).

The implementation of the project not only improved environmental quality but also increased farmers' income. According to a survey in 2009, from 2000 to 2009, the rate of shelterbelts for protecting farmland and pasture increased from 3.5% to 14.8%, and the surface dust release decreased by 19%; the concentration of air negative oxygen ions in the Beijing region doubled, and the cumulative number of days with the air quality equal to or better than Level 2 was 285 days, increasing by 110 days from the start of the project; the income of farmers increased significantly, which was higher than the national average in the same period (Wang 2011a). Although the Beijing-Tianjin Sand Source Control Project has achieved periodic results, it is arduous and long-term; thus, it is necessary to continue to consolidate achievements and further utilize science and technology.

12.2.3 Other Important Desertification Control and Ecological Construction Projects

There are two other key national desertification control and ecological construction projects that occurred over the first 10 years of the twenty-first century.

1. The returning farmland to forest/grass project was mainly implemented in areas with increased land desertification, soil erosion, and grassland degradation due to vegetation deterioration caused by firewood overharvesting, over-reclamation, overgrazing, and steep slope-farming, among other reasons. During the 10-year period, 1.4×10^6 ha of desertified land was controlled, and the ability of the area to resist wind and sand hazards was enhanced by returning farmland to forest/grass and restoring and increasing grassland vegetation (Wang 2011a).
2. The grassland desertification control project aimed to protect existing grasslands by fencing and rotational grazing and to promote conservation by restoration, such as artificial plantings, aerial sowing, and grassland improvements. In alpine regions, river source ecosystems should be restored and protected mainly through measures of returning grazing land to grassland and grass plantings; in areas with better light, heat, and water conditions, grass-crop rotation should be carried out to increase the development of high-yield and high-quality artificial pastures. During the project period, control of desertified land occurred on 12.79×10^6 ha (Wang 2011a).

12.3 Successful Models of Aeolian Desertification Control and Sand Industry in Typical Regions

The sand industry, a growing area in future desert science development, is an agricultural knowledge-intensive industry, and it involves the development of desertification control engineering. The intention of the sand industry is to coordinate the relationship between ecological environmental construction and economic development of sandy areas to fully utilize resources and advantages, avoid disadvantages, and establish a good feedback relationship between the ecological environment and eco-social sustainable development. Successful desertification control and development patterns under several typical natural conditions in China are summarized based on the activities in the agricultural areas of China for more than 50 years.

12.3.1 The Hotan Oasis in Xinjiang: Representing Efforts to Combat Desertification and a Sand Industry in an Oasis in an Extremely Arid Region

The Hotan oasis is located on the southwestern edge of the Taklimakan Desert. The southern part of the oasis contains a sand gravel plain in front of the Kunlun Mountains, and the northern part is directly connected to moving sand dunes. There is an average annual precipitation of only 34.8 mm and evaporation up to 2564 mm. The main mechanism to combat oasis desertification is to set up protection systems and stabilize the oasis ecosystem by implementing the following approaches (Wang 2011a):

1. Constructing water conservancy infrastructure. An integrated irrigation system was constructed to make full use of the water resources of the Yurungkax River and the Karakax River. Eighty percent of farmlands have been equipped with small agricultural water conservancy facilities. The utilization coefficient of the canal system increased from 0.35 to 0.40, and the irrigation quota decreased from 1800–2250 m^3/ha to 900–1050 m^3/ha.
2. Establishing a complete desertification control system with oases at the center. To protect natural vegetation through enclosures in the semifixed sand dune areas outside the oasis, a protective belt was established by planting shrubs and grasses through flooding irrigation. This approach not only prevents the occurrence of blown sand but also creates advantageous conditions for the development of animal husbandry. In the project area, a windbreak forest belt that was 358 km long and 10–300 m wide was built around the oasis, and a field protection network with a narrow forest belt and small grid was built inside the oasis. The total forest coverage of the oasis has reached 40.2%. On the farmlands, many

kinds of fruit trees were selected and arranged to carry out intercropping with grain crops or cotton.

3. Implementing additional measures. On the periphery of the oasis, in addition to enclosures for vegetation restoration, some other measures were applied to stabilize shifting sand dunes and to expand arable land through flooding in regions with suitable conditions, such as flattening isolated sand dunes and afforesting interdune lowlands for those that continuously distribute moving sand dunes.

This model to combat desertification has improved the oasis climate environment and produced evident economic benefits. For example, the wind speed has been reduced by 25%, and the sand content in the sand flow was reduced by 40–60% on the farmland compared with that in open areas (Zhu et al. 1989). In comparison to in the late 1970s, at the end of the twentieth century, the total grain, cotton, and vegetable cooking oil production in Hotan County increased 1.17-fold, 1.1-fold, and 2.31-fold, respectively; the grain yield per unit increased 3.3-fold, and the per capita income increased by 7.5-fold (Team for RDIX 1989).

12.3.2 The Pingchuan Oasis in Gansu: Representing Efforts to Combat Desertification and the Sand Industry in an Oasis in an Arid Region

The Pingchuan oasis in Linze County of Gansu Province, situated on the northern bank of the Heihe River in the middle Hexi Corridor, has a long and narrow shape. Its northern part is close to dense moving sand dunes and gobi. The annual precipitation is 117 mm, and there is a prevailing northwest wind. Aeolian desertification in the oasis is serious due to the original fixed shrub-coppice dunes being reactivated because of vegetation destruction, resulting in most cultivated lands being abandoned and the oasis retreating southward for 200–500 m (Wang 2011a). Therefore, establishing a complete protection system along the northern side of the oasis is an important measure.

First, using the long and narrow interdune lowland terrain along the northern side of the oasis, a sand-prevention forest belt with a width of 10–50 m was constructed through irrigation with spare cropland water. The forest belt contains *Populus gansuensis* and *Elaeagnus angustifolia*. The former has a remarkable wind-prevention effect and is mostly planted in areas with underlying soil layers, and the latter has abundant branches and leaves as well as a good ability to block wind and sand and is suitable for infertile soil layers.

Second, a 300 m × 500 m farmland shelter forest network was also built inside the oasis. The tree species were mainly *Populus gansuensis*, *P. nigra var. thevestina*, *Salix matsudana,* and *Ulmus pumila.*

For the dune and interdune lowlands outside the sand-prevention forest belt, sand-blocking barriers consisting of clay or reed were first set up, and then, different *sand-fixation plants such as Haloxylon ammodendron, Hedysarum scoparium, Caragana korshinskii*, and *Tamarix chinensis were planted under protection of the sand-blocking barriers.* To further prevent the external sand source, an enclosure belt with a width of 800–1000 m for vegetation recovery was planned outside the protection system. In winter, the farmland irrigation spare water is diverted into the enclosed area to accelerate the restoration of the vegetation.

Thus, an oasis-centered protection system was formed, using the combination effect of "blocking blown sand, fixing surface moving sand, and enclosing sand land for vegetation recovery" from the edge of the oasis to the periphery (Wang 2011a). Sheltered by the integrated protection system, the cultivated land in the oasis affected by desertification decreased from 17.8% to 0.4%, and the amount of agricultural and forestry lands in the sandy areas increased from 6.1% to 43%; in addition, the per capita income has increased by 153.6%.

12.3.3 The Naiman Banner in Inner Mongolia: Representing Efforts to Combat Desertification and the Sand Industry in an Interlaced Agropastoral Zone in a Semiarid Region

Naiman Banner is located in the middle of the Horqin Sandy Land and belongs to the interlaced area of agriculture and animal husbandry in the semiarid zone, where the average annual precipitation is 352.1 mm, the annual number of gale days is 21 days, and sandstorm days total 26 days. The original landscape is sparsely forested grassland developed on a terrain of sand dunes interlaced with bottomlands. The vegetation coverage varies regionally according to human activity intensity, which is at approximately 5–50%. As a result of excessive cultivation, overgrazing, and firewood harvesting, aeolian desertification has developed rapidly, and desertified land accounted for 20% of the total area in the late 1950s, 53% of the total area in the mid-1970s, and 77.6% of the total area in the late 1980s. Vegetation is commonly seen retrogressively developing from sparsely forested grassland to sandy vegetation. It is one of the severely desertified areas in northern China.

The model used to combat desertification and implement the sand industry is different according to different landscapes and land use characteristics in a region (Wang 2011a):

1. For wide interdune bottomlands mainly used for agriculture, the following measures have been implemented: (a) reestablishing artificial sand-fixing vegetation belts vertical to the main wind direction, where 3–5 years later, a relatively stable sandy shrub vegetation system formed; (b) contracting with each household, setting up fences to protect the vegetation on semifixed and fixed sand dunes, strictly prohibiting grazing and firewood collection, and taking supplementary measures such as sowing grasses; (c) adjusting the land use structures, returning desertification-influenced croplands to grassland, strengthening the

capital construction of farmlands with good soil and water conditions, increasing high-yield crop planting, and developing cash crops; and (d) adjusting the livestock structure, increasing the number of house-feeding animals, and reducing the number of grazing animals.

This desertification control model was adopted in Yaoledianzi village in the middle of Naiman Banner. After 10 years of desertification control, the desertified land area decreased from 77% to only 13% of the total area; the vegetation coverage increased from less than 15% to more than 30%; the proportion of dry cropland to cultivated land decreased from 78.8% to 17.3%, while the share of irrigated land increased from 21.1% to 82.7%; the per unit area yield of grain increased from 675 kg/ha to 5130 kg/ha, and the per capita income increased by 126.3%.

2. For fixed and semifixed sand dune lands mainly used for animal husbandry, each ecological household considered as a unit was contracted as responsible for desertification control of its adjacent land, and the tasks mainly included (a) flattening the interdune lowland near the homestead to develop irrigated agriculture with groundwater (water table generally at a depth of 3–4 m), where if conditions permit, then rice cultivation in film-bottomed sandy land could be adopted, using this as the base to further develop cash crops, and (b) setting up enclosures around the base for future rangeland use, that is, to build tree and shrub forest belts near the base, to plant sand-fixing plants on sand-shifting lands, and to enable grass vegetation to naturally develop or to supplement by sowing forage seeds.

   Thus, many small eco-economic areas combining agriculture, forest, and animal husbandry were formed, with areas of 0.15–0.25 km^2 each. This small green eco-economic area not only solved the economic issues of each household but also prevented the expansion of land desertification. The model was adopted in Baiyintala village in Naiman Banner. Five years later, the vegetation coverage in the enclosed area increased from 5–15% to 60–70%; the per capita annual income increased from 400 to 1500 RMB yuan, with an increase of 75% over the 10-year period.

3. For slightly undulating desertified lands, the countermeasures implemented were the following: (a) changed from extensive rainfed farming with grain as the main crop and adjusted the ratio of agriculture to forestry and to animal husbandry from 6:2:2 to 2:4:4; (b) constructed farmland and rangeland sheltering forest networks at a large scale and enclosed the marginal areas for vegetation restoration; (c) developed irrigation agriculture in sheltered grids with good soil and water conditions; and (d) developed animal husbandry forage bases in grids on fixed and semifixed sand dunes on slightly undulating terrain.

   This model was adopted in Huanghuatala village in the south of Naiman Banner, where desertified land originally accounted for 83% of the land area. After the model, more than 500 shelterbelts of small ecosystem grids were built. Approximately 85% of the desertified lands have been managed. The agriculture and animal husbandry economies developed harmoniously, and the total grain production, forage yield, and per capita annual income increased 3-fold, 1.5-fold, and -fold, respectively.

4. For greatly undulating regions mainly with dense moving sand dunes, the following measures were implemented: (a) shrubs were planted on moving sand dunes, under the premise of establishing a sand-blocking tree forest belt in front of them, and evergreen trees were planted within the shrubs; (b) in some open interdune lowlands, groundwater was exploited to irrigate crops by applying rice cultivation in film-bottomed sandy land; and (c) natural enclosures combined with artificial supplementary grass sowing were the most basic form of land desertification control in the interlaced semiarid agriculture and animal husbandry area.

   This model is well demonstrated in Bagapoli village in Naiman Banner. In the approximately 10 years from 1975 to 1985, vegetation coverage increased from 20% to 71%, and the total biomass dry weight increased from 1089.0 kg/ha to 7675.5 kg/ha; the organic matter content in the 0–50 cm soil layer increased from 0.045% to 0.537%, and the grass yield increased 1.5-fold. The per capita annual income increased two-fold.

12.3.4 The Zhanggutai Region of Liaoning Province: Representing Efforts to Combat Desertification and the Sand Industry in a Loess Hilly Area in a Transition Zone Between Semiarid and Semihumid Regions

The Zhanggutai area, situated at the southeastern margin of the Horqin Sandy Land and in the agricultural and pastoral interlaced zone, has a temperate continental monsoon climate. Historically, the Zhanggutai region had dense forests and lush pastures. However, driven by immediate interests, the vegetation was seriously damaged and became a large area of sandy desertified land due to unsustainable human activities such as forestland reclamation, overharvesting of firewood, overgrazing, and exploiting natural resources. On the other hand, because the land surface in the Zhanggutai region originally consisted of sandy sediments, the soil mechanism is unstable; the monsoon climate increases the effect of wind erosion when the surface is exposed during periods of farmland cultivation in winter and spring, which aggravates the aeolian desertification of the lands. This is a typical area of land aeolian desertification in northern China; 41.2% of the area has moving or semimoving sand dunes, and 58.8% of the area has fixed dunes (Su et al. 2006).

The main mode of desertification control in this area is to restore vegetation mainly by afforestation. Measures taken include the following (Zhao 2013): (a) strengthening the conservation and sustainable exploitation of existing forest resources and establishing a long-term mechanism of harmony between man and nature; (b) promoting afforestation steadily; creating an integrated desertification control system combining trees, shrubs and grasses; and converting sandy cultivated land to forest and grass in a planned way; (c) establishing and improving protective measures for natural restoration of surface vegetation in hill regions; (d) constructing nature reserves to improve microclimates, fanning the effect out; (e) nationally exploiting and utilizing water resources, ensuring ecological water use in desertification areas; and (f) strengthening management and protection of forests according to laws and comprehensively consolidating the achievements of desertification control.

After a group of young science and technology researchers from all over the country established the Sand-Fixing Afforestation Experimental Station in Zhanggutai in 1952, they successfully introduced and built the first windbreak and sand-fixing forest of *Pinus sylvestris* in China in 1955. They then continued to carry out a series of studies on the construction techniques of mixed forests, farmland/grassland shelterbelts, and sandy orchards, as well as on forest conservation techniques. After more than 50 years of efforts by several generations, the forest cover in the Zhanggutai region has increased from 6% to 67%, and the total grain output increased tenfold (Zhu 2014). The Zhanggutai area, developing toward the integration of forestry, agriculture, animal husbandry, and tourism, is now a national sandy land forest park.

12.4 Successful Protection System for Traffic Projects Related to the National Economy

Shifting sand is one of the serious contributors to desertification in China. In the vast desert areas in northern China, transportation facilities, as important infrastructure and the leading industry of the national economy, often suffer from moving sand damage. The establishment of a protection system along traffic lines has always been the key topic of scientific researchers, for which several generations have implemented painstaking efforts and have achieved a series of successes in different regions. These successful experiences can be used for reference in combating desertification in areas with similar conditions.

12.4.1 The Protection System for the Baotou-Lanzhou Railway: The Shapotou Section as a Typical Case in the Moving Sand Dune Region

The Baotou-Lanzhou Railway, formally opened to traffic in 1958, is a pioneering work in the history of railways in China because it crossed the Tengger Desert six times, with a cumulative length of 40 km through the desert. In particular, it crossed a region of undulating latticed crescent dunes with a relative height of 10–20 m at the

Shapotou section in the southeastern edge of the desert. The Shapotou Desert Experimental Research Station, established in 1956, has always provided theoretical and technical support for preventing the railway from moving sand damage. Experimental research in the Shapotou region over many years indicated that the railway sand-prevention system essentially achieved the purpose of long-term biological sand fixation and eco-environmental improvement (Wang 2011a):

1. Primary vegetation formed, and its protection function improved. In comparison to in bare sand dune areas, in the fixed sand area with vegetation coverage at 20–30%, the wind speed was reduced by 55% (compared with that at a 20 cm height over the moving sand dune area), the ground roughness increased 957.7-fold, and the near-surface sand flux was only 21.7%.
2. Microclimatic environmental improvements occurred due to the existence of aboveground sparse vegetation and plant roots in soil. Vegetation allows solar radiation to transmit but slows airflow so that solar and surface radiative heat is not easy to emit; thus, heat conduction and heat storage capacity are relatively weak in soils with plant roots.
3. Biological soil crust formed, and soil developed. Wind tunnel experiments have shown that gray or gray-black crust can prevent soil erosion under a wind speed of 25.6 m/s. The contents of physical clay particles and organic matter in the soil crust and lower substratum increased significantly in comparison with those in the moving sand dune area. The minimum water-holding capacity of this thin sandy soil layer was as high as approximately 20%, and the thin sandy soil layer intercepted rain water four times more than that of bare dune sands. The relatively fertile wet topsoil created a favorable habitat for the reproduction of annual plants and cryptogams in the rainy season. These plants were well developed and in turn beneficial to biological crust formation and soil-forming processes.
4. Based on artificial vegetation, biodiversity increased, and herbs, lower plants, and microbes multiplied in large numbers, forming a naturalized multifunctional ecosystem more stable than artificial vegetation. The sand land ecosystem composed of animals, plants, and the environment has a tendency to evolve from an azonal geographical environment (moving sand) to a zonal geographical environment (steppe desert).

The successful experiences in moving sand fixation for the Baotou-Lanzhou Railway have been extended in the whole country and even in developing countries. The comprehensive techniques of the railway protection system are detailed in Chap. 10 and will not be repeated here.

12.4.2 The Protection System for the Lanzhou-Xinjiang Railway: The Yumen Section as a Typical Case in the Gobi Region

The Lanzhou-Xinjiang Railway, opened to traffic in 1962, is an important traffic line leading from northwestern China to all parts of the country. It includes an important part of the New Eurasian Continental Bridge, with a total length of 1892 km, of which 1100 km runs in the vast Gobi Desert. Most of the gobi surface is covered by a Quaternary gravel layer, and there is no sufficient sand source; thus, the wind-sand flow is mostly unsaturated. More than 80% of the sands in wind-sand flows are distributed within a height of 200 cm, far more than the 10 cm over bare sand dunes (Wu and Ling 1965; Yin 1989; Zou et al. 1995). The gobi surface is rough and often cemented with mud or calcium; the threshold of the wind speed for sand movement is 6–7 m/s, slightly higher than that in desert areas. The blown sand damage in the gobi area can be classified into three categories: wind erosion, sand accumulation, and sand abrasion. The establishment of railway forest sand control systems is an economical, reliable, and long-term effective method to control sand (Feng et al. 1995). The research project "Sand-combating System in Yumen section of Lanzhou–Xinjiang Railway" is a successful demonstration of irrigation afforestation in the gobi region with good water conditions. This project involved forest belts with different widths and different numbers of strips, which were arranged in parallel or nearly parallel to the railway line (Fig. 12.2). The main purpose of this system was to reduce the wind speed and cut off the sand supply (Qi 1996).

Erecting a vertical sand barrier on the windward side of sand-prevention belts is an effective measure to prevent sand accumulation. The distance between a sand barrier and forest belt depends on the strength of the wind-sand flow, generally 15–20 m; the height of the sand barrier should not be less than 1.5 m; and its direction needs to be consistent with the direction of the forest belt.

Forest belts are the basis of the railway protection system; the width, configuration, and tree composition of forest belts directly affect the function of the system.

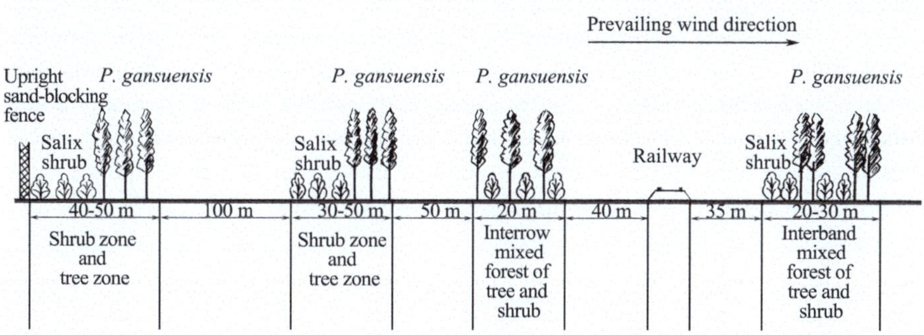

Fig. 12.2 Cross section of a complete sand-prevention system for the Yumen section of the Lanzhou-Xinjiang Railway (Wang 2011b)

The setting of forest belts follows a principle of adapting measures according to local conditions, defending against harm and emphasizing prevention over treatment. A forest belt is mainly used to prevent outside wind-sand flow and can be widened appropriately to control local sand. The distance between forest belts is most effective when the main afforestation tree species in the forest belt grow to their normal heights. Tree species selected for building forest belts should have the following properties: (a) substantial ground coverage and long life spans; (b) strong root sprouting, fast-growing and uniformly distributed roots, and large root amplitude; and (c) tolerance to atmospheric and physiological droughts as well as to soil infertility and strong wind resistance. The forest belts should include at least 3/5 to 2/3 of sandy or xerophytic shrubs and semishrubs that are easy to propagate, fast growing, and strong self-renewing. The afforestation method with nonnative soil naturally deposited in excavated ditches has been proven to be beneficial to the survival and growth of tree species. The stable water content in guest soil can be increased by 54–60%, and the total nitrogen content can be increased by 62.5% (Wang 2011a). This approach can promote early canopy closing of stands. In addition, irrigation and coppicing in time can promote the rejuvenation of trees.

The function of railway sand-prevention systems to reduce wind speed is closely related to their own structure (forest belt height, width, and sparseness), wind velocity, and other meteorological conditions. The higher the forest belt, the greater the protective distance is; a forest belt that is too wide or too narrow will reduce its effect, and the optimum width of a forest belt is 25 to 8 m; the best effect is when the porosity of the forest belt is at 0.3–0.4.

The completely established railway forest protection system showed obvious eco-environmental effects. Wind-sand damage to railways can be reduced after 1 year and can be essentially eliminated within 2–3 years. These protection systems have reduced wind speeds by 20.2–30.5% within the distance of 0–30 H (H, height of forest belt) behind the forest belt, with the best reduction at 47.0% at 10–15 H; in comparison with that in an open field, the forest protection system can increase air temperatures by 0.6–2.8 °C (at 0.5–2.0 m in height) in spring and autumn seasons and decrease it by 1.2–1.5 °C in summer. In summer, these systems can also reduce the daily maximum ground surface temperature by 3–5 °C and the minimum by 4–6 °C; it can reduce evaporation by 38.7–71.8% and increase air relative humidity by 5.1–6.8%; under the forest, fine composition of less than 0.1 mm in the 0–5 cm soil layer increased by 6.0–3.7% or two- to sixfold than that in the open field; the soil bulk density decreased by 0.06–0.11 g/cm^3, and the porosity increased by 1.6–4.1% (Wang 2011a). In addition, a large amount of plant litter and decomposed roots participate in fertility cycling, supplying nutrients for plants.

12.4.3 The Protection System for the Tarim Desert Highway: A Typical Case in an Extremely Arid Region

The Taklimakan Desert is the largest desert in China and the second largest desert in the world. The Tarim Desert Highway runs through the Taklimakan Desert from north to south, with a total length of 522 km. The Tarim Desert Highway, with its whole-line sand-prevention system, has set a Guinness World Record. The idea of sand prevention is taken into consideration not only in the maintenance of highways but also in early route selection, linear and cross-section design, and construction.

In establishing the highway sand-prevention system, it followed the two-step strategy of "first engineering then biological" measures and the implementation method of "engineering measures ensuring highway operations as a basis, chemical measures as assistance, and ecological rehabilitation and restoration as a goal" (Wang 2011a). A comprehensive highway sand-prevention system (Fig. 12.3) has been designed based on a series of concentrated studies: field observations of meteorological and climatic conditions, studies of blown sand movement laws, wind tunnel simulation tests of different desert landforms and highway directions, field tests of various sand-prevention methods (mechanical, chemical, and biological), observations of sand-prevention benefits of the system, investigations of all kinds of sand-fixation material resources, and studies on development and breeding of sand-fixation plants.

To increase the development rate of oil fields, the highway was partially opened to traffic while being built, and a highway protection system of "sand blocking in combination with sand fixing," using sand-blocking fences and straw-checkerboard sand barriers as the principal measures, was simultaneously established to ensure smooth operation of the highway. This combination system initially included the front sand-blocking belt, the bare sand-accumulation belt, the semihidden straw-checkerboard sand barrier belt, and the highway slope protection belt. As of September 1995, when the highway ran through the Taklimakan Desert, 800 km of sand-blocking fences had been erected, and 42×10^6 m^2 of checkerboard sand barriers had been laid (Wang 2011a).

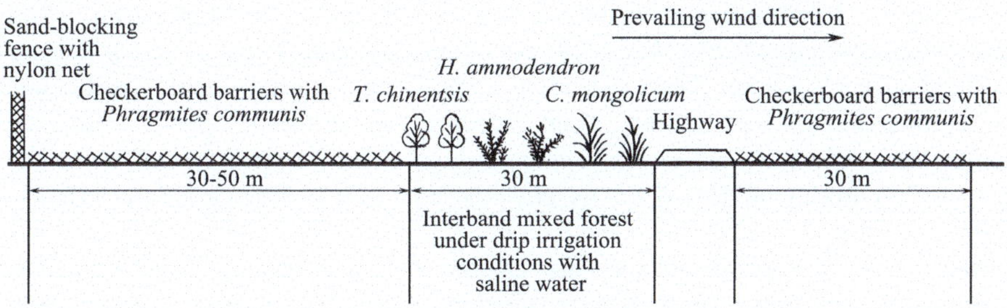

Fig. 12.3 Cross section of a complete sand-prevention system along the Tarim Desert Highway (Wang 2011b)

To ensure that the sand-blocking engineering performed its sand-prevention function well with a maximum time duration, it was necessary to adopt a technology of fixed fences combined with movable fences to strengthen the front sand-blocking effect according to local terrain and sand damage conditions (Feng and Qian 1986; Dong et al. 1998). To improve the sand-fixation efficiency of straw-checkerboard sand barriers, it is also necessary to improve the design and construction of sand barriers at different sites; for example, it is advisable to set up a 0.5 m × 0.5 m scale of checkerboard sand barriers in an upwind landform region, 2 m × 2 m in a leeward region, and 1 m × 1 m or 1 m × 2 m in a smooth, open, and flat sandy area.

In continuously or intermittently distributed flat, open sandy areas, where sand damage mainly occurs from blown sand and low dune movements, the shoulder slope of a highway could be solidified by using synthetic chemical materials. There is a certain distance on both sides of the highway that can be solidified after clearing obstacles (obstacles include plants and shrub-coppice dunes, and the solidified width is far smaller than the cleared width); thus, a smooth section can be created, and the wind-sand flows can smoothly transmit over the road. Such measures can effectively control sand damage to highways and save costs.

The width of a protection belt should be determined flexibly according to the local wind strength, cross-section sand flux, dune distribution and moving speed, as well as the angle between the highway direction and the main wind direction (Feng and Qian 1986). Although factors affecting the reasonable protection width are complex, many different empirical models to obtain in situ optimum widths have been proposed by Chinese scholars, and they have significant reference value (Wang and Chen 1997).

Biological sand fixation is a fundamental solution to control sand damage to desert highways. However, plant habitats in the Taklimakan Desert are extremely harsh, so irrigation is a prerequisite for plant survival. First, reasonable irrigation should be implemented using local salty groundwater to ensure the normal growth of plants; second, plant species suitable for local climate conditions and water sources and planting methods are key to the success of biological sand fixation.

Taking into consideration the adaptability of plants to arid environments, the stability of plant growth, and salinity adaptability to soil salt accumulation under different irrigation methods, three tree species, i.e., *Tamarix chinensis* (*Tamarix taklamakanensis*, *Tamarix austromongolica*, and *Tamarix gansuensis*), *Calligonum arborescens* (*Hedysarum scoparium* and *Calligonum caput-medusae*), and *Haloxylon ammodendron* (*Haloxylon persicum* and *Haloxylon aphyllum*), were selected as sand-fixation plants cultivated in large areas through field experiments. In comparison to introduced plant species, as domestic plants, *Tamarix taklamakanensis* and *Calligonum caput-medusae* have better adaptability to the growing environment. Based on its physiological and ecological characteristics, *Tamarix taklamakanensis* was planted on the outermost side in the upwind direction because of its strong resistance to sand erosion and sand burying, fewer branches, and greater sparseness; *Hedysarum scoparium*, being fast growing and relatively drought tolerant, was planted in the middle dunes; *Tamarix austromongolica*, which grows fast, has many branches and tall trees and was planted near the highway.

Haloxylon ammodendron grows slowly in the early stage and has bright green leaves; it was also planted near the highway. All these species have developed considerable tolerance for aridity, wind, salinity, and limited nitrogen. In terms of plant growth performance, the method of border irrigation is better than furrow irrigation and drip irrigation; in terms of physiological adaptation, the stress resistance of the three tree species under drip irrigation was the highest among the three irrigation patterns.

In October 2006, the project "Ecological Forest Engineering for the Tarim Desert Highway" received approval, soon after the Tarim Desert Highway was completed (in autumn 2005). A forest belt of 436 km was built along the desert highway, with more than 20 million shrubs planted, covering an area of 3128 ha; there were 111 newly drilled irrigation wells, 430 km UPVC main water supply pipelines, 25.63 km of 10 kV HV line, and many other service facilities (TOB 2007).

The sand-blocking effects of high fences and the sand-fixation effect of straw-checkerboard barriers are outstanding. Sand-blocking fences rapidly reduced the energy of wind-sand flow and changed saturated or unsaturated wind-sand flows into supersaturated flows so that some sand grains were intercepted. The checkerboard sand barriers changed the structure of sand flow fields near the ground surface, which decreased the wind speeds rapidly at a 2 cm height, and the sand-carrying capacity of flow and the sand flux decreased sharply at a 10 cm height. The shrub sand-fixation belt constructed along the Tarim Desert Highway is a tight-structure forest. Sand blocking and sand fixation are its main functions, and the resulting improvements in the local microclimate will also be beneficial to the permanent role of the system. Considering a shrub forest belt 3 years after its planting as an example (Wang et al. 2003; Wang 2011a, b), the wind speeds at a 2 m height within the checkerboard barrier 20 m from the forest, 2 m from the forest, and at site in the forest decreased by 21.05%, 32.69%, and 77.68%, respectively, compared with that in an open field (50 m to the forest belt), and wind speeds at a 0.5 m height decreased by 44.67%, 93.72%, and 98.23%, respectively; the shrub cover was over 90%, and the wind speed in the forest belt was often less than the minimum threshold of the anemometer, which completely prevented the local wind erosion. In comparison with that in an open field (50 m to forest belt), the sand flux at 2 m and 15 m behind the forest belt decreased by 95.33% and 85.82%, respectively, but the sand flux at 2 m to the front of forest belt decreased by 26.10%, indicating that the shrub forest belt had a greater sand-preventing effect than that of checkerboard sand barriers. Overall, the establishment of a shrub sand-fixing forest belt not only provided protection for the Tarim Desert Highway but also improved the local microclimates, resulting in the reestablishment of some natural plants, an increase in birds and other animals, and the creation of a stable and harmonious eco-environment.

12.5 Conclusion

The Chinese government has always attached great importance to combating aeolian desertification for the protection of national structures, improvement in the eco-environment and improvement in living standards of the rural population, and has incorporated combating aeolian desertification as a basic state policy in the National Economic and Social Development Plan. In the long history of combating aeolian desertification, some acceptable and practical techniques, successful demonstrations, and valuable experiences as well as extension service models have been developed at community, local, and national levels, for instance, the techniques for dune fixation and sand stabilization by using biological and mechanical measures; the techniques of aerial seeding to restore vegetation on sand-shifting lands; the establishment of sheltering forest networks on farmlands; the construction of national ecological barriers; sustainable rotational grazing and determination of rangeland quality and carrying capacity; approaches to develop fodder farms and high-efficiency grazing lands and the development of eco-farms as a means of improving vegetation cover; rice cultivation in film-bottomed sandy land; and other successful experiences. Combating aeolian desertification is a long-term and arduous task as well as a grand effort in systematic engineering, which requires the integration of knowledge among various disciplines and the mobilization of all social forces to participate in and make use of advanced science and technology to promote aeolian desertification control.

References

Dong ZB, Chen GT, Yan CZ (1998) Sand dunes movement near the Tarim desert oil-transportation highway. J Desert Res 18(4):328–333 (in Chinese with English abstract)

Feng LC, Qian ZY (1986) On the sheltering principle of straw checkerboard sand barriers and the calculation formula of effective protective width. In: The Industrial Aerodynamics Professional Committee of China Aerodynamics Society. In: Proceedings of the second national conference on wind engineering and industrial aerodynamics (in Chinese with English abstract)

Feng LC, Lu J, Di YQ (1995) A Summary of research on railway wind-sand hazards and their prevention and control in desert region. In: Proceedings of the Taklimakan Desert international scientific congress: Desert past, present and future. Arid Zone Research (Suppl), pp 475–487 (in Chinese with English abstract)

Planning Office for Construction of "Three North" Shelter-forest System (Three North Office), Northwest Institute of Forestry Investigation Planning and Design (NW Institute) (1990) The atlas of the construction of the shelter-forest system in northwest, north and northeast China. Xi'an Map Publishing House, China (in Chinese)

Qi YZ (1996) Study on the prevention system from wind-sand flow in railway construction in desert and semi-desert areas. J Desert Res 16(3):266–274 (in Chinese with English abstract)

Su F, Guo C, Xu Q et al (2006) Growth status of Pinus sylvestris var. mongolicas plantion for sand fixation in wind erosion area. J Northwest For Univ 21(04):65–68 (in Chinese)

The Comprehensive Study Team for Resources Development in Xinjiang, CAS (Team for RDIX), (1989) Study on ecology and environment in Xinjiang. Science Press, Beijing (in Chinese)

The Tarim Oilfield Branch of China National Petroleum Corporation (TOB) (2007) Chronicle of events in green corridor construction— an ecological engineering of shelter forest belt across the Taklimakan desert known as "sea of death". Xinjiang People's Publishing House, Urumqi (in Chinese)

The "Three North" Shelter-forest System Construction Bureau (Three North Bureau) (1993) Plan for construction of the protection forest system in northwest, north and northeast China. Ningxia Publishing House, Yinchuan (in Chinese)

Wang T (2011a) Sand control engineering in China. Science Press, Beijing (in Chinese)

Wang T (2011b) Deserts and aeolian desertification in China. ELSEVIER in Amsterdam and Science Press, Beijing

Wang XM, Chen GT (1997) Evaluation of effectiveness and discussion of reasonable width for the mechanical sand control system along the Tarim desert highway. J Arid Land Res Environ, 11(1):28–35 (in Chinese with English abstract)

Wang J, Sang CQ, He Q (2003) Analysis of microclimate on different underlying surfaces at middle section of the Tarim desert highway. J Desert Res 23(5):577–580 (in Chinese with English abstract)

Wu Z, Ling YQ (1965) A preliminary study on some rules of wind-sand movement and prevention of wind-sand hazards. Research on Sand Control (No. 7). Science Press, Beijing (in Chinese)

Yin YS (1989) Study on wind and sand flow in gale area in gravel desert. J Desert Res 9(4):27–36 (in Chinese with English abstract)

Zhao YM (2013) Current situation and countermeasures of land desertification in Zhangwu County, Liaoning Province. Protection Forest Science and Technology, (7) (in Chinese with English abstract)

Zhu YX (2014) Study on sand fixation effect of *Pinus sylvestris* plantation in Zhanggutai sandy land, Liaoning Province, Theory Horiz (5):62–64 (in Chinese with English abstract)

Zhu ZD, Liu S, Di XM (1989) Sandy desertification and its control in China. Science Press, Beijing (in Chinese)

Zou XY, Dong GR, Wang ZL (1995) Some characteristics of wind and sand flow in Gobi region. J Desert Res 15(4):368–373 (in Chinese with English abstract)

Chapter 13
Implementation of Measures to Combat Aeolian Desertification in Mongolia

Tao Wang, Mandakh Nyamtseren, and Jing Pan

Abstract In Mongolia, there are several major anthropogenic factors that have a serious impact on the aeolian desertification process, including animal husbandry and related overgrazing, inappropriate transformation of rangelands into croplands, deforestation mainly for household purposes, inadequate rehabilitation of lands affected by mining activities, and poor planning efforts related to settlement and road construction, with animal husbandry having the highest pressure on land over the years. As a party to the United Nations Convention to Combat Desertification, Mongolia has developed United Nations Convention on Combating Desertification-National Action Plans (UNCCD-NAPs) to combat aeolian desertification, and under this comprehensive framework, many national and international projects with assistance from international organizations and donor countries have been implemented to tackle aeolian desertification. Some success stories have emerged related to the implementation of projects to combat aeolian desertification in Mongolia, and the

---

T. Wang (✉)
Northwest Institute of Eco-Environment and Resources, Chinese Academy of Sciences, Lanzhou, Gansu, China

Key Laboratory of Desert and Desertification, Chinese Academy of Sciences, Lanzhou, Gansu, China

Drylands Salinization Research Station, Northwest Institute of Eco-Environment and Resources, Chinese Academy of Sciences, Lanzhou, Gansu, China
e-mail: wangtao@lzb.ac.cn

M. Nyamtseren
The Institute of Geography and Geoecology, Mongolia Academy of Sciences, Ulaanbaatar, Mongolia

J. Pan
Northwest Institute of Eco-Environment and Resources, Chinese Academy of Sciences, Lanzhou, Gansu, China

Drylands Salinization Research Station, Northwest Institute of Eco-Environment and Resources, Chinese Academy of Sciences, Lanzhou, Gansu, China

Key Laboratory of Desert and Desertification, Chinese Academy of Sciences, Lanzhou, Gansu, China

T. Wang et al. (eds.), *Combating Aeolian Desertification in Northeast Asia*, Ecological Research Monographs, https://doi.org/10.1007/978-981-16-9028-0_13

future focus of UNCCD-NAPs in Mongolia should be on a sustainable mechanism to enable the interaction of agriculture, animal husbandry, forestry, and industry.

Keywords Desertification combating · Project · Implementation · Demonstration

13.1 Introduction

Based on the definition of desertification in the United Nations Conference on Environment and Development (UNCED) (1992), approximately 90% of the Mongolian territory is considered vulnerable to land degradation and aeolian desertification. However, 75% of Mongolian land is rangelands, including grasslands, shrublands, sandy lands, and semideserts. Vast territory and low population density on the one hand and the harsh climate with the highest amplitude of diurnal, monthly, and seasonal temperatures and relatively low precipitation on the other hand have predetermined the development of an extensive grazing pattern in the country. In terms of other land use types, urban land occupies 5.2% of the territory, croplands occupy 6.4%, and mining occurs in ~2.2% of the territory (Gazriin hariltsaa, geodezi, zuragzuin gazar 2019). According to the latest assessment of land degradation and desertification, approximately 76.8% of the land in Mongolia is considered degraded, of which 22.1% is severely degraded and needs immediate measures on restoration and rehabilitation. Among the major factors that have had the strongest influence on grassland degradation and aeolian desertification over the years, anthropogenic activities and animal husbandry have been the main causes of land degradation in the country (Hudulmur and Tsogtbaatar 2010; Tuvshintogtokh and Ariungerel 2013). Improving rangeland management, therefore, should be one of the main solutions for tackling degradation and desertification in Mongolia (Bedunah and Schmidt 2004; Nixson and Walters 2006; Sekiyama et al. 2014).

There are different reasons for the desertification of grasslands in Mongolia, and desertification is closely linked with the increased overgrazing that occurred after the country's socioeconomic transformation, which started in the late 1990s. The other major reasons for aeolian desertification, in addition to climate change, which is highly pronounced in the steppe regions of the country, are disturbance from mining activities, uncontrolled roads and transportation networks (especially dirt roads), inappropriate transformation of grasslands into croplands, deforestation, and forest degradation.

This section will discuss recent progress toward addressing land degradation and desertification, focusing on aspects such as policy, legislation, projects, programs, best practices, and future projections.

13.2 Policy and Legislation

Until 1996, Mongolia had no policies or actions focused on land degradation and aeolian desertification, although the national assessment of desertification was conducted in 1992 jointly with the Turkmen Institute of Desert Study. Even though this first-ever assessment considered only the Gobi region of the country, it highlighted several problems related to land use practices, especially the high concentration of livestock in one place (Kharin and Tsolmon 1994). With adoption of the United Nations Convention on Combating Desertification (UNCCD) in 1996, Mongolia agreed to mainstream policies addressing desertification. The first National Action Plan to Combat Desertification and Drought was developed in 1996 with technical support from the United Nations Environment Programme (UNEP), United Nations Development Programme (UNDP), and Economic and Social Commission for Asia and the Pacific (ESCAP). In conjunction with implementing this plan, the National Committee to Combat Desertification was established under the Ministry of Environment. The first National Action Plan (NAP) of Mongolia to combat desertification developed a project list with strict milestones and practical objectives. In total, 11 project concepts were developed, and of these, 3–5 projects were supported by donor countries and international organizations. The advantages of this first NAP were in its concreteness, robustness, and management focus. However, because not all proposed projects were implemented, several gaps were defined and addressed during the development of the second NAP, which was developed in 2003 (Mandakh et al. 2016). In the following years, NAP development followed the general guidelines defined by the UNCCD in connection with its strategies. In total, the Mongolian government changed its NAP, which until 2015 was as both a legal and policy document, three times, in 1996, 2003, and 2010 (Banzragch and Enkhbold 2010; Dorj et al. 2013). The current NAP to combat desertification aims to (1) strengthen institutional capacity; (2) improve the legal and policy framework; (3) enhance science, technology, and knowledge integration; (4) increase advocacy to raise awareness and increase education; and (5) intensify concrete actions at the grassroots level and increase investment (Jamsran et al. 2015). The advantage of this NAP is in its practical approaches to solve complex problems, and it is more action-oriented than previous versions; thus, in comparison to earlier versions, this version is people-centered. For the first time in 20 years, Mongolia aimed to separate legal, policy, and institutional issues related to combating desertification, which created the space for involving broader communities, especially from non-environmental sectors that have impacts on land.

To support the implementation of the UNCCD-NAP, improve cross-sectoral involvement, and address other development issues, the government of Mongolia has been initiating and implementing several national programs, e.g., the National Green Belt program, Mongolian Livestock program, Millennium Road program, and Poverty Eradication program. The implementation of these cross-sectoral programs is supported by the national budget, donor countries, and international organizations.

In addition to these nationwide programs, the government of Mongolia and Parliament, with support from donor countries and international organizations and the impact of the global trends, have paid substantial attention to improving its environmental legislation and enhancing policies toward achieving SDGs and conserving the environment. During the past 30 years, Mongolian environmental legislation has drastically improved; e.g., the concepts of community-based natural resource management have become increasingly important and been included in several laws and policies as major mechanisms for sustaining environmental resources and their flows. The importance of restoration and support for the natural regeneration of native ecosystems has become a cornerstone of all resource-, land-, and environment-related sectoral policies. With the legislation related to the conservation offset concept, all extractive industries, road construction activities, and large construction projects are urged to implement actions to avoid, reduce, or restore any impacts to nature. Such changes and integration of environmental requirements into non-environmental sectors may have a positive impact in the long term.

In addition, in 2012, the Parliament and government of Mongolia made a "revolutionary" decision by approving the Law on Soil Protection and Desertification Prevention. The Soil Protection and Desertification Prevention Law provides a framework of actions dedicated to preventing, restoring, and rehabilitating degraded land (NCCD 2018). Compared to other environmental laws, the current law has the advantage of defining different response levels to address land degradation and desertification, which is novel to Mongolian environmental legislation. The law stresses that preventing and combating desertification and land degradation should be addressed jointly with other economic sectors because they are at the core of the problem. Although much attention is given to pasture use problems, the law pronounces that all actions should follow a sustainable land management approach. As the law is new and no monitoring efforts or evaluations exist yet to judge the pros and cons of this law, it is expected that several amendments must be added, especially considering the new approaches and practices that are emerging with targets to achieve land degradation neutrality for the country.

Furthermore, land degradation and desertification problems are highlighted in several national development policies, e.g., Sustainable Development Vision 2050 and the Green Development Policy (Lee & Ahn 2016). According to these nationwide plans, Mongolia aims to reverse degradation by 10–30%, increase forest coverage by as much as to 8–9%, and expand areas of protected lands to ensure sustainability and resilience in the future.

13.3 Projects and Programs

Although the government of Mongolia is paying considerable attention to land degradation and desertification problems, its human, technical, and financial capacity is not enough to address all aforementioned goals and targets. Therefore, the country is taking a significant step by initiating bilateral and multilateral cooperation and mechanisms at regional and global levels. Several projects and programs have been implemented in Mongolia with the support of donor countries and international organizations (MNEM 1997; UNCCD 2006; Darima 2020). Below, the projects that have had a significant impact are highlighted:

- Sustainable land management to combat *desertification* in *Zamyn Uud* Sum (1997–2000). Supported by the Global Environment Facility-Small Grants Programme (GEF-SGP).
- Rangeland Management Project in Mongolia's Gobi Desert (1995–2006). Implemented by the German Agency for Technical Cooperation.
- Combating Desertification in Asia (2001–2005). Supported by the Asian Development Bank, UNCCD Secretariat.
- Combat Desertification by Improving Pasture Management (2000–2003). Supported by the German Agency for Technical Cooperation.
- Sustainable Grassland Management (2002–2007). Implemented by the GEF, UNDP, the Government of Netherlands.
- Prevention of Dust and Sand Storms in North-East Asia (2003–2005). Supported by the Asian Development Bank, Government of Japan, and Global Environment Facility.
- Mongolian Pasture – Green Gold (2004–2007). Supported by the Swiss Development Cooperation Agency.
- GL-CRSP Gobi Forage Project (2004–2008). Implemented by USAID.
- Let's Make Mongolia Green (2004–2009). Initiated and implemented by the International Rotary Organization, Government of Mongolia.
- Korea-Mongolian Green Belt (2005–2035). Implemented by the Korea Forest Service, Republic of Korea.
- Coping with Desertification in Mongolia Project (2005–2015). Funded by the Swiss Development Cooperation (SDC).
- Sustainable Land Management to Combat Desertification in Mongolia (2008–2012). Financed by the Global Environment Facility, Netherlands, government in-kind contributions.
- Prevention and Control of Dust and Sand Storms in North East Asia (2002–2011). Financed by the UN Economic and Social Commission for Asia and the Pacific (ESCAP), among others.

The projects supporting the implementation of national strategies to combat desertification can be classified into the following three categories:

13.3.1 Sustainable Land Management Projects

Sustainable land management projects are principally for protecting and restoring ecosystems and essential ecosystem services that are key to reducing poverty in Mongolia, strengthening the environment for sustainable land management by building capacities in appropriate government institutions and user groups, and demonstrating good practices through on-the-ground interventions that are integrated with national economic and social development policies. The following projects are included in this category: (1) UNDP-supported Sustainable Land Management to Combat Desertification, (2) SDC-supported Coping with Desertification, (3) Sustainable Grassland Management, and (4) the Green Gold projects. All abovementioned projects involved implementing activities at all household, local, regional, and national levels by leveraging national policies and adopting new concepts and models, specifically for land use and land management. In particular, sustainable grassland management projects in Mongolia were intended to increase the welfare of herding families through sustainable management of grasslands by strengthening and formalizing the existing customary herder community institutions and strengthening linkages between them and formal governance structures and the private sector. In addition to these objectives, some of the projects included creating a competitive and cost-effective market environment, improving livestock productivity by supporting genetic diversity, and enhancing the value chain and all other market-oriented models. This category includes projects such as the UNDP-led Sustainable Grassland Management project, which was a pilot for all preceding projects and programs; the SDC-supported Mongolian Pasture – Green Gold project; the USA-supported Mercy Corp Grassland project; etc. These projects successfully implemented and adapted the community-based natural resource concept in Mongolia, even though there were some conflicts with the land tenure system that small models established by these projects, indicating that such a governance system may contribute greatly to addressing both the environmental and socioeconomic problems facing the country.

13.3.2 Sustainable Livelihoods and Rural Poverty Reduction Projects

All these projects aim to reduce vulnerability and achieve secure and sustainable livelihoods by targeting poor and vulnerable near-poor households and individuals nationwide. Projects such as Pastoral Risk Management (PRM), Sustainable Grassland Management, and Green Gold addressed using grasslands wisely, and these projects directed a great amount of their actions to livelihood and alternative income generation. The majority of the projects have implemented the following activities under this objective: (1) risk forecasting, preparedness, and response planning; (2) pasture-land tenure, management, and use; (3) implementation of good practices in terms of pastoral livelihood improvement and alternative income generation; and (4) institutionalization of pasture management and risk reduction. The aforementioned efforts and many other local initiatives have addressed livelihood and poverty with several impactful solutions.

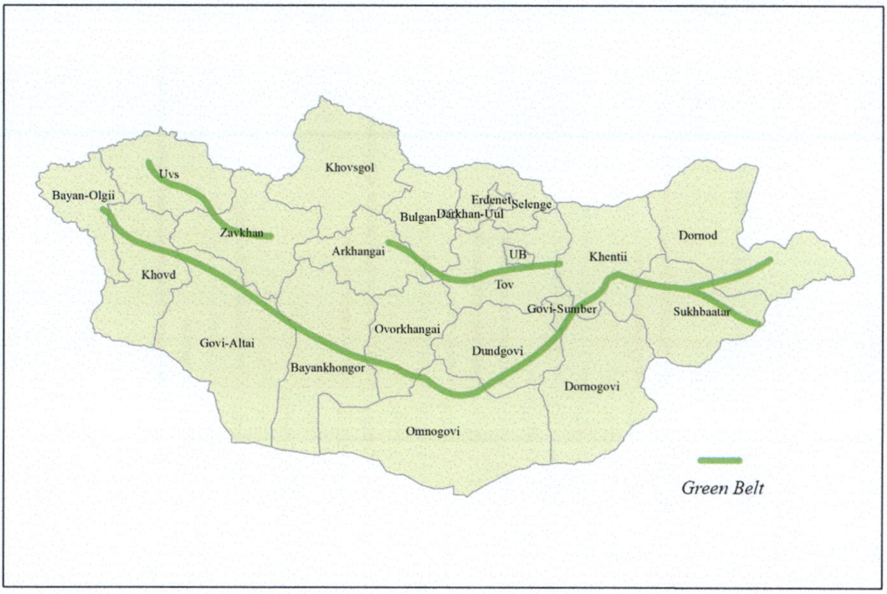

Fig. 13.1 Spatial plan of the National Green Belt program in Mongolia (revised according to Khaulenbek and Avirmed 2005)

13.3.3 Afforestation Projects

With the ratification of the UNCCD, the government paid attention to the development of country-specific approaches to fight sand encroachment and prevent degradation and desertification. Reforestation and afforestation are currently considered the core activities for implementing national strategies. The first large-scale afforestation project was implemented by the Institute of Geoecology (former name) in 1997–2000 with the support of the GEF-SGP. As a result of this project, a total of 200 ha of a green strip was established that prevented the soum center from dust storms and sand encroachment. In addition to tree plantations, 400 ha of an additional buffer zone was fenced, and the natural regeneration of vegetation occurred (Khaulenbek and Mandakh 2004; Schoberg 2008). This project laid the foundation for the development of the national Green Belt program.

The Green Belt program was adopted by Government Resolution # 44 in 2005. The program proposed establishing a forested belt with a total length of 3700 km to cover 200,000 ha of territory. The Green Belt has main and subforest strips and was developed in three phases: phase 1, established 26,692 ha of forest strips during 2005–2015; phase 2, established 39943.8 ha of forest strips in 2015 and will establish until 2025; and phase 3, established 66,573 ha of forest strips in 2025 and will establish until 2035 (Fig. 13.1).

The aim of this program is to establish a "green" zone along the natural borders between the steppe and semidesert to create favorable environmental conditions that prevent further degradation of steppe ecosystems, reduce risks associated with dust

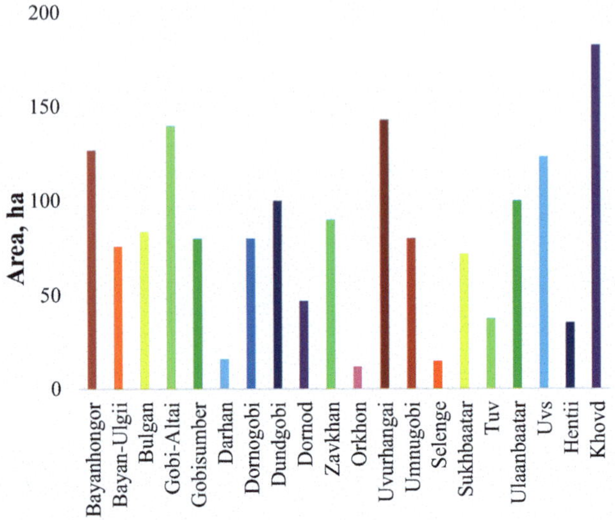

Fig. 13.2 Area of afforestation, by aimag

and sandstorms, and create soil conditions suitable for further expansion of forests. The national program has its own dedicated fund, which is sourced from the national budget and supported by donor countries, international organizations, regional funds, and public donations. The program approaches are defined as follows:

1. Overall management should use the government of Mongolia's institutional strategy.
2. The program should adhere to a participatory approach as much as possible.
3. The program should prioritize projects/initiatives seeking multiple benefits for both the environment and socioeconomy.
4. The program should involve different economic sectors, e.g., road and transportation, agriculture, mining, and urban development (Mongol Ulsiin Zasgiin Gazar 2005).

To date, the National Green Belt program has been implemented in 82 sums within 20 aimags, and a total of 1640.4 ha of land have been afforested (Banzragch 2014; Fig. 13.2).

According to the Ministry of Environment and Tourism (MET), since its implementation, 2127.44 million MNT investments have been made from the state budget (Fig. 13.3). Unfortunately, the allocation of the budget for the Green Belt program has been decreasing yearly, and the reason for this decrease is linked to very diverse issues. First, because the project relies on private-public partnerships (PPPs), which are a relatively new concept, few were prepared for such changes. Second, during the period of Green Belt program initiation and implementation, there were at least two significant economic collapses in the global market that impacted state decisions on how to prioritize investments from the budget. Third, starting in 2005, many small

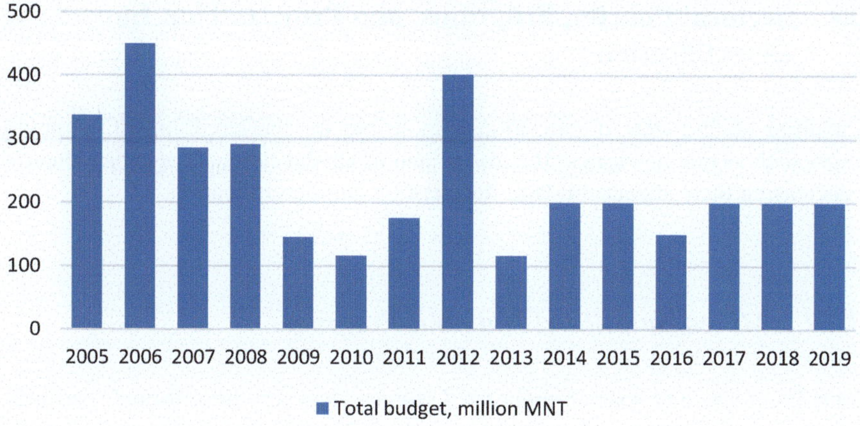

Fig. 13.3 Total funds allocated to the Green Belt program from the state budget (Baigali Orchin Ayalal Juulchlaliin Yam 2019)

and large international initiatives began funding the Green Belt directly or by implementing independent projects under a common goal.

As mentioned above, the National Green Belt program initiative of the government of Mongolia is favored by UNCCD, ESCAP, and different countries. The government of Korea, Korean NGOs, and volunteers are actively participating in efforts to reach the goals and targets of this program. However, an evaluation of the National Green Belt program showed that 15–20% of the program has been implemented. Such relatively low implementation can be explained first by the harsh biophysical settings of the Mongolian drylands. Second, given the predominantly nomadic lifestyle in the country, implementing this program requires time, technology, and a solid knowledge base.

Parallel to the Green Belt program, the government of Mongolia has also developed several policies related to forests and woodlands. One of them was a Government Resolution prohibiting the cutting and collecting of saxaul and other dryland trees for firewood in 2006, and this effort positively impacted the state of saxaul growth in the Gobi region of the country.

In conclusion, the government of Mongolia, in collaboration with the international community, is eager to succeed in the field of combating desertification. In addition, all the projects and activities have tried to integrate lessons learned from preceding projects. Even though, in addition to afforestation, there is a lack of different actions focused on on-the-ground implementation, a positive aspect is that the country has started to consider the socioeconomic side of the problem, thus acting on two fronts. This scenario might result in better solutions in the near future, especially in regard to achieving multiple benefits through nature-based solutions related to land use.

13.4 Sectoral Policies, Programs, and Projects That Impact Desertification

In addition to the state organization responsible for the environment, economic sectors such as mining, agriculture, roads, and urban development also pay attention to decreasing their impacts on land degradation and desertification.

13.4.1 Millennium Road National Program

Mongolia, a vast and sparsely populated country, has very little formal road infrastructure, and the total road length in Mongolia is 11,100 km, of which only 11.9% is asphalt-paved, 12.5% is gravel-paved, and the remaining 75.6% is not paved (Chimgee and Uranbileg 2001; UNESCAP 2001; Keshkamat et al. 2013). Vehicle tracks in the dirt are one of the serious environmental problems in Mongolia's fragile grassland ecosystems. The vehicle tracks in the dirt destroy vegetation and cause soil erosion, resulting in approximately 8000–10,000 km^2 of desertified land across the country (Dregne 1983; MNEM 1997; Belnap 2002; Keshkamat et al. 2013). Li et al. (2006) noted that revegetation of such damaged areas due to vehicle tracks requires approximately 10–15 years after the track has ceased to be used. Moreover, the pioneer plants in natural revegetated tracks are invasive weed species and not endemically edible forage species, which are socioeconomically important bases for livestock herding.

Since 2005, the Mongolian government, through the support of the Asian Development Bank and the World Bank, has begun building a formal system of paved roads across the country as a means of social and economic connectivity for its people. Mongolia has proposed a "Millennium Road" project to eliminate vehicle tracks as a factor of land desertification, but the proposed paved road system also can link the key cities and resources in Mongolia (Fig. 13.4).

Although paving essential arterial roads throughout Mongolia is undoubtedly an ideal solution, without foregoing future related efforts, the adoption of low-investment and low-risk mitigation measures to reduce or control dirt vehicle track corridor widths is currently necessary.

The asphalt ring roads around towns, which were designed following a user-centric approach, could serve people within towns and help reduce the land desertification caused by these dirt tracks. Other similar cost-effective actions could also be developed. The creation, shift, and modification of asphalt roads could then be verified based on nomadic practices and seasonal forage situations and used to adjust them as needed.

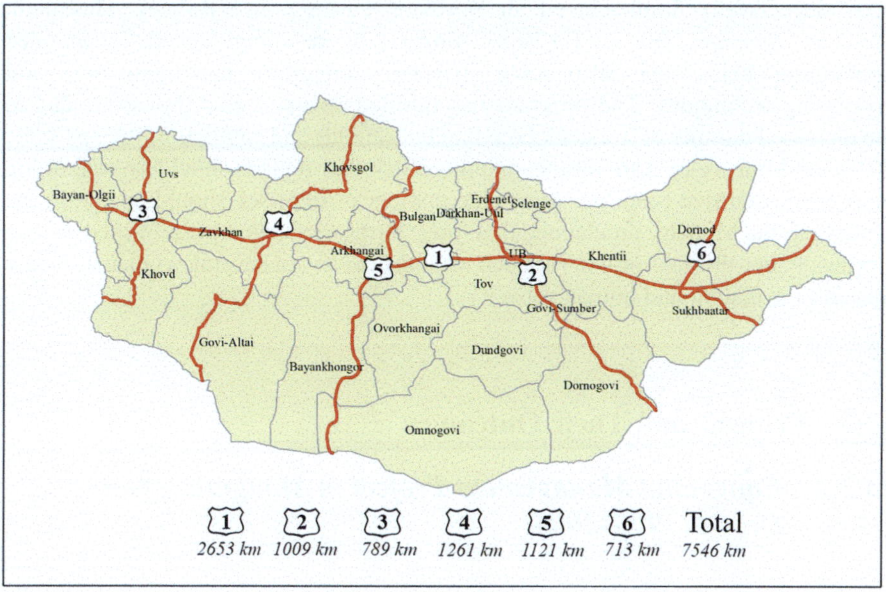

Fig. 13.4 Millennium Road program in Mongolia (revised based on Ukhnai 2005)

13.4.2 Renewable Energy to Reduce Fuelwood Collection

Renewable energy is widely available in Mongolia and could be considered a key solution for providing energy supplies to remote areas, particularly small villages and settlements, including traditional households in arid and semiarid zones. Despite its high installation costs, renewable energy has very low operating costs and significant advantages, such as lower greenhouse gas emissions, toxicity levels, and particulate emissions. For residents in the Gobi areas in Mongolia, solar and wind energy are good alternatives for reducing the vegetation destruction resulting from gathering fuelwood. The government of Mongolia has announced a project of "100 thousand solar Ger (Mongolian traditional house)" to improve the electricity supply to remote rural area and avoid cutting desert bushes for household use. This effort is another positive step in minimizing triggers of sand-dust storms and ultimately preventing desertification overall.

13.4.3 Sustainable Agriculture

To provide healthy and nutritious food to the population as well as to address overgrazing and create green jobs, the government of Mongolia has developed several policy steps focused on sustainable agriculture. The Atar 5 program has been successfully implemented in northern agri-regions (Bulgan, Huvsugul, Zavkhan, Selenge, Tuv, and Khentii aimags) by restoring recently abandoned agricultural fields and introducing irrigated agriculture and innovative soil processing techniques. The program has fulfilled its main goal; however, due to the increased number of livestock breeding households and conflicts of interest since 2018, the Ministry of Agriculture announced that agri-regions should favor farming more than extensive pasture use, and the number of livestock should be cut. In the same year, the Ministry proclaimed that forest strips needed to be established around croplands, but these have not yet been established due to several problems mainly related to irrigation and human capacity.

13.5 Projects and Their Outcomes

13.5.1 Rangeland Management Project in Mongolia's Gobi Desert (1995–2006)

The project was designed to provide both environmental conservation and livelihood benefits. It has been reported that the project resulted in great increases in household incomes and increases in rangeland productivity, and the success of this project was mainly attributed to the incorporation of community-based management into the design (Leisher et al. 2012). The project created community-based organizations (CBOs) to improve pasture management, develop alternative livelihoods, and strengthen cooperation among local communities and district governments. Rangeland management improvement involved the coordination of access to rangelands for all participants, improvements in water sources for livestock, and development of specific winter grazing areas for CBO members. Every district had a community organizer who supported the development of the CBO, was responsible for organizing and encouraging the communities, and acted as a liaison with local government, resource agencies, and the rest of the project team. By the end of the project in 2006, the project covered an area of 13.5 million ha, comprising 12 districts across 3 provinces, and there were 83 CBOs, involving 1175 households, or approximately 14% of the households in the project area (Dorj et al. 2013).

13.5.2 GL-CRSP Gobi Forage Project (2004–2008)

The Gobi Forage project under the Global Livestock Collaborative Research Support Program (http://glcrsp.ucdavis.edu), described as a cutting-edge project, established forage monitoring and forecasting services that regularly provide pastoral communities, policy makers, and administrators in charge of agriculture and rural development with valuable information through forage resource maps for the future

30 and 60 days. The successes of the project were partly derived from the project's ability to successfully carry out four complementary activities tailored to local conditions: (1) adopting specific technology to measure forage quantity; (2) conducting detailed field measurements of forage quality; (3) effectively extending accurate forage distributing information; and (4) linking information with herder alliances. This seminal method (technology) applied in Mongolia has resulted in dramatic effects. The high-quality predictive ability of the project technology and the effective dissemination of its information were sufficient to convince people to use the forage prediction maps who were originally suspicious of their accuracy. Approximately 93% of government officials use Gobi forage products to make recommendations on livestock movements.

The beneficiaries of this project were not just livestock farmers. In Mongolia, almost all important short-term population movements are related to forage conditions. Gobi forage products can assist in managing pastoral migration, greatly improving the natural resource management and institutional functioning. One provincial governor described how the forage information system helped him manage the influx of approximately 50,000 herders and their families from neighboring drought-stricken provinces, avoiding conflicts over pasture resources.

13.5.3 Coping with Desertification in Mongolia (2004–2011)

The objective of this project was to support Mongolia on effective coordination and implementation of obligations toward coping with desertification and promoting sustainable livelihoods in arid and semiarid areas. The project resulted in four different outcomes: (1) The NCCD is effectively coordinating and reporting on desertification issues relevant to Mongolia; (2) local communities are empowered in the sustainable management of natural resources and diversification of livelihoods to address desertification; (3) Mongolia's youth and the general public are aware of the country's environmental challenges and have basic knowledge about the threats of desertification; and (4) a database with appropriate information, technologies, approaches, and tools to cope with desertification in Mongolia was established and disseminated to support project implementation, scaling up at the grassroots level and informed decision-making at the policy level (https://www.eda.admin.ch). Within a framework of knowledge management, Mongolia first assessed and mapped the extent of desertification and land degradation using a spatial model, remote sensing, and GIS. The project supported the transfer of the WOCAT approach to documenting and reporting best practices through a framework in which 30 technologies and approaches were documented and disseminated in the form of books, brochures, and flyers. Another great impact of this project was the implementation and experimentation of watershed management in the Buyant River basin.

13.6 Conclusion

Aeolian desertification, as a form of land degradation and desertification, clearly threatens Mongolian lands. Although this process is associated with natural and biophysical factors, the exacerbation of aeolian desertification in Mongolia, unlike in other parts of the world, is closely related to the human-environment nexus. Over the past 30 years, Mongolia has developed policies and legislation integrating both lessons learned and international trends; however, there are still more efforts needed to be able to succeed in achieving defined goals and targets. The environmental sectors of the country are also young compared with other economic sectors as is the current state of the art in nature conservation, especially in terms of combating desertification based on classical notions of protecting land. The projects and programs implemented in this period showed that there are tremendous possibilities for preventing desertification and land degradation, and the key is working toward public engagement and raising awareness. Although a number of projects have been implemented as part of public-private partnerships, livelihoods, and diversification of income sources, it is still important to pay attention to how these activities can benefit nature; i.e., the use of novel concepts related to ecosystem services and nature-based solutions need to be translated and implemented on the ground.

References

Baigali Orchin Ayalal Juulchlaliin Yam (2019) Mongol ornii baigali orchnii tuluv baidaliin tailan, Ulaanbaatar

Banzragch T (2014) The evaluation of the National Green Wall programme implementation. MNE Workshop, Ulaanbaatar

Banzragch, T, Enkhbold S (2010) New Policy on Combating Desertification in Mongolia. Institute for Global Environmental Strategies (IGES), June 17–18

Bedunah DJ, Schmidt SM (2004) Pastoralism and protected area management in Mongolia's Gobi Gurvansaikhan National Park. Dev Change 35:167–191

Belnap J (2002) Impacts of off-road vehicles on nitrogen cycles in biological soil crusts: resistance in different U.S. deserts. J Arid Environ 52:155–165

Chimgee P, Uranbileg S (2001) Development of spatial data infrastructure in Mongolia and its application to the 'Millennium Road' Project. International conference on spatial information for sustainable development, Nairobi, Kenya, 2–5 October 2001

Darima D (2020) The socio-economic consequences of the desertification processes in Mongolia. Conference paper, E3S web of conferences 164:11001. https://doi.org/10.1051/e3sconf/202016411001

Dorj O, Enkhbold M, Lkhamyanjin S et al (2013) Mongolia: country features, the main causes of desertification and remediation efforts. In: Ali Heshmati G, Squires VR (eds) Combating desertification in Asia, Africa and the Middle East (proven practices). Springer

Dregne HE (1983) Soil and soil formation in arid regions. In: Webb RH, Wilshire HG (eds) Environmental effects of off-road vehicles: impacts and management in arid regions. Springer, New York, pp 15–30

Gazriin hariltsaa, geodezi, zuragzuin gazar (2019) Mongol ulsiin gazriin negdmel sangiin tailan. BHBYA, Ulaanbaatar

Hudulmur S, Tsogtbaatar J (eds) (2010) Desertification atlas of Mongolia. Admon, Ulaanbaatar
Jamsran T, Doljin D, Nyamtseren M et al (2015) Issues related to policy and measures to combat desertification in Mongolia. In: Combating drought, land degradation and desertification for poverty reduction and sustainable development. UNCCD 3rd scientific conference, 9–12 March 2015, Cancun Mexico
Keshkamat SS, Tsendbazar NE, Zuidgeest MHP et al (2013) Understanding transportation-caused rangeland damage in Mongolia. J Environ Manag 114:433–444
Kharin NG, Tsolmon P (1994) Vliyanie socialno-economicheskih factorov na razvitie opustinivaniya Mongolii. Problemi osvoeniya pustini 6:17–23
Khaulenbek A, Avirmed A (2005) Nogoon herem ecotrassiin undeslel. Baigali Orchnii Yam, Ulaanbaatar
Khaulenbek A, Mandakh N (2004) Tuv suurin gazriin elsjilt, hamgaalah agra zam. In: Mongol ornii oig ashiglah, nuhun sergeeh asuudluud, Ulaanbaatar, pp 160–170
Lee D, Ahn G (2016) A way forward to sustainable international forestry cooperation: a case study of the 'greenbelt plantation project in Mongolia'. J Rural Develop/Nongchon-Gyeongje 39:143
Leisher C, Hess S, Boucher TM et al (2012) Measuring the impacts of community-based grasslands management in Mongolia's Gobi. PLoS One 7(2):e30991
Li SG, Sujimura MAT, Ugimoto ATS et al (2006) Natural recovery of steppe vegetation on vehicle tracks in Central Mongolia. J Biosci 31:85–93
Mandakh N, Dash D, Khaulanbek A et al (2016) Tsuljilttei temtsekh arga khemjee, technologi. Admo, Ulaanbaatar. 268 pp
Mongol Ulsiin Zasgiin Gazar (2005) Nogoon Herem Undesnii Hutulbur. Accessed from www.legalinfo.mn on 23 Dec 2020
NCCD (2018) National report on voluntary target setting to achieve land degradation neutrality in Mongolia. http://www.unccd.int
Nixson F, Walters B (2006) Privatization, income distribution, and poverty: the Mongolian experience. World Dev 34(9):1557–1579
Schoberg D (2008) A tree grows in Mongolia. The Rotarian 187(4):53–57
Sekiyama A, Takeuchi W, Shimada S (2014) Detection of grassland degradation using MODIS data in Mongolia. J Arid Land Stud 24:175–178
The Ministry of Nature and Environment of Mongolia (MNEM) (1997) National plan of action to combat desertification in Mongolia. Ulaanbaatar
Tuvshintogtokh I, Ariungerel D (2013) Degradation of Mongolian grassland vegetation under overgrazing by livestock and its recovery by protection from livestock grazing//The Mongolian Ecosystem Network. Springer, pp 115–130
Ukhnai B (2005) Road transportation and tourism in Mongolia. Ulaanbaatar
UNCCD (2006) Mongolia National Report to CRIC5. Ulaanbaatar
UNCED (1992) Earth summit Agenda 21: programme of action for sustainable development. United Nations Department for Public Information, New York
UNESCAP (United Nations Economic and Social Commission for Asia and the Pacific) (2001) The road networks connecting China, Kazakhstan, Mongolia, the Russian Federation and the Korean Peninsula (doc HC412 ST ESCAP 2173). United Nations, New York

Chapter 14
Sustainable Land Management to Combat Aeolian Desertification

Atsushi Tsunekawa

Abstract The activities implemented by the international community to combat desertification can be traced back to the United Nations Conference on Desertification (UNCOD) held in Nairobi, Kenya, in 1977. Since this conference, the international goal has been "combating desertification," which means entirely preventing desertification. However, the new concept of "land degradation neutrality" (LDN) was included in 15.3 of the Sustainable Development Goals (SDGs) that were adopted in 2015, and the goal of today's international community is to achieve LDN. Sustainable land management (SLM) plays an essential role in addressing desertification and land degradation. SLM refers to technologies and approaches that enable sustainable land production, livelihood improvement, and environmental conservation through appropriate soil and water management, and various SLM projects are being implemented worldwide. Future challenges for SLM relate to linking land management activities to socioeconomic empowerment of people, exploring exit strategies from development aid and introducing diverse sources of funding, and responding to climate change in terms of both adaptation and mitigation.

Keywords United Nations Convention to Combat Desertification (UNCCD) · Combat desertification · Land degradation neutrality (LDN) · Sustainable Development Goals (SDGs) · Sustainable land management · Soil and water conservation

14.1 Introduction

The activities implemented by the international community to combat desertification can be traced back to the United Nations Conference on Desertification (UNCOD) held in Nairobi, Kenya, in 1977. This conference adopted a Plan of Action to

A. Tsunekawa (✉)
Arid Land Research Center, Tottori University, Tottori, Japan
e-mail: tsunekawa@tottori-u.ac.jp

T. Wang et al. (eds.), *Combating Aeolian Desertification in Northeast Asia*, Ecological Research Monographs, https://doi.org/10.1007/978-981-16-9028-0_14

Combat Desertification (PACD) that nations and international communities could implement "to prevent and to arrest the advance of desertification and, where possible, to reclaim desertified land for productive use."[1] Since this conference, the international goal has been to combat desertification to prevent the advancement of desertification. However, land degradation neutrality (LDN), which was included in 15.3 of the Sustainable Development Goals (SDGs) that were adopted in 2015, is a new concept that calls for avoiding further land degradation and restoring degraded land. The goal of today's international community is to realize the net land restoration (=(restored land area) – (degraded land area)) by avoiding further land degradation and restoring degraded land, that is, to achieve land degradation neutrality (Safriel 2017).

Sustainable land management (SLM) plays an essential role in addressing desertification and land degradation. SLM refers to technologies and approaches that enable sustainable land production, livelihood improvement, and environmental conservation through appropriate soil and water management. A variety of SLM projects are being implemented around the world.[2]

This chapter summarizes the international efforts led by the UNCCD from the past to the present to cope with desertification, explains the significance of LDN and its monitoring methods, and determines the current status and future challenges of SLM as it relates to combating Aeolian desertification and achieving LDN.

14.2 United Nations Convention to Combat Desertification (UNCCD)

The UNCCD[3] is officially named the "United Nations Convention to Combat Desertification in Those Countries Experiencing Serious Drought and/or Desertification, Particularly in Africa" and is an international treaty that stipulates that affected countries that are party to the Convention prepare and implement national action programs and developed countries that are party to the Convention provide substantial financial resources and other forms of support to assist the affected developing countries.

The Convention was adopted in Paris, France, on June 17, 1994, and it entered into force on December 26, 1996, following the agreement to establish an intergovernmental negotiating committee to prepare the Convention at the United Nations Conference on Environment and Development (UNCED, abbreviated Earth Summit or Rio Summit) held in Rio de Janeiro, Brazil, in 1992. This Convention, the United Nations Framework Convention on Climate Change (UNFCCC), and the

---

[1] http://www.ciesin.org/docs/002-478/002-478.html.

[2] https://www.wocat.net/en/.

[3] https://www.unccd.int/convention/about-convention.

Fig. 14.1 UNCCD COP 14 held in New Delhi, India (Photo: September 3, 2019, by Atsushi Tsunekawa)

Convention on Biological Diversity (CBD), which were established in the wake of the Rio Summit, are referred to as the "Rio Conventions."

The Convention consists of a preamble, 40 articles in the main text, and 5 regional implementation annexes for the 5 regions affected by desertification and drought: Africa, Asia, Latin America and the Caribbean, the northern Mediterranean, and Central and Eastern Europe.

In Article 1 of the main text, "desertification" is defined as a "reduction or loss, in arid, semi-arid and dry sub-humid areas, of the biological or economic productivity and complexity of rainfed cropland, irrigated cropland, or range, pasture, forest and woodlands resulting from land uses or from a process or combination of processes, including processes arising from human activities and habitation patterns, such as (i) soil erosion caused by wind and/or water; (ii) deterioration of the physical, chemical and biological or economic properties of soil; and (iii) long-term loss of natural vegetation."

The Conference of the Parties (COP) is the supreme body of the Convention (Article 22), which has met every 2 years since 2001 and held its 14th session in September 2019 in New Delhi, India (Fig. 14.1). The Committee on Science and Technology (CST) is a subsidiary body of the Convention under Article 24 of the Convention, and the Committee for the Review of the Implementation of the Convention (CRIC) is responsible for the periodic review of the implementation of the Convention and was established by the decision of the Fifth Conference of the Parties (COP5/Decision 1).

As of April 2021, 197 countries have ratified the UNCCD. One of the features of the UNCCD is that it stipulates the obligations of affected party countries and those of developed party countries separately. That is, UNCCD provides the general obligations in Article 4, obligations of affected party countries in Article 5, and obligations of developed party countries in Article 6. In effect, every country that is party to the Convention is obliged to submit periodic country reports to the COP through the Secretariat (Article 26), and affected party countries are obliged to formulate national action programs to combat desertification and mitigate the effects of drought (Article 10).

14.3 From "Combating Desertification" to "Land Degradation Neutrality"

The activities implemented by the international community to combat desertification can be traced back to the United Nations Conference on Desertification (UNCOD) held in Nairobi, Kenya, in 1977. This conference adopted a Plan of Action to Combat Desertification (PACD) that nations and international entities could implement "to prevent and to arrest the advance of desertification and, where possible, to reclaim desertified land for productive use." Since this conference, the international goal has been to combat desertification to prevent the advancement of desertification.

However, the aim of completely stopping desertification and reducing the area of land degradation to zero has not been achieved. For example, the Millennium Ecosystem Assessment report (MA 2005) reported that 10–20% of the world's drylands are degraded, and Luc Gnacadja, who became Executive Secretary of the UNCCD in 2007, launched an initiative to revitalize the UNCCD. Building on the introduction of offset mechanisms by the Convention on Biological Diversity (CBD) and the United Nations Framework Convention on Climate Change (UNFCCC), the UNCCD Executive Secretary envisioned adapting the offset principles to address land degradation globally, and a "zero net rate of land degradation (ZNLD)" that incorporated these offset principles was proposed (Safriel 2017).

Thus, "combat desertification" as a goal was renewed to realize the net land restoration (=(restored land area) – (degraded land area)) by avoiding further land degradation and restoring degraded land, that is, to achieve land degradation neutrality (Fig. 14.2), which is a new goal of the international community.

In June 2012, the United Nations Conference on Sustainable Development[4] was held in Rio de Janeiro, Brazil. The conference, also known as the Rio+20 Summit, was held with the aim of reviewing and following up on the 20 years of activity since the UN Conference on Environment and Development that was held in Rio de Janeiro in 1992. On the final day of the conference, the outcome document, "The Future We Want,"[5] was adopted. Desertification, land degradation, and drought (DLDD) were included in Articles 205–209 of the document as one of the

[4] https://sustainabledevelopment.un.org/rio20.

[5] https://sustainabledevelopment.un.org/futurewewant.html.

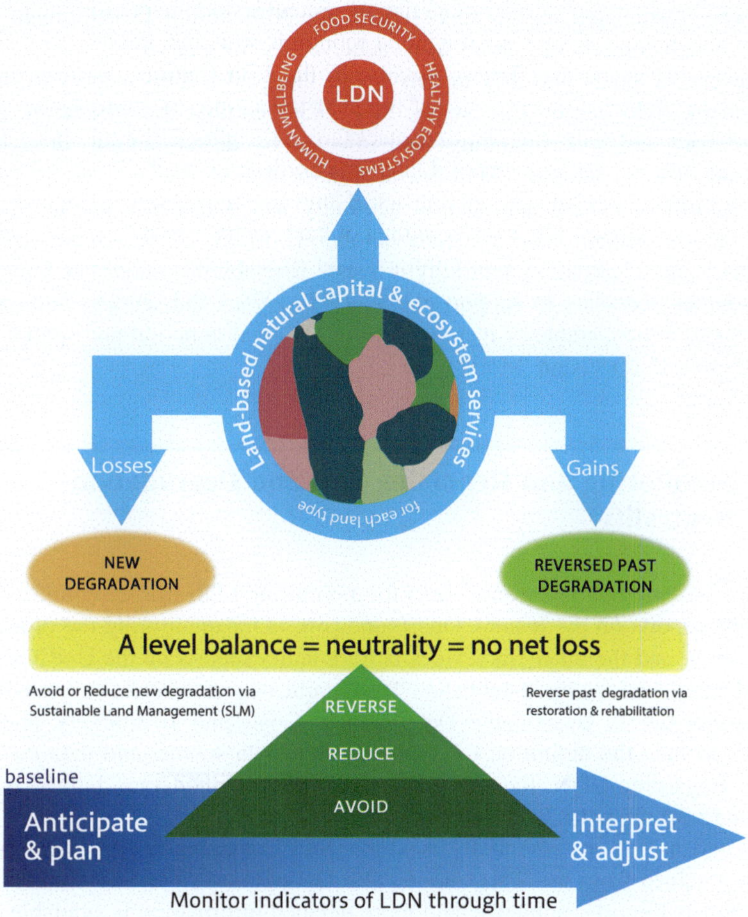

Fig. 14.2 The key elements of the scientific conceptual framework for LDN and their interrelationships (cited from Orr et al. 2017 with modification)

26 thematic areas and cross-sectoral issues for action and follow-up. Article 206 calls for achieving a "land degradation neutral world."

At the United Nations Sustainable Development Summit[6] on September 25, 2015, more than 150 world leaders adopted the new 2030 Agenda for Sustainable Development,[7] including the Sustainable Development Goals (SDGs). The

[6] https://sustainabledevelopment.un.org/post2015/summit.

[7] https://sdgs.un.org/2030agenda.

SDGs were formulated as the successor to the Millennium Development Goals (MDGs) which consists of 8 goals to be achieved by the year 2015.[8]

There are 17 goals and 169 targets under the goals. With respect to dryland development, SDG target 15.3 states: "By 2030, combat desertification, restore degraded land and soil, including land affected by desertification, drought and floods, and strive to achieve a land degradation-neutral world."

The definition of land degradation neutrality was determined at the UNCCD/COP12 held in Ankara, Turkey, from October 12 to 23, 2015, and the definition states that "land degradation neutrality is a state whereby the amount and quality of land resources necessary to support ecosystem functions and services and enhance food security remain stable or increase within specified temporal and spatial scales and ecosystems" (Decision 3/COP.12[9]).

14.4 Monitoring and Reporting on Land Degradation Neutrality

There are two international frameworks for monitoring LDN in each country and for submitting reports on the results. One framework is for monitoring the progress of the 17 SDGs, and the other framework is for country reports on the UNCCD.

The United Nations has established "SDG indicator metadata" for the indicators that measure the 17 Sustainable Development Goals and 169 targets, and these metadata include the definition of each indicator, data source, and data collection method. Regarding LDN, the specific measurement methods are defined as SDG indicator 15.3.1 (Fig. 14.3).

SDG indicator 15.3.1 essentially uses the UNCCD country report process, and the data are obtained from the 2018 report. The UN publishes the Sustainable Development Goals Report[10] annually, and more detailed information is available on the Internet. For LDN, the statistical annex of 2020[11] includes supplementary information (United Nations Economic and Social Council 2020), and indicator 15.3.1 includes the table "Proportion of land that is degraded over total land area by regions" that is current as of 2015 (Table 14.1). The table notes state that regional data are based on the country-level data that were in the 2018 UNCCD national reports submitted by 123 countries and that the estimates were prepared by UNCCD based on global data sources.

On the other hand, the UNCCD party countries are required under the Convention to provide, through the UNCCD Secretariat, reports on measures undertaken to

[8] https://www.un.org/millenniumgoals/.

[9] https://www.unccd.int/sites/default/files/sessions/documents/2019-08/3COP12_0.pdf.

[10] https://www.un.org/sustainabledevelopment/progress-report/.

[11] https://unstats.un.org/sdgs/files/report/2020/secretary-general-sdg-report-2020%2D%2DStatistical-Annex.pdf.

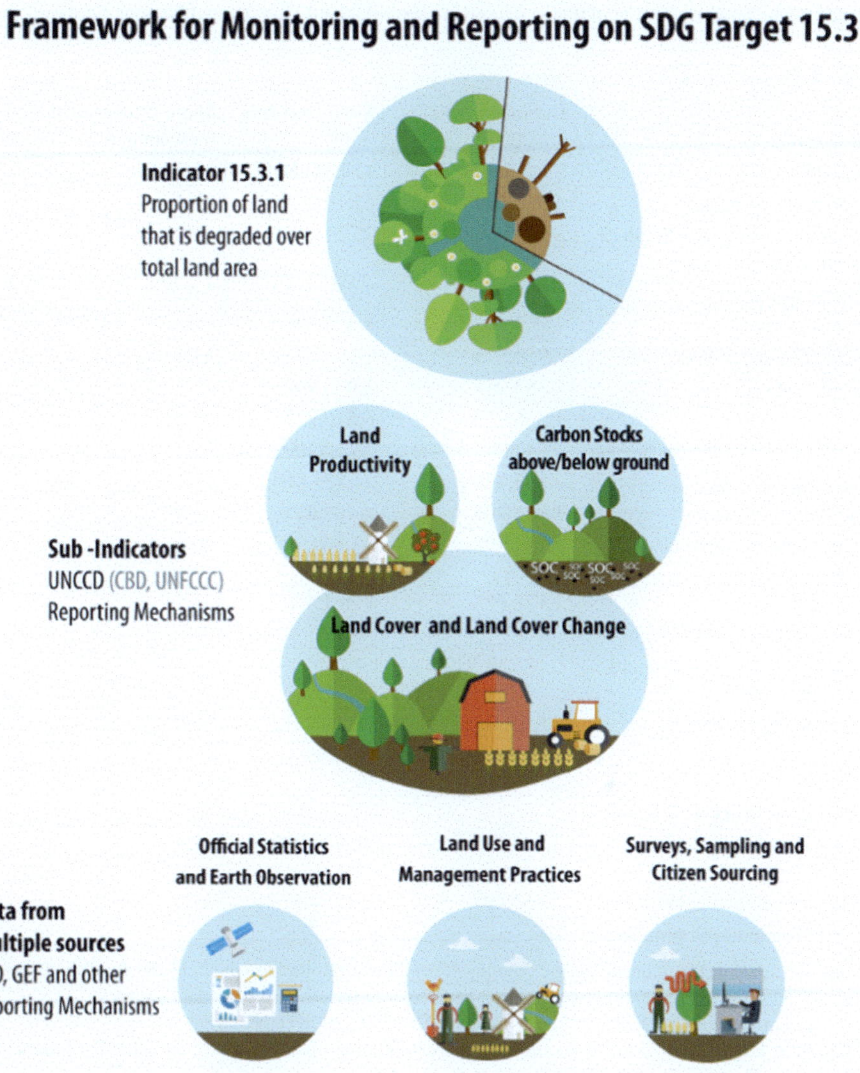

Fig. 14.3 Framework for monitoring and reporting on land degradation neutrality (The Global Mechanism of the UNCCD 2016)

implement the Convention. This is in accordance with Article 26 and the decisions of the Conference of the Parties (COP), particularly decision 11/COP.1. Accordingly, the UNCCD party countries were required to submit reports to the Secretariat once every 2 years prior to 2017 and once every 4 years starting from the 2017–2018 reporting cycle, and the UNCCD uses submitted Convention country reports to monitor LDN. To determine the progress toward achieving land degradation neutrality, the initial status is quantified and considered the baseline of land degradation, and then, the balance between the area of "gains" (significant positive changes/

Table 14.1 Proportion of land that is degraded over total land area (United Nations Economic and Social Council 2020)

Regions	2015 Degraded land area (km[a])	Total land area (km[a])	Share of degraded land (percentage)
World	23,962,509	119,681,858	20.0
Sub-Saharan Africa	4,950,699	22,107,557	22.4
Northern Africa and Western Asia	847,523	12,211,454	6.9
Northern Africa	432,119	7,720,758	5.6
Western Asia	415,405	4,490,696	9.3
Central and southern Asia	2,950,693	10,557,737	27.9
Central Asia	1,383,958	3,940,962	35.1
Southern Asia	1,566,735	6,616,775	23.7
Eastern and southeastern Asia	3,942,095	16,140,899	24.4
Eastern Asia	2,888,889	11,731,466	24.6
Southeastern Asia	1,053,205	4,409,432	23.9
Latin America and the Caribbean[b]	5,257,898	19,809,979	26.5
Oceania[a]	2,978,078	8,391,420	35.5
Australia and New Zealand	...	...	...
Oceania (exc. Australia and New Zealand)	...	...	...
Europe and Northern America[c]	3,035,523	30,462,812	10.0
Europe	...	...	...
Northern America	...	...	...
Landlocked developing countries	3,835,033	16,734,270	22.9
Least developed countries	3,453,498	20,354,573	17.0
Small island developing states	...	...	...

Note: Regional data are based on the country-level data submitted in UNCCD 2018 national reports from 123 countries and estimates prepared by UNCCD based on global data sources
Source: United Nations Convention to Combat Desertification (UNCCD)
[a]Including Papua New Guinea, Australia, and New Zealand but excluding the islands of Oceania
[b]Excluding the islands of the Caribbean
[c]Excluding the USA and Switzerland

improvements) and "losses" (significant negative changes/degradation) is measured against the baseline for each land type at the end of LDN implementation[12] (Fig. 14.4). The indicators and their associated metrics (in parentheses) are already used for UNCCD reporting and for the sustainable development goals (SDGs) related to land cover (land cover change), land productivity (net primary productivity, NPP), and carbon stocks (soil organic carbon, SOC).

[12]https://knowledge.unccd.int/ldn/ldn-monitoring.

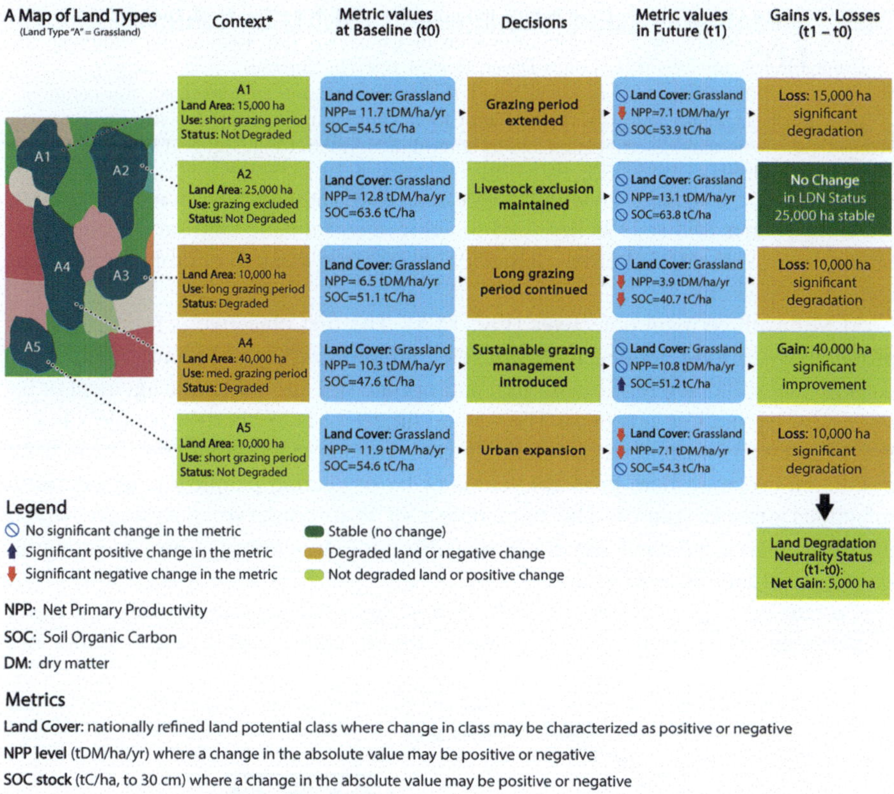

Fig. 14.4 A hypothetical example showing how LDN status is monitored on the basis of changes in the value of the metrics, using the one-out, all-out approach applied to each land unit (cited from Orr et al. 2017 with modification)

As of April 2021, the most recent report is the 2017–2018 report,[13] and the reports from each country are available on the Internet, and the reports include information on indicators SO1–1: Trends in land cover, SO1–2: Trends in land productivity or functioning of the land, and SO1–3: Trends in carbon stocks above and below ground, and SO1: Proportion of land that is degraded over total land area (SDG 15.3.1). However, the methods used to measure these indicators vary from country to country. Therefore, great care must be taken when comparing and interpreting these data across countries.

[13] https://prais.unccd.int/unccd/reports.

14.5 Sustainable Land Management (SLM) to Achieve LDN

In recent years, the concept of sustainable land management has played a central role in efforts to address desertification and land degradation and achieve land degradation neutrality. Sustainable land management (SLM) is a concept that encompasses the technologies and approaches used to achieve sustainable land production, livelihood improvement, and environmental conservation through appropriate soil and water management.

SLM has three main objectives (Fig. 14.5). The first objective is to increase the productivity of food, fodder, fuelwood, etc., which is achieved by improving water use efficiency and soil fertility through appropriate soil and water management or by improving agronomic practices. The second objective is to improve people's livelihoods and well-being, for example, by increasing income through conversion from traditional crops to more profitable crops or by protecting people's health through infectious disease control. The third objective is to conserve ecosystems, which involves preventing land degradation, conserving biodiversity, and maintaining proper water and material cycles.

Current good practices in drylands include not only appropriate soil and water management technologies directly related to desertification and land degradation but

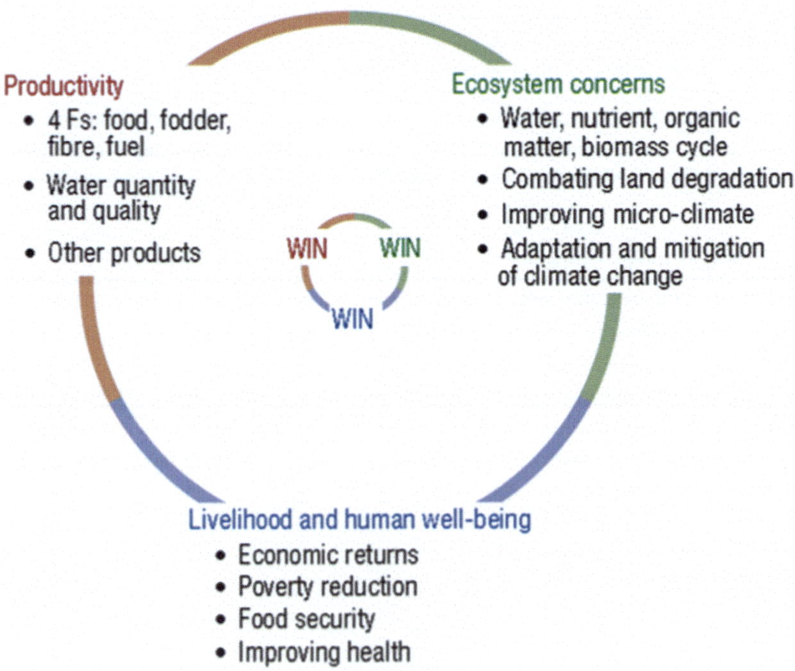

Fig. 14.5 Principles of the best SLM practices (Liniger et al. 2011)

also capacity building for local people, promotion of voluntary efforts, institutional frameworks, generation of the necessary funding for implementing SLM projects, and broad networks of stakeholders and relevant organizations. From our past experience, in addition to individual technologies, we have gradually begun to understand the importance of the methods and approaches for disseminating these technologies, for example, the process for encouraging local people to voluntarily use the technologies and approaches for providing institutional and financial support for such activities. Therefore, in the following, individual elemental technologies are described as SLM technologies, and the methods, mechanisms, and devices used to introduce, implement, and disseminate them are described as SLM approaches.

14.5.1 SLM Technologies

The importance of local and traditional knowledge or indigenous technologies that have been developed over time to combat desertification has been recognized. The SLM technologies used in a given area mainly depend on how the land is used, and these technologies are introduced to address the challenges the land faces (Fig. 14.6).

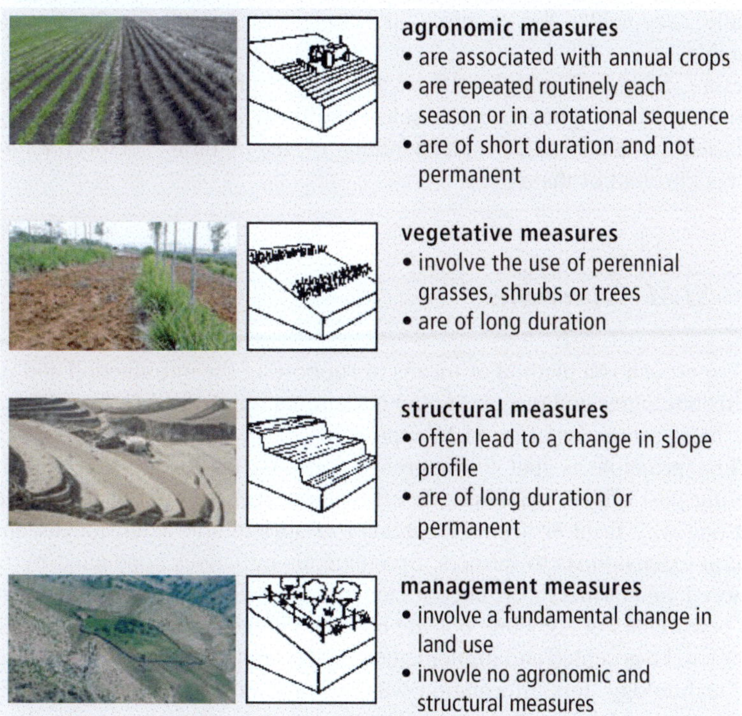

Fig. 14.6 Categories of SLM measures by WOCAT (Harari et al. 2017)

More specifically, to combat Aeolian desertification, technologies to appropriately manage land surface conditions are important. The magnitude of Aeolian desertification can be explained through wind erosivity and land surface erodibility; wind erosivity is the capacity of wind to cause dust emission and is often expressed by wind speed; land surface erodibility is the susceptibility of land to wind erosion. Therefore, measures to alleviate Aeolian desertification are based on two aspects: how to reduce wind erosivity and how to enhance resistance of land surfaces to erodibility. For arable land, managing residues, mulching, composting, and improving compost and manure are used to prevent land surfaces from being exposed to directly blowing wind. For grazing land, it is necessary to mitigate overgrazing through proper livestock and grazing pressure management, such as herd management, changes in stocking rates, or timing of use (Duniway et al. 2019), so that the grass cover can protect the land surface from blowing wind. For plantations and afforested areas, SLM technologies include the introduction of wind breaks and agroforestry, which can reduce the wind speed, eventually mitigating wind erosivity.

The following conditions are required for SLM technology in the target area: first, the technology should be affordable, and the technology should be able to be introduced and maintained without a heavy economic burden to farmers. Second, the technology should be easy to use and accessible to the farmers who are going to use it. Third, the technology should be environmentally sound and not impair biodiversity or impose a burden on the environment. Fourth, the technology should be socially acceptable; for example, it should not be religiously or culturally unacceptable.

Therefore, SLM technologies vary according to the target region and people's requirements, and it is necessary to select the most appropriate technology for a region based on a thorough understanding of the natural, socioeconomic, and cultural background of the region.

14.5.2 SLM Approaches

An SLM approach is a method or means of supporting the introduction and practice of SLM technologies and promoting the dissemination of SLM technologies. There are three basic principles of an SLM approach.

The first principle is that the approach must be participatory (collaborative). Considering past efforts to address desertification, technology transfer was carried out in a one-way fashion, from researchers to agricultural extensionists and then agricultural extensionists to farmers. Spontaneous initiatives conducted by farmers were ignored, and farmers were encouraged to implement technologies developed by external researchers or someone outside of their villages. There was no way for the technology to be accepted and implemented on the land. Even if there was a problem with the technology, this information was not provided to the researchers, and no improvement could be implemented. On the other hand, the recently developed participatory approach puts the farmers at the center of the process, emphasizes bottom-up involvement from the farmers, and encourages the engagement of farmers by giving them responsibility. Empowerment, capacity building, research, dissemination, and organization are the key elements of this approach.

The second principle is that the approach must be integrated. In the past, SLM projects often had a single objective, such as increasing crop yield, and aimed to introduce a single type of technology. However, in recent years, there has been a trend toward implementing an integrated approach that aims to improve the environment, economy, and society of a village in a comprehensive and integrated manner, such as by increasing the productivity of the land, improving the livelihood of farmers, and empowering vulnerable people such as women. Examples of integrated approaches include integrated watershed management and integrated rural development.

The third principle is that the approach must involve a partnership through which diverse stakeholders participate. It is important to establish a framework where the various actors involved in the SLM project, i.e., government officials, NGOs, farmers, landowners, researchers, mass media, and aid agencies, can cooperate with each other while sharing roles.

The significance and role of SLM were recognized at the United Nations Conference on Environment and Development (Earth Summit) held in 1992. Since that time, SLM has been developed in various parts of the world depending on the land uses and other aspects of the local natural environment in those areas.

SLM is further expected to play a major role to realize land degradation neutrality. Moreover, given the current situation caused by the COVID-19 pandemic, the importance of maintaining food security, especially in rural villages, is being increasingly recognized due to the lockdown of cities and restrictions on the transportation of people and goods. SLM is also even more important given these issues.

Given this context, the following three challenges facing SLM are addressed:

First, it is necessary to link SLM activities with the socioeconomic empowerment of local people. Most current SLM projects are expensive and often carried out by local laborers (sometimes free labor) from the local village where the project is being implemented, which raises questions about the economic viability, sustainability, and autonomy of SLM. In addition, donors are considering exit strategies to avoid continually providing financial aid; thus, the issue of how to end this aid-heavy system needs to be addressed. Thus, the first challenge for SLM is how to link SLM to improving the livelihoods of local people so that they will be able to implement SLM autonomously and sustainably. It is also important to establish a mechanism to address the empowerment of socially vulnerable and people that live in poverty, such as women and landless youth, through their participation in SLM activities, such as acquiring skills to improve their livelihoods, increasing their cash income through the sale of their products, and establishing an innovation platform where empowerment can be discussed among a community.

Second, it is expected that countries and funding entities will explore exit strategies so that they do not continually provide aid; thus, diverse sources of funding need to be identified. Achieving the global goal of land degradation neutrality by 2030 (SDG target 15.3) requires mobilizing large amounts of financial resources. Public and philanthropic resources alone will not be enough, as recognized in the Addis Ababa Action Agenda.[14] Thus, various new financial instruments and intermediaries are currently being explored to catalyze private capital to achieve land degradation neutrality (LDN). For example, in relation to the UNCCD, an LDN fund was officially launched at the COP 13 held in Ordos, China, which is the first investment instrument to leverage public money to raise private capital for SLM projects. The LDN fund will invest in financially viable private projects focused on land rehabilitation and sustainable land management, and the fund is expected to provide environmental and socioeconomic benefits as well as financial returns.

Third, it is crucial to respond to climate change through both adaptation and mitigation. The IPCC Special Report on Climate Change and Land (2019) shows that better land management can contribute to addressing climate change, and SLM can prevent and reduce land degradation, maintain land productivity, and sometimes reverse the adverse impacts of climate change on land degradation. SLM contributes to climate change mitigation through, for example, the sequestration of soil organic carbon on cropland. On the other hand, SLM practices can also play an important role in climate change adaptation by coping with changing climate conditions in the future. For example, more runoff accumulation and improved sediment runoff functions will be required to cope with intensified precipitation under a changed climate in the future. Therefore, this type of innovative SLM that contributes to both adaptation and mitigation of climate change is a climate-smart SLM, and we are currently developing its methodology and trying to implement the concept in practice.

14.6 Case Study of an SLM Project That Addresses Aeolian Desertification in Northeast Asia

This section introduces "Sustainable Land Management for Combating Desertification in Mongolia" as an example of an SLM project to address Aeolian desertification in Northeast Asia. Funded by the UNDP, the government of the Netherlands, and the Swiss Agency for Development and Cooperation (SDC), this project began in January 2008 and ended on December 31, 2012.

The overall goal of the project was to ensure that pastures, agricultural areas, forests, and other terrestrial land use types are productive and sustainable and that ecosystem services and functions that are essential to improving livelihoods and reducing poverty are protected and enhanced, as well as to build institutional capacities within government, research agencies, organizations, individual personnel, and local communities and demonstrate good practices of sustainable land

[14] https://sustainabledevelopment.un.org/index.php?page=view&type=400&nr=2051&menu=35.

management (SLM) in line with national economic and social development policies (Swenson and Erdenebileg 2012).

The project worked extensively in the 4 aimags of Tuv, Sükhbaatar, Dornogovi, and Uvurkhangai, concentrating on 13 target soums. Land management personnel in the soums were trained in land use planning and mapping. Seventy-four different trainings were conducted, including training on pasture management, traditional rotational grazing practices, and planting of trees and shrubs. In addition, material support was provided for the restoration of wells and springs, and fencing was provided for hay production and vegetable-growing areas.

The project supported the formation of 109 herder communities (HGs) and 13 forest user groups (FUGs) for a total of 122 resource user groups. These groups consisted of 1236 households with 2781 members from the 13 target soums in the 4 target aimags. The 74 different trainings included training on a wide range of SLM activities, desertification control and livelihood enhancement occurred, and exchanges and sharing of experiences with pastoralists and forest user groups were conducted. Over the 5 years of the project, a total of 8605 local participants from the soum government and pastoralist groups were trained in land management planning, vegetable growing, tree planting, producing better milk products, wood product processing, water harvesting, producing brick fuel, increasing energy efficiency, constructing barriers against sand movement, hay making, managing pastures, controlling rodents (Brandt's vole), teaching herder group leadership, and other related training. This approach laid a solid foundation for knowledge sharing and ongoing activities to sustain the SLM activities of the local soum and herder communities even after the completion of the project.

The project also actively supported the restoration of water springs and shallow hand-dug wells and worked with local soum officers and herder groups to determine the best ways to restore and protect these water sources. New deep wells were dug, and many of deep wells were rehabilitated. Three of the deep wells were equipped with solar-powered pumps, allowing them to pump water without relying on a motorized engine, thus reducing the cost of benzene while protecting the environment. Ponds were also installed and rehabilitated early in the project.

Fenced areas were created for vegetative cover and hay production to protect and provide winter forage production for livestock. Trees and shrubs such as larch, tamarisk, elm, and saxaul were initiated and planted in the targeted soum areas during the project period. In addition, approximately one million hectares of grassland were protected through traditional rotational grazing practices.

As summarized above, the project worked at the "local level, on the ground," while building the capacity of government land management agencies and academic institutions and helping improve policies and laws. As mentioned in Sect. 14.5, individual SLM technologies such as livestock fencing, installing wells, and tree planting were introduced, but there was also an emphasis on various SLM approaches such as training and forming herder groups to ensure the other technologies and efforts could be maintained. The importance of combining SLM technologies and approaches in an appropriate manner is exemplified by this project.

14.7 Conclusion

The current, common global goal to combat desertification is to achieve land degradation neutrality by 2030, as set out in SDG 15.3. The progress of each country toward achieving its LDN targets is reported in the country reports submitted to the UNCCD and the annual Sustainable Development Goals Report of the United Nations. Great expectations have been placed SLM in terms of achieving LDN, and various SLM projects have been implemented worldwide based on the land uses and agricultural practices of the project sites.

Future challenges for SLM are to link SLM activities with the socioeconomic empowerment of local people, explore exit strategies to end development aid for SLM, identify diverse sources of funding, and respond to climate change in terms of both adaptation and mitigation through the establishment of a new climate-smart SLM methodology.

References

Duniway MC, Pfennigwerth AA, Fick SE, Nauman TW, Belnap J, Barger NN (2019) Wind erosion and dust from US drylands: a review of causes, consequences, and solutions in a changing world. Ecosphere 10:e02650. https://doi.org/10.1002/ecs2.2650

Harari N, Gavilano A, Liniger HP (2017) where people and their land are safer: a compendium of good practices in disaster risk reduction. Centre for Development and Environment (CDE), University of Bern, and Swiss NGO Disaster Risk Reduction (DRR) Platform, with Bern Open Publishing, Bern and Lucerne

IPCC (2019) Climate Change and Land: an IPCC special report on climate change, desertification, land degradation, sustainable land management, food security, and greenhouse gas fluxes in terrestrial ecosystems [Shukla PR, Skea J, Calvo Buendia E, Masson-Delmotte V, Pörtner H-O, Roberts DC, Zhai P, Slade R, Connors S, van Diemen R, Ferrat M, Haughey E, Luz S, Neogi S, Pathak M, Petzold J, Portugal Pereira J, Vyas P, Huntley E, Kissick K, Belkacemi M, Malley J (eds)]

Liniger HP, Mekdaschi Studer R, Hauert C, Gurtner M (2011) Sustainable land management in practice – guidelines and best practices for sub-Saharan Africa. WOCAT and FAO, TerrAfrica

MA (2005) Millennium ecosystem assessment: ecosystems and human Well-being: synthesis. Island Press, Washington, DC. 137 p. Available from: http://www.maweb.org/en/Products.aspx?

Orr BJ, Cowie AL, Castillo Sanchez VM, Chasek P, Crossman ND, Erlewein A, Louwagie G, Maron M, Metternicht GI, Minelli S, Tengberg AE, Walter S, Welton S (2017) Scientific conceptual framework for land degradation neutrality. A report of the science-policy interface. United Nations Convention to Combat Desertification (UNCCD), Bonn

Safriel U (2017) Land Degradation Neutrality (LDN) in drylands and beyond – where has it come from and where does it go. Silva Fennica 51(1B):1650. https://doi.org/10.14214/sf.1650

United Nations Economic and Social Council (2020) Progress towards the sustainable development goals, report of the secretary-general, supplementary information. E/2020/57*. https://unstats.un.org/sdgs/files/report/2020/secretary-general-sdg-report-2020%2D%2DStatistical-Annex.pdf

The Global Mechanism of the UNCCD (2016) Land degradation neutrality: the target setting programme. Bonn, p 20

Swenson K, Erdenebileg S (2012) Sustainable land management for combating desertification in Mongolia—MON/08/301: Report of the Final Evaluation. 52 p. Available from: https://erc.undp.org/evaluation/documents/download/8165

Index

T. Wang et al. (eds.), *Combating Aeolian Desertification in Northeast Asia*, Ecological Research Monographs, https://doi.org/10.1007/978-981-16-9028-0

C

D

The manufacturer's authorised representative in the EU is Springer Nature Customer Service Centre GmbH, Europaplatz 3, 69115 Heidelberg, Germany. If you have any concerns regarding our products, please contact ProductSafety@springernature.com

Printed and bound by CPI Group (UK) Ltd, Croydon, CR0 4YY
30/09/2026
02225241-0005